T0180606

Multiscale Processes in the Earth's Magnetosphere: From Interball to Cluster

NATO Science Series

A Series presenting the results of scientific meetings supported under the NATO Science Programme.

The Series is published by IOS Press, Amsterdam, and Kluwer Academic Publishers in conjunction with the NATO Scientific Affairs Division

Sub-Series

I. **Life and Behavioural Sciences**	IOS Press
II. **Mathematics, Physics and Chemistry**	Kluwer Academic Publishers
III. **Computer and Systems Science**	IOS Press
IV. **Earth and Environmental Sciences**	Kluwer Academic Publishers
V. **Science and Technology Policy**	IOS Press

The NATO Science Series continues the series of books published formerly as the NATO ASI Series.

The NATO Science Programme offers support for collaboration in civil science between scientists of countries of the Euro-Atlantic Partnership Council. The types of scientific meeting generally supported are "Advanced Study Institutes" and "Advanced Research Workshops", although other types of meeting are supported from time to time. The NATO Science Series collects together the results of these meetings. The meetings are co-organized bij scientists from NATO countries and scientists from NATO's Partner countries – countries of the CIS and Central and Eastern Europe.

Advanced Study Institutes are high-level tutorial courses offering in-depth study of latest advances in a field.

Advanced Research Workshops are expert meetings aimed at critical assessment of a field, and identification of directions for future action.

As a consequence of the restructuring of the NATO Science Programme in 1999, the NATO Science Series has been re-organised and there are currently Five Sub-series as noted above. Please consult the following web sites for information on previous volumes published in the Series, as well as details of earlier Sub-series.

http://www.nato.int/science
http://www.wkap.nl
http://www.iospress.nl
http://www.wtv-books.de/nato-pco.htm

Series II: Mathematics, Physics and Chemistry – Vol. 178

Multiscale Processes in the Earth's Magnetosphere: From Interball to Cluster

edited by

Jean-André Sauvaud

Centre d'Etude Spatiale des Rayonnements,
Centre National de la Recherche Scientifique, Université Paul Sabatier,
Toulouse, France

and

Zdeněk Němeček

Faculty of Mathematics and Physics,
Charles University,
Prague, Czech Republic

Kluwer Academic Publishers

Dordrecht / Boston / London

Published in cooperation with NATO Scientific Affairs Division

Proceedings of the NATO Advanced Research Workshop on
Multiscale Processes in the Earth's Magnetosphere: From Interball to Cluster
Prague, Czech Republic
9–12 September 2003

A C.I.P. Catalogue record for this book is available from the Library of Congress.

ISBN 1-4020-2767-2 (PB)
ISBN 1-4020-2766-4 (HB)
ISBN 1-4020-2768-0 (e-book)

Published by Kluwer Academic Publishers,
P.O. Box 17, 3300 AA Dordrecht, The Netherlands.

Sold and distributed in North, Central and South America
by Kluwer Academic Publishers,
101 Philip Drive, Norwell, MA 02061, U.S.A.

In all other countries, sold and distributed
by Kluwer Academic Publishers,
P.O. Box 322, 3300 AH Dordrecht, The Netherlands.

Printed on acid-free paper

This book is dedicated to the
memory of Yuri Galperin,
pioneer of space
investigations.

Contents

Contributing Authors

Alexander Z. Bochev
Solar—Terrestrial Influences Laboratory
Bulgarian Academy of Sciences (BAS)
1113 Sofia, Bulgaria

Joseph E. Borovsky
Space and Atmospheric Science Group
Los Alamos National Laboratory
Los Alamos, New Mexico 87545, USA

Jiasheng Chen
Center for Space Physics
Boston University
725 Commonwealth Avenue, Boston, Massachusetts 02215, USA

C. Philippe Escoubet
European Space Agency
European Space Research & Technology Centre
Solar and Solar-Terrestrial Missions Division
Keplerlaan 1, 2200 AG Noordwijk, The Netherlands

Galina I. Korotova
Institute of Terrestrial Magnetism, Ionosphere and Radiowave Propagation
Russian Academy of Sciences
142190, Troitsk, Moscow Region, Russia

Zdeněk Němeček
Department of Electronics and Vacuum Physics
Faculty of Mathematics and Physics
Charles University
V Holešovičkách 2, 180 00 Prague 8, Czech Republic

Jolene S. Pickett
Department of Physics and Astronomy
The University of Iowa
Iowa City, Iowa 52242, USA

John D. Richardson
Massachusetts Institute of Technology 37-655
Cambridge, Massachusetts 02139, USA

Christopher T. Russell
University of California
405 Hilgard Ave
1567 Los Angeles, California 90095, USA

Jean-André Sauvaud
Centre d'Etude Spatiale des Rayonnements
Centre National de la Recherche Scientifique/Universite Paul Sabatier
9, ave. du Colonel Roche — Boite postale 4346 31028 Toulouse, France

Victor A. Sergeev
Institute of Physics
St. Petersburg State University
198904 St. Petersburg, Russia

David G. Sibeck
NASA Goddard Space Flight Center
Greenbelt, Maryland 20771, USA

Adam Szabó
NASA Goddard Space Flight Center
Greenbelt, Maryland 20771, USA

Jana Šafránková
Department of Electronics and Vacuum Physics
Faculty of Mathematics and Physics
Charles University
V Holešovičkách 2, 180 00 Prague 8, Czech Republic

Simon Wing
Applied Physics Laboratory
The Johns Hopkins University
11100 Johns Hopkins Road, Laurel, Maryland 20723-6099, USA

Georgy N. Zastenker
Space Research Institute
Russian Academy of Sciences
Profsoyuznaya Str. 84/32, 117997, Moscow, Russia

Lev M. Zelenyi
Space Research Institute
Russian Academy of Sciences
Profsoyuznaya Str. 84/32, 117997, Moscow, Russia

Preface

The past forty years of space research have seen a substantial improvement in our understanding of the Earth's magnetosphere and its coupling with the solar wind and interplanetary magnetic field (IMF). The magnetospheric structure has been mapped and major processes determining this structure have been defined. However, the picture obtained is too often static. We know how the magnetosphere forms via the interaction of the solar wind and IMF with the Earth's magnetic field. We can describe the steady state for various upstream conditions but do not really understand the dynamic processes leading from one state to another. The main difficulty is that the magnetosphere is a complicated system with many time constants ranging from fractions of a second to days and the system rarely attains a steady state. Two decades ago, it became clear that further progress would require multi-point measurements. Since then, two multi-spacecraft missions have been launched — INTERBALL in 1995/96 and CLUSTER II in 2000. The objectives of these missions differed but were complementary: While CLUSTER is adapted to meso-scale processes, INTERBALL observed larger spatial and temporal scales.

However, the number of papers taking advantage of both missions simultaneously is rather small. Thus, one aim of the workshop "Multiscale processes in the Earth's magnetosphere: From INTERBALL to CLUSTER" hosted by Charles University in Prague, Czech Republic in September 2003 was to bring the communities connected with these projects together to promote a deeper cooperation. The leaders of projects presented summaries of the achievements made by their investigations and demonstrated the special capabilities of these missions to fulfill particular requirements. Other key speakers emphasized the importance of multipoint measurements for the research in their particular areas. The second aspect of the meeting was to stress the importance of the solar wind input on magnetospheric processes.

In course of the above workshop, 21 invited or solicited lectures, 14 oral contributions, and 18 posters were presented and 17 of these presentations were chosen for publication in the volume of the NATO Science Series which you are now reading. We hope that this volume brings not only a summary of INTERBALL and CLUSTER achievements but that it will serve as a useful aid for planning of further investigations and preparation of new multisatellite missions.

We gratefully acknowledge the funds provided by the NATO Scientific Affair Division for this workshop. We would like to express our thanks to all participants for their contributions to the success of the workshop and all the authors who submitted their manuscripts for publication in this volume. We acknowledge with thanks the effort of numerous reviewers who helped us to improve the readability and scientific quality of all contributions, namely: Elizaveta E. Antonova, Daniel Berdichevsky, Natalia L. Borodkova, Mohammed Boudjada, Patrick Canu, James Chen, Giuseppe Consolini, Charles Farrugia, Althanasios Geranios, Chaosong Huang, Christian Jacquey, Alan J. Lazarus, Janet Luhmann, Volt N. Lutsenko, Jan Merka, Karim Mezaine, Patrick T. Newell, Steven M. Petrinec, Anatoli A. Petrukovich, Tai Phan, Lubomir Prech, Patricia Reiff, John D. Richardson, Jana Safrankova, Victor A. Sergeev, James A. Slavin, Charles W. Smith, Paul Song, Yan Song, Marek Vandas, Shinichi Watari, Georgy N. Zastenker, Eftyhia Zesta. Last but not the least, we would like to thank Jana Safrankova for careful organization of the reviewing process and Jiri Pavlu for extended technical assistance in preparation of the final manuscript.

JEAN-ANDRÉ SAUVAUD

ZDĚNĚK NĚMEČEK

PROPAGATION AND EVOLUTION OF ICMES IN THE SOLAR WIND

John D. Richardson, Ying Liu, and John W. Belcher
Massachusetts Institute of Technology, Cambridge, MA, USA

Abstract   Interplanetary coronal mass ejections (ICMEs) evolve as they propagate outward from the Sun. They interact with and eventually equilibrate with the ambient solar wind. One difficulty in studying this evolution is that ICMEs have no unique set of identifying characteristics, so boundaries of the ICMEs are difficult to identify. Two characteristics present in some ICMEs but generally not present in the ambient solar wind, high helium/proton density ratios and low temperature/speed ratios, are used to identify ICMEs. We search the Helios 1 and 2, WIND, ACE, and Ulysses data for ICMEs with these characteristics and use them to study the radial evolution of ICMEs. We find that the magnetic field magnitude and density decrease faster in ICMEs than in the ambient solar wind, but the temperature decreases more slowly than in the ambient solar wind. Since we also find that ICMEs expand in radial width with distance, the protons within ICMEs must be heated. Scale sizes for He structures are smaller than for proton structures within ICMEs.

Key words: ICME; solar wind.

1.   INTRODUCTION

Coronal mass ejections (CMEs) are eruptions of matter from the Sun into interplanetary space. A CME may result in the ejection of 10^{16} g of matter with a broad range of speeds, up to at least 2000 km/s (e.g., Lepping et al., 2001). The ejected material forms an interplanetary coronal mass ejection (ICME) in the solar wind. The ICMEs that collide with Earth often produce large effects in Earth's magnetosphere; almost all of the largest geomagnetic storms result from ICMEs (Gosling, 1993).

A centerpiece of space weather research is the forecasting of ICMEs and their magnetospheric effects. Solar observations have been used to detect Earthward CMEs (i.e., Zhao and Webb, 2003). Spacecraft monitors such

J. –A. Sauvaud and Z. Němeček (eds.),
Multiscale Processes in the Earth's Magnetosphere:From Interball to Cluster, 1-14.
© 2004 *Kluwer Academic Publishers. Printed in the Netherlands.*

as ACE and SOHO at the L1 Lagrange point provide warnings of incoming ICMEs 30-60 minutes upstream of Earth. Proxies such as the field direction within magnetic clouds (Chen, 1996) and shocks (Jurac et al., 2002) can be used to predict geomagnetic storms. In the future, STEREO is designed to remotely sense the propagation of CMEs in the solar corona which produce Earthward-propagating ICMEs.

For long-range, several day in advance forecasting based on solar observations to succeed, we need to understand better how ICMEs evolve in the solar wind. For the shorter-term, 30-60 minute forecasting based on L1 observations, the effect of radial evolution between L1 and Earth is likely small, but since L1 monitors are often hundreds of R_E from the Earth-Sun line, we need to understand the spatial extent of CME material perpendicular to the solar wind flow. Variations in the magnetic field magnitude, plasma bulk speed, and plasma density have larger scale lengths in solar wind which causes large geomagnetic disturbances than in the typical solar wind (Jurac and Richardson, 2001), but length-scales of variations within the ICME material have not been studied.

This paper reviews the radial evolution and spatial scales of ICMEs in the solar wind. One of the difficulties in studying these subjects is that it can be difficult to identify ICMEs and their boundaries. We discuss two criteria which may be sufficient to identify ICMEs, enhanced helium abundances and lower than expected temperatures for a given solar wind speed. We use a list of ICMEs produced using these criteria which spans radial distances from 0.3 to 5.4 AU to described the radial changes in ICMEs. We also compare the helium abundances observe by ACE and WIND to determine the scale size of enhanced helium events perpendicular to the solar wind flow.

2.   IDENTIFICATION OF ICMES

ICMEs have many identifying characteristics (see reviews by Gosling, 1990; Neugebauer and Goldstein, 1997; Gosling, 1997); among them are

1 The temperature is lower than normal for the observed solar wind speed.

2 The fluctuation level of the magnetic field magnitude is small.

3 The ratio of the He to H density is larger than normal but strongly fluctuating (e.g., Berdichevsky et al., 2002).

4 Energetic protons and cosmic rays stream along the magnetic field.

5 Bi-directional electrons are observed, moving in both directions along the magnetic field indicating that the field lines are closed loops connected to the Sun or to themselves.

6 Enhancements in minor ions such as Fe and higher charge states of heavy ions such as Fe and O (Burlaga et al., 2001; Lepri et al., 2001).

7 Depressed energetic particle intensities known as Forbush decreases resulting from the increased magnetic field.

8 Slowly rotating magnetic fields are signatures of a subset of ICMEs known as magnetic clouds.

9 A preceding shock formed by faster CME material encountering slower solar wind when solar counterpart observations or remote radio sensing indicate the presence of ejecta.

The problem is that, although ICMEs may have some of these characteristics, few ICMEs have all of them and many have a small subset of them. In addition, these characteristics may not persist across the entire ICME but may come and go within one ICME.

Various lists of ICMEs have been developed based on various of these criteria. We refer in this paper to the list developed by Cane and Richardson (2003) (hereafter referred to as C&R) using data from 1996 through 2002 which covered the period from solar minimum to solar maximum. The starting point for this list is the criterion that the observed temperature, T_{obs}, be less than 0.5 of the temperature expected for the observed solar wind speed, T_{exp} (Richardson and Cane, 1995). The expected temperatures for a given speed are taken from Lopez and Freeman (1986). In addition to the temperature criterion, C&R used shocks, Forbush decreases, and energetic particle signatures (but not helium abundance or bi-directional electrons) to identify ICMEs.

Figure 1 shows 12 days of WIND data from 1999. The hatched area shows the time of an ICME from the C&R list. The ICME region lasts about two days, has a low temperature given the relatively high speed, follows a shock, and has little magnetic field variation. Although this criterion was not part of the C&R ICME search criteria, the ICME also has enhanced He abundances. This example is clearly consistent with the definition of an ICME, although some ambiguity exists as to the location of the trailing edge. But on day 260 the helium abundance also increases for about a day coincident with a small temperature decrease. Magnetic fluctuations through most of this time period are small. Based on the He abundance, this event is also likely an ICME, although the boundaries of the event are not obvious.

The hatched ICME in Figure 1 shows two other important features of ICMEs. The speed decreases across the ICME, so that the forward edge moves faster than the trailing edge. This speed difference results from expansion of the ICME. The magnetic field (and thus magnetic pressure) is high within the first half of the ICME; many ICMEs have a larger internal pressure than does the

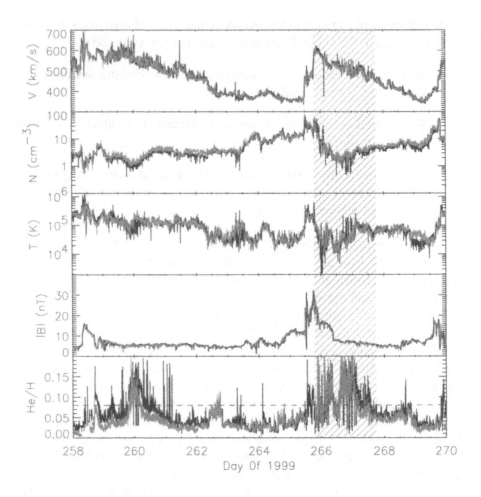

Figure 1. Solar wind speed, density, temperature, magnetic field magnitude, and helium/proton density ratios from WIND (black) and ACE (gray). The hatched region is a ICME identified by C&R. The dashed line in the bottom panel shows the 0.08 helium to proton density ratio considered sufficient to identify ICME plasma.

ambient solar wind, also leading to expansion of the ICME as it moves outward.

The examples in Figure 1 show the difficulties inherent in the study of ICMEs; namely determining if they are present and when they start and end. In a few cases when ICMEs have apparently been larger in angular extent than the spacecraft separation, the same ICME has been observed at widely separated radial distances. For example, the Bastille day 2001 CME occurred when Earth and Voyager 2 were at nearly identical heliolongitudes and was observed at 1 AU and 63 AU (Wang et al., 2001). But since radial alignments of spacecraft are rare, we want to look at ICME evolution on a statistical basis. We choose two ICME characteristics that are thought to be sufficient (but not necessary) to identify ICMEs; 1) $T_{obs}/T_{exp} < 0.5$ and 2) $N_{He}/N_H > 8\%$. This method makes the implicit assumption that ICMEs with these characteristics are typical of ICME plasma.

We identify the times in the Helios 1 and 2, WIND, ACE, and Ulysses data that both these criteria are met. Helios 1 and 2 operated from 1976 to 1980 and 1976 to 1985, respectively, at distances of 0.3 to 1 AU. WIND was launched in late 1994 and is near 1 AU. ACE was launched in 1997 and orbits Earth's L1 Lagrange point. Ulysses was launched in 1990 and orbits over the solar poles at 1.3 - 5.4 AU (although most ICMEs detected are at low latitudes). Since the temperature decreases with distance, we normalized to 1 AU assuming a R^{-1} dependence (Totten et al., 1995; see also Steinitz and Eyni, 1981; Lopez and Freeman, 1995) before applying the temperature criteria. The N_{He}/N_H ratio should be independent of distance which makes it a useful tracer of ICMEs (Paularena et al., 2001). The complete list of ICMEs we identified using these criteria is given by Liu et al. (2004).

3.   RADIAL EVOLUTION OF ICMES

Figure 2 shows the distribution of ICMEs at 1 AU from ACE and WIND over a solar cycle using the above criteria. The solar cycle dependence is as expected, with less ICMEs at solar minimum than at solar maximum. We note that adherence to these criteria results in about 50% less ICMEs than in the C&R list. Also note that not all the magnetic clouds identified by the WIND magnetic field instrument (http://lepmfi.gsfc.nasa.gov/mfi/mag_cloud_pub1.html) are on our list, reinforcing the uncertainty inherent in choosing ICMEs.

Figure 3 shows the average solar wind density, speed, temperature, and magnetic field magnitude within ICME plasma. The circles show data from Helios 1 and 2, the triangles from WIND and ACE, and the squares from Ulysses. The solid lines show the best fit of a power law to the data. The top panel shows that the density profile which best fits the data is $N(r) = 6.2 \, R^{-2.3}$. The average density (normalized to 1 AU) in the ICMEs of 6.2 cm^{-3} is slightly less than the 7 cm^{-3} in the background solar wind, consistent with previous results (Crooker et al., 2000). The decrease with R is faster than in the background solar wind; as expected, the solar wind density as a whole decreases as R^{-2} out to 70 AU (Richardson et al., 2003). The more rapid decrease of density within ICMEs is due to the expansion of the ICMEs, as discussed above.

The second panel of Figure 3 shows the average speed within the ICMEs. The fit line shows that the average speed is about 450 km/s and does not change

with distance. This is comparable to the average speed of all solar wind near
Earth, 440 km/s. The variations of the ICME speeds decrease with distance.

The third panel shows the temperature in ICMEs. The best fit is $T(r) =$
$3.5 \times 10^4 R^{-0.3}$K. The average solar wind temperature at 1 AU is about 9.5×10^4
K, so the ICME temperature is well below this (as expected given that one of
the ICME identification criteria is low temperature). The small R dependence
of T with distance was unexpected. The temperature of the background solar wind in the inner heliosphere decreases as R^{-1} (Totten et al., 1995). Since
ICMEs expand with distance, the temperature in ICMEs should decrease faster
than in the background solar wind due to adiabatic cooling; instead it decreases
less quickly. This result implies that significant heating of the protons in the
ICMEs takes place, more than in the normal solar wind. ICMEs are often associated with streaming electrons and high heat flux; some of this energy may
couple to the protons.

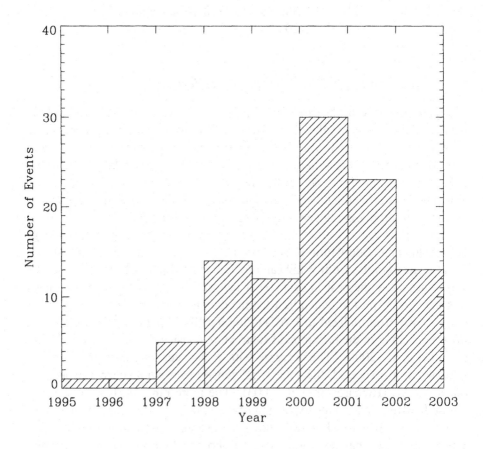

Figure 2.   Number of ICMEs at 1 AU as a function of time.

The bottom panel shows the magnetic field magnitude within the ICMEs. The fit to the data gives $B(r)=7.4\ R^{-1.4}$ nT. For an ideal Parker spiral, the radial component of B decreases as R^{-2} and the tangential field as R^{-1}. The average magnetic field in the solar wind at 1 AU is 6.3 nT, so B is larger within

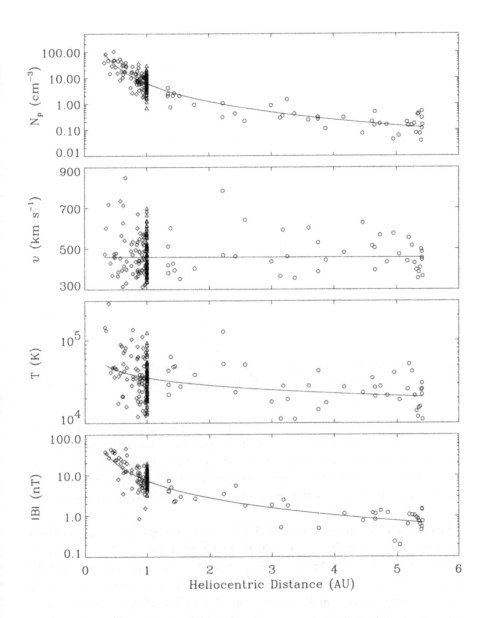

Figure 3.   Average values of the density, speed, temperature, and magnetic field magnitude within ICMEs as a function of distance. Also shown are power law fits to each parameter.

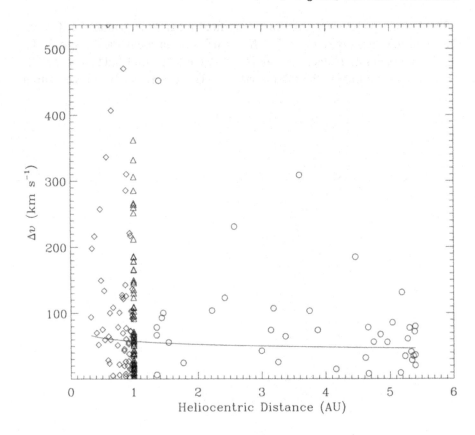

Figure 4. The speed difference, Δv, across the ICMEs as a function of distance. Diamonds show Helios 1 and 2 data, triangles show ACE and WIND data, and circles show Ulysses data.

ICMEs. The higher B often results in a higher internal pressure within ICMEs which contributes to their expansion. The best power law fit to the magnetic field magnitude observations in the low-latitude background solar wind gives a $R^{-1.1}$ decrease. Thus the magnetic field within ICMEs decreases faster than that in the solar wind as a whole, again consistent with ICMEs expanding with distance.

One of the characteristics of an ICME is that the speed of the leading edge is generally greater than the speed of the trailing edge, which results in a dynamic expansion of the ICME. Figure 4 shows how Δv, the difference between the speeds on the leading and trailing edges, changes with distance. The average value of Δv only has a small change, from about 65 to 45 km/s, but the scatter decreases quickly with distance and few ICMEs beyond 5 AU have large Δv.

Figure 5 shows the radial width of ICMEs as a function of distance, where the width is determined by multiplying the time it takes an ICME to pass the

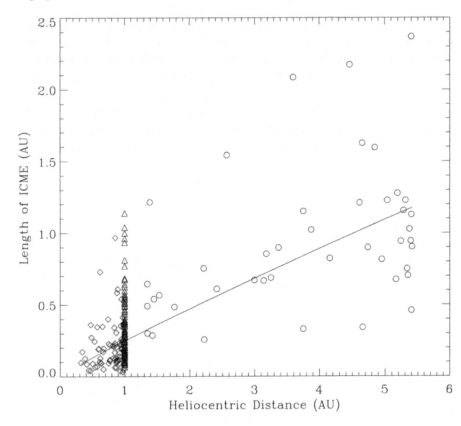

Figure 5. The radial width of the observed ICMEs as a function of distance. Diamonds show Helios 1 and 2 data, triangles show ACE and WIND data, and circles show Ulysses data.

spacecraft by the average speed of the ICME. The average ICME length increases from 0.25 AU at 1 AU to about 1 AU at 5 AU, a factor of four increase. Thus the expansion of ICMEs inferred from observations is a measurable, and significant, effect.

4.   SPATIAL SCALES

The space weather program resulted in numerous studies of the scale lengths of plasma and magnetic field features in the solar wind. Plasma features have scale lengths of order 100 R_E while magnetic field scale lengths are tens of R_E (Paularena et al., 1998; Zastenker et al., 1998; Richardson and Paularena, 2001). Scale lengths were longer for geoeffective solar wind features (Jurac and Richardson, 2001).

The regions of He enhancement are thought to be prominence material which has its origin lower on the solar surface. The He enhancements are often patchy and variable, but it has not been clear whether these are temporal or spatial variations. Since WIND and ACE provide He data near Earth, we can investigate the scale sizes of the He enhancements. As in previous work, we look at six hour segments of data from two spacecraft. The data are time-shifted using the observed solar wind speed to account for the radial separation of the spacecraft. We then perform correlations on the data as a function of lag.

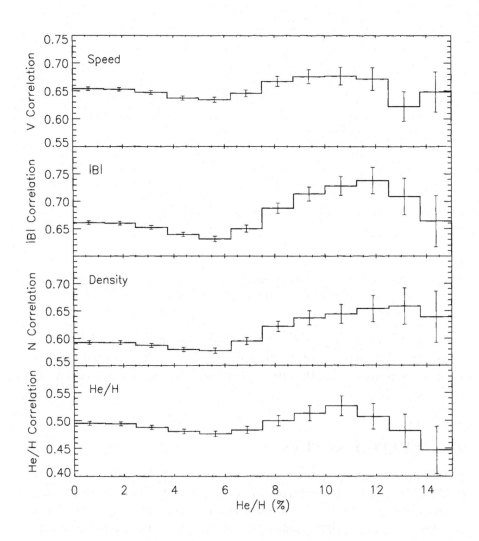

Figure 6.   Correlations as a function of the He/H ratio.

Figure 6 shows correlations as a function of the He/H ratio. We did not specifically separate out ICME regions in this part of the study, but most of the high He/H regions are likely ICMEs. The top 3 panels show that the speed, density, and B correlations are better when He/H ratios are greater than about 5%, consistent with the Jurac and Richardson (2001) results. The bottom panel shows the He/H correlations; these correlations are not significantly better for higher He/H ratios.

To determine scale lengths of the solar wind, we look at correlations as a function of spacecraft separation perpendicular to the solar wind flow. To maintain meaningful statistics, we divide the data into times when the He/H ratio is less than 4% and times it is greater than 5%.

Figure 7 shows the correlations of density and He/H ratio as a function of Y_{GSE}-separation of the spacecraft for these two cases. The density correlations are fairly constant out to separations of about 220 R_E for the low He case and 250 R_E for the large He case. For the low He case, the He/H ratio correlations are very similar to those for the density. This implies the source regions of the protons and He vary similarly. For the case where He/H is greater than 5%, which should be predominately ICME plasma, the scale length of the He/H correlations is much smaller than for the proton density, with a decrease in correlations at about 140 R_E. Thus the He seems to be generated by a small region of a much bigger ICME source structure.

5.   SUMMARY

We investigated the radial evolution and spatial scales of ICMEs. We used the temperature/speed ratio and the He/H ratio criteria to identify ICMEs in spacecraft at positions from 0.3 to 5.5 AU from the Sun. We then investigated ICME evolution in a statistical sense and find that ICMEs are about a factor of 4 larger in radial width at 1 than at 5 AU. The density and magnetic field magnitude within ICMEs decrease faster than those in the background solar wind. These data are interpreted as indicating that ICMEs expand with distance out to at least 5 AU. The temperature decreases less fast in ICMEs than in the solar wind, opposite to expectations for a radially expanding (and thus adiabatically cooling) structure, which implies that the ICME plasma is heated significantly more than the background solar wind. The spatial scales of He perpendicular to the solar wind flow are similar to that of the density in normal solar wind, but are about half the length scales of protons in the ICMEs.

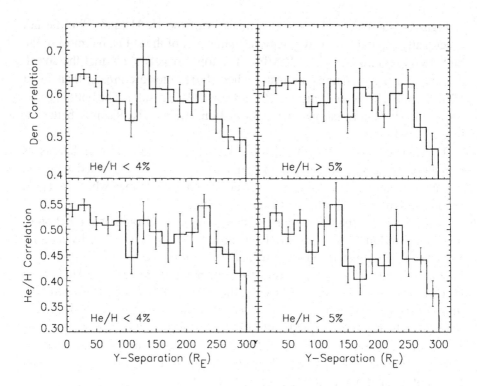

Figure 7.   Correlations as a function of the Y_{GSE}-separation of the spacecraft for He/H < 4% and He/H > 5%.

ACKNOWLEDGMENTS

We are grateful to the ACE science center for providing the ACE plasma and magnetic field data, to the ESA Ulysses web page for the Ulysses data, to J. Kasper and A. Lazarus for the WIND alpha data, and to R. Schwenn for the Helios data. The work at MIT was supported NASA grant NAG-10915 and by NSF grants ATM-9810589, ATM-0203723 and ATM-0207775.

REFERENCES

Berdichevsky, D. B., Farrugia, C. J., Thompson, B. J., Lepping, R. P., Reames, D. V, Kaiser, M. L., Steinberg, J. T., Plunkett, S. P., and Michels, D. J., 2002, Halo-coronal mass ejections near the 23rd solar minimum: Lift-off, inner heliosphere, and in situ signatures, *Ann. Geophys.* **20**(7):891–916.

Burlaga, L. F., Skoug, R. M., Smith, C. W., Webb, D. F., Zurbuchen, T. H., Reinard, A., 2001, Fast ejecta during the ascending phase of solar cycle 23: ACE observations, 1998–1999 *J. Geophys. Res.* **106**(A10):20957–20978.

Cane, H. V., and Richardson, I. G., 2003, Interplanetary coronal mass ejections in the near-Earth solar wind during 1996–2002, *J. Geophys. Res.* **108**(A4):1156, doi: 10.1029/2002JA009817.

Chen, J., 1996, Theory of prominence eruption and propagation: Interplanetary consequences, *J. Geophys. Res.* **101**(A12),27499–27519.

Crooker, N. U., Shodhan, S., Gosling, J. T., Simmerer, J., Lepping, R. P., Steinberg,J. T., and Kahler, S. W., 2000, Density extremes in the solar wind, *Geophys. Res. Lett.* **27**(23):3769–3772.

Gosling, J. T., 1990, Coronal mass ejections and magnetic flux ropes in interplanetary space, in: *Physics of Magnetic Flux Ropes*, C. T. Russell, E. R. Priest, and L. C. Lee, eds., Geophysical Monograph 58, American Geophysical Union, pp. 343.

Gosling, J. T., 1997, Coronal mass ejections: An overview, in: *Coronal Mass Ejections*, N. Crooker, J. A. Joselyn, and J. Feynman, eds., Geophysical Monograph 89, American Geophysical Union, pp. 9.

Gosling, J. T., 1993, The solar flare myth, *J. Geophys. Res.* **98**(A11):18937–18949.

Jurac, S., and Richardson, J. D., 2001, The dependence of plasma and magnetic field correlations in the solar wind on geomagnetic activity, *J. Geophys. Res.* **106**(A12):29195–29206.

Jurac, S., Kasper, J. C., Richardson, J. D., and Lazarus, A. J., 2002, Geomagnetic disturbances and their relationship to interplanetary shock parameters, *Geophys. Res. Lett.* **29**(10):1463, doi: 10.1029/2001GL014034.

Lepping, R. P., Berdichevsky, D. B., Burlaga, L. F., Lazarus, A. J., Kasper, J., Desch, M. D., Wu, C.-C., Reames, D. V., Singer, H. J., Smith, C. W., and Ackerson, K. L., 2001, The Bastille day magnetic clouds and upstream Shocks: Near-earth interplanetary observations, *Solar Phys.* **204**(1–2):287–305.

Lepri, S. T., Zurbuchen, T. H., Fisk, L. A., Richardson, I. G., Cane, H. V., and Gloeckler, G., 2001, Iron charge distribution as an identifier of interplanetary coronal mass ejections, *J. Geophys. Res.* **106**(A12):29231–29238.

Liu, Y., Richardson, J. D., and Belcher, J. W., 2004, A Statistical Study of the Properties of Interplanetary Coronal Mass Ejections from 0.3 to 5.4 AU, *Plan. Sp. Sci.*, in press.

Lopez, R. E., and Freeman, J. W., 1986, Solar wind proton temperature-velocity relationship *J. Geophys. Res.*, **91**(A2):1701–1705.

Neugebauer, M., and Goldstein, R., 1997, Particle and field signatures of coronal mass ejections in the solar wind, in: *Coronal Mass Ejections*, N. Crooker, J. A. Joselyn, and J. Feynman, eds., American Geophysical Union, Washington, pp. 245.

Paularena, K. I., Zastenker, G. N., Lazarus, A. J., and Dalin, P. A., 1998, Solar wind plasma correlations between IMP 8, INTERBALL-1 and WIND, *J. Geophys. Res.* **103**(A7):14601–14617.

Paularena, K. I., Wang, C., von Steiger, R., and Heber, B., 2001, An ICME observed by Voyager 2 at 58 AU and by Ulysses at 5 AU, *Geophys. Res. Lett.* **28**(14):2755–2758.

Richardson, I. G., Cane, H. V, 1995, Regions of abnormally low proton temperature in the solar wind (1965–1991) and their association with ejecta, *J. Geophys. Res.* **100**(A12):23397–23412.

Richardson, J. D., and Paularena, K. I., 2001, Plasma and magnetic field correlations in the solar wind, *J. Geophys. Res.* **106**(A1):239–251.

Richardson, J. D., Wang, C., and Burlaga, L. F., 2003, The solar wind in the outer heliosphere, *Adv. Sp. Res.* in press.

Steinitz, R., and Eyni, M., 1981, Competing effects determining the proton temperature gradient in the solar wind, in: *Solar Wind Four*, H. Rosenbauer, ed., MPI Katlenburg-Lindau and MPI Garching, MPAE-W-100-81-31, pp. 164–167.

Totten, T. L., Freeman, J. W., and Arya, S., 1995, An empirical determination of the polytropic index for the free-streaming solar wind using Helios 1 data, *J. Geophys. Res.* **100**(A1):13–18.

Wang, C., Richardson, J. D., and Burlaga, L., 2001, Propagation of the Bastille Day 2000 CME shock in the outer heliosphere, *Solar Phys.* **204**(1–2):413–423.

Zastenker, G. N., Dalin, P. A., Lazarus, A. J., and Paularena, K. I., 1998, Comparison of solar wind parameters measured simultaneously by several spacecraft, *Cosmic. Res.* **36**(3):214–225 (Transl. from *Kosmicheskie issledovaniya* **36**(3):228–240, *(in Russian)*).

Zhao, X. P., and Webb, D. F., 2003, Source regions and storm effectiveness of frontside full halo coronal mass ejections, *J. Geophys. Res.* **108**(A6):1234, doi: 10.1029/2002JA009606.

THE SOLAR WIND INTERACTION WITH PLANETARY MAGNETOSPHERES

C. T. Russell[1], X. Blanco-Cano[2], N. Omidi[3], J. Raeder[4], and Y. L. Wang[5]

[1]*University of California, Los Angeles, CA 90095-1567;* [2]*Ciudad Universitaria, Coyoacan D.F. 04510, Mexico;* [3]*University of California, San Diego, CA 92093-0407;* [4]*University of New Hampshire, Durham, NH 03824-3525;* [5]*Los Alamos National Laboratory, Los Alamos, NM 87545*

Abstract: The solar wind interaction with planetary magnetospheres is a multifarious topic of which our understanding continues to grow as we obtain more detailed observations and more capable numerical codes. We attempt to explain how the system functions by examining the output of models of increasing sophistication. A gasdynamic numerical model produces a standing bow shock in front of a fixed impenetrable obstacle. The post-shock flow is heated and deflected but no plasma depletion layer is formed in the subsolar region contrary to observations. If magnetic forces are included, then a self-consistent obstacle size can be produced and plasma depletion extends all the way to the subsolar region. While a standing slow mode wave has been reported in the subsolar region, it appears that such a wave is not essential to the formation of a subsolar plasma depletion layer. Both the gasdynamic and magnetohydrodynamic models are self-similar. They do not change with the size of the obstacle. However, in the real solar wind interaction we expect that the relative scale size of ion motion and the radius of the obstacle will change the nature of the interactions. Hybrid simulations allow this multiscale coupling to be explored and shrinking the size of the obstacle relative to the gyroradius enhances the role of kinetic processes. Phenomena such as upstream ions, plasma sheet formation, and reconnection can be found in surprisingly tiny magnetospheres. Finally, we contrast how the magnetospheres of the Earth and Jupiter are powered. In the former case the solar wind interaction is very important and the latter case much less so.

Key words: magnetosphere; solar wind interaction; gasdynamic simulation; magneto-hydrodynamic simulation; hybrid simulation; Earth; Jupiter.

15

J. –A. Sauvaud and Z. Němeček (eds.),
Multiscale Processes in the Earth's Magnetosphere:From Interball to Cluster, 15-35.
© 2004 *Kluwer Academic Publishers. Printed in the Netherlands.*

1.   INTRODUCTION

Measurements of the interaction of the solar wind with planetary magnetospheres have been gathered for over 40 years and theoretical models predate the *in situ* observations by another three decades. More recently numerical simulations have proven themselves to be a most useful adjunct to theory and observation. Simulations have continued to improve in capability both as numerical techniques and computational capabilities have improved. Our understanding of the solar wind interaction with planetary magnetospheres is now at a very sophisticated level. In fact entire books have been written about this subject. Rather than attempt to review all of what we presently know about the solar wind interaction, this paper is intended to be a tutorial stressing topics concerning the external interaction but introducing in the section on hybrid simulations the full panoply of magnetospheric processes. It closes with a discussion of what provides the power for magnetospheric processes using Jupiter and the Earth to illustrate two very different means of providing this power. The second topic requires resort to observations as the system is too complex to be convincing simulated at present while the first topic can be now best discussed through the device of examining simulations of differing approximations. Again, the reader should not expect a complete review of the topic in these pages. These topics were selected based on the authors' perceptions about what are the contemporary, key issues and where there might be some confusion due to the plethora of models and data.

The solar wind interaction with planetary magnetospheres has been treated numerically since the pioneering work of J. R. Spreiter and colleagues in the 1960's (see Spreiter et al., 1966). Such models were extremely useful in understanding the observed properties of the solar wind interaction with the Earth and have been used to make important inferences about the location of reconnection sites on the magnetopause for example (Luhmann et al., 1984a). However, the gas dynamic model is limited in what it can tell us about the behavior of the magnetosheath. Thus with the advent of increased computer power, numerical magneto-hydrodynamic models (e.g., Raeder, 2003) have become the state of the art technique for understanding the solar wind interaction with the Earth. These models allow us to predict more realistic behavior for the plasma but such models do not produce kinetic phenomena, such as upstream waves, or a radiation belt. To address kinetic effects and especially to show how cross-scale coupling can produce a hierarchy of magnetospheres, we need to run hybrid codes in which ion motion is followed and electrons are treated as a massless fluid (Winske and Omidi, 1996). At present obstacles as large as that of Mercury can be treated at least in two dimensions. Such solutions (Omidi et al., 2003) begin to produce the rich variety of plasma phenomena that we see in our space measurements. The models we use herein consider only the solar-wind-magnetosphere interaction and do not properly consider the physics of the upper atmosphere and ionosphere or other non-solar wind plasma sources. To illustrate how these other effects may alter the dynamics of a planetary magnetosphere, in general we must at present rely on observations.

In the next section we examine the three-dimensional gasdynamic simulation results of Spreiter et al. (1966) to show what phenomena can be produced with this simple physics of the compressional sound wave in the absence of magnetic forces. We then probe the results of three-dimensional MHD simulations to see how the solution is modified by inclusion of the magnetic forces. Then we examine the need for the reported standing, slow-mode wave (Song et al., 1992) in the formation of a plasma depletion layer in the subsolar region. Next we move on to the hybrid simulation to reveal the addition physical processes that occur in a plasma controlled by motion on the ion gyro scale. We can enhance the effect of the gyro scale processes by shrinking the size of the obstacle. We find that a hierarchy of magnetospheres is produced as the scale size of the obstacle increases from sub-ion-inertial-length scales to sizes of the order of Mercury. Finally, we contrast the dynamics of the magnetosphere of the Earth that are dominated by the solar wind with those of Jupiter's magnetosphere whose dynamical processes are dominated by internal mass-loading by Io. In the latter case we have not reached the state where simulations can be used with confidence and merely for our discussion on observations.

2.   THE GASDYNAMIC MODEL

In the gasdynamic model the only wave present is the compressional sound wave. The gas is isotropic and the thermal distribution maxwellian. The magnetic field is convected with the flow. The major success of the gasdynamic model is that it produces a standing bow shock around a fixed obstacle with a magnetosheath flow between the shock and the obstacle. We illustrate first in Figure 1 how deflection around an obstacle occurs and why there is a standing bow shock in supersonic flow. A pressure gradient is needed to produce the force that deflects the flow around an obstacle. As illustrated in the top panel if the flow is subsonic, the thermal or static pressure well exceeds the dynamic pressure associated with the flow of the plasma, gas or fluid. Then a sufficiently strong pressure gradient can form and the necessary flow deflection can take place. If the flow is highly supersonic as in the lower panel, the proper pressure gradient cannot form without the formation of a shock. The shock converts flow energy into thermal energy allowing a proper gradient for deflection to be produced. This is shown in the lower panel of Figure 1.

The location of the shock is not arbitrary. The shock compresses the flow so that it is denser than the preshocked flow. The shock front sits at a distance from the obstacle that allows the shocked flow to move around the obstacle. Thus the shock distance depends on the radius of curvature of the obstacle and the compression ratio of the flow. The compression ratio depends on the Mach number of the flow relative to the obstacle and on the polytropic index of the equation of state of the gas. The location of the shock moves to infinity as the Mach number approaches unity or the radius of curvature of the obstacle goes to infinity. We note that the gasdynamic model uses a fixed obstacle size. There is no way it could treat a compressible magnetic obstacle. Also there is no change in the solution with increasing size of the obstacle.

Figure 2 shows the flow lines, bow shock and obstacle of the Spreiter et al. (1966) solution. Also shown are convected magnetic field lines. In this model the magnetic field exerts no force on the flow. The shaded region is the foreshock region and the magnetosheath behind the quasi-parallel shock.

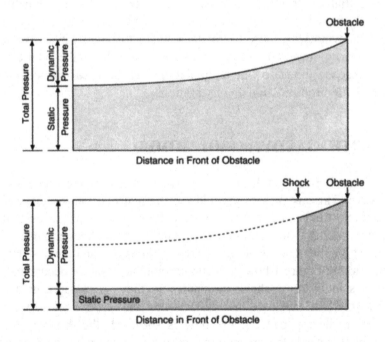

Figure 1. Flow deflection in subsonic (top) and supersonic (bottom) situations. When the static pressure well exceeds the dynamic pressure (i.e. in subsonic flow) a pressure gradient can provide the needed force to deflect the flow. When the dynamic pressure exceeds the static pressure (i.e. in supersonic flow) the needed pressure gradient is produced with the aid of a standing shock that heats the flow.

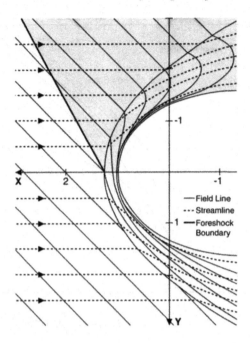

Figure 2. The flow deflection and convected magnetic field in the Spreiter et al. (1966) gas-dynamic calculation. The empirically derived foreshock region is shaded but cannot be directly derived from the code.

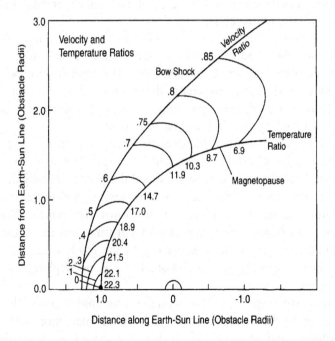

Figure 3. The velocity and temperature ratios in the gasdynamic solution.

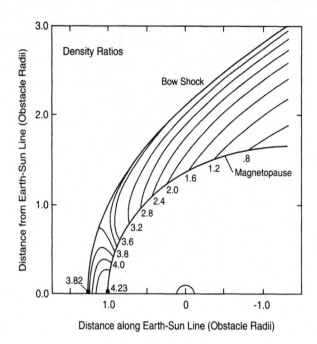

Figure 4. The number density ratio in the gasdynamic solution.

These regions are observed to be associated with certain angles ($\leq 45°$) between the interplanetary magnetic field and shock normal but are not produced explicitly by the code.

Figure 3 shows the iso-contours of velocity and temperature ratios (relative to the solar wind values) for the gasdynamic model. In gasdynamics the two ratios have the same iso-contours. The temperature peaks at the subsolar point and decreases along the flanks while the velocity ratio increases from zero at the subsolar point. In short, the shocked solar wind is expanding and cooling as it flows around the obstacle.

Figure 4 shows contours of the density. There is a strong compression in the subsolar region where the temperature is high. This also contributes to the deflecting pressure gradient. As expected the plasma density decreases as the flow moves away from the subsolar region but there is no near-magnetopause depletion of plasma until down the flanks. The highest densities are just behind the shock at all local times except in the subsolar region. While away from the subsolar region the gasdynamic solution does produce a near-magnetopause depletion, there is no depletion seen in the subsolar region.

In closing, we emphasize how valuable this simple model has been in teaching us about the behavior of the solar wind interaction with planetary magnetospheres and ionospheres. It has been used to determine where particles would go that leaked from the magnetosphere (Luhmann et al.,

1984b) and where antiparallel merging on the magnetopause would occur (Luhmann et al., 1984a). It provided rapid, realistic magnetosheath and shock properties well before computers were able to provide equivalent MHD calculations enabling early space measurements to be rationally interpreted.

3. THE MAGNETOHYDRODYNAMIC MODEL

The addition of magnetic forces in the magnetohydrodynamic model allows the obstacle to be a compressible magnetic dipole field and produces a more realistic magnetosheath structure (e.g., Wang et al., 2003a). While

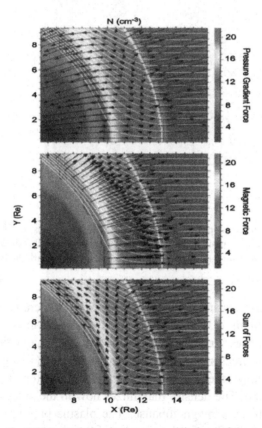

Figure 5. Forces in the MHD simulation in the equatorial plane of the magnetosheath drawn with arrows on a color contour background of the plasma density. The top panel shows the plasma pressure gradient force. The middle panel shows the magnetic force. The bottom panel shows the sum of the two forces. The magnetic field is northward perpendicular to the plane of the figure. Upstream solar wind density is 5 cm^{-3}. Stream lines are shown in white; the last closed field line is pink.

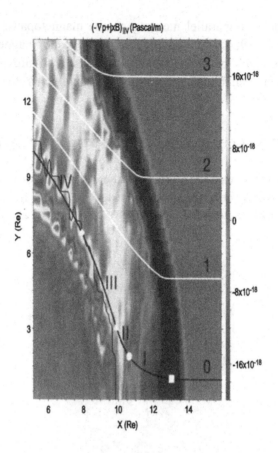

Figure 6. Net MHD force ($-\nabla p + \mathbf{j} \times \mathbf{B}$) resolved along streamlines, i.e. $\parallel$V. Units of color bar are pascals/m. Color contours show net force. Numbers along the streamline marked '0' indicate regions with different stress features. The white symbols mark the transition from one regime to another.

anisotropic plasma pressure can be treated in MHD generally models assume isotropic maxwellian distributions as in the gasdynamic case. Even so important differences arise. Figure 5 shows the magnetic and plasma pressure gradient forces separately and then combined (Wang et al., 2003b). The top panel shows the pressure gradient force on top of a color-coded density background. The density maximizes now in the outer magnetosheath and not at the subsolar magnetopause. The plasma pressure gradient force acts to slow down the flow in the vicinity of the shock so the arrows are all outward there. Where the density is dropping the pressure gradient reverses and the force pushes in toward the magnetopause while in the gasdynamic case the force is away from the magnetopause everywhere. The magnetic force provides the missing force and adds a twist of its own. The middle panel of Figure 5 shows the magnetic force and it is "outward" everywhere.

It adds to the deceleration of the flow near the shock and overcomes the inward force of the plasma pressure gradient in the inner magnetosheath. The sum of the two forces in the bottom panel show that plasma decelerates everywhere along the subsolar streamline but that the force turns to parallel to magnetopause at an increasingly greater distance from the magnetopause as one moves away from the subsolar point. Thus a plasma depletion layer is formed all the way to the subsolar point.

In Figure 6 we look at the force with color contours rather than arrows. Here the force is resolved along the streamlines. The deep blue color shows that the flow is decelerated at and behind the shock. The region of deceleration extends close to the magnetopause in the subsolar region but elsewhere there is a thick acceleration region near the magnetopause that carries material away from this region and creating a plasma depletion layer. We do not display here, but do recommend also using the N/B ratio that is a measure of field line stretching. This parameter can be used for an alternate definition of the plasma depletion layer.

There are three propagating waves in a magnetized plasma: fast, intermediate and slow. To create an arbitrary perturbation in the plasma requires all three waves to be produced. Observations (Song et al., 1990; 1992) suggest that a slow-mode, standing shock wave is seen in the subsolar region near the plasma depletion layer so it is reasonable to ask if such a standing shock is seen in the simulations. Figure 7 shows the phase and group velocity of the slow mode wave when the Alfven speed equals 80% of the sound speed. The slow mode wave does not propagate perpendicular to the magnetic field. The group velocity plot shows that the energy of the wave is strongly guided by the magnetic field.

We need to demonstrate that slow-mode waves are supported by the MHD simulations. To do this we create a plasma box and perturb it with a pressure pulse as shown in Figure 8 (Wang et al., 2003c). Initially there is no flow in this plasma box. A fast mode wave has a positively correlated magnetic field and pressure. A slow mode wave has a negatively correlated magnetic field and plasma pressure. We can see from the lower two panels of Figure 8 that the code does resolve both the fast and slow waves equally well. Nevertheless, in none of the MHD runs to date (Wu, 1992; Ogino, 1992; Lyon, 1994; Siscoe et al., 2002; Wang et al., 2003c) has evidence been observed for a standing slow mode shock, leaving some mystery about the physical cause of the features observed by Song et al. (1990; 1992).

In this paper we have discussed only one application of the MHD model, to the Earth's magnetosheath and stressed the physics near the subsolar magnetopause. In fact the MHD model has been used for a wide range of problems including comets where the obstacle is provided by mass-loading when the expanding cometary atmosphere is ionized (e.g. Fedder et al., 1986).

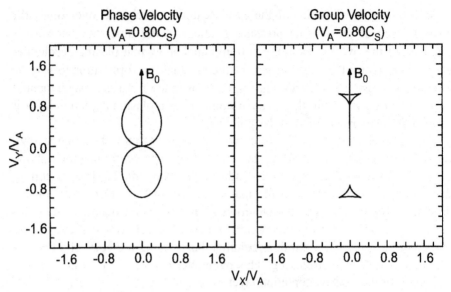

Figure 7. Phase velocity and group velocity of the slow mode wave when the Alfven speed is 80% of the sound speed. Direction of the magnetic field is upward.

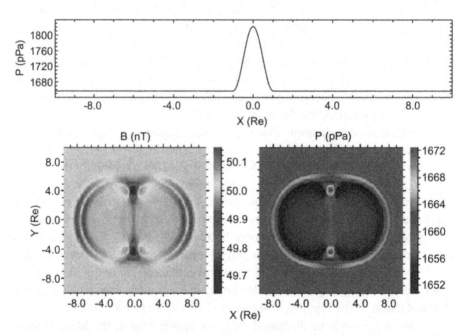

Figure 8. Verification that the MHD code can produce a slow mode wave. The pressure pulse in the top panel was introduced in the center of a square grid of stationary plasma. The two panels below show the magnetic field strength and pressure expected for the fast mode (positively correlated B and P) and slow mode (anticorrelated B and P).

The MHD code, like the gasdynamics code, has proven invaluable in interpreting space measurements (see e.g. Fedder et al., 1997; Russell et al., 1998a). Most recently it has been used to explore how the size of the reconnection region on the dayside magnetopause might change with dipole tilt (Russell et al., 2003).

While the gasdynamic solution is strictly self-similar because there is no physics that depends on absolute scale, some MHD models have introduced absolute scales in two ways. In parameterizing resistivity they may use a length scale that is absolute and in treating the inner boundary condition, the coupling to the ionosphere or planet, a length scale may be introduced. However, we expect these introductions of scale size to have relatively minor effects on the solution and the MHD models will be nearly self-similar. It is clear, though, that kinetic processes are important in magnetospheres and we must go beyond the MHD models. When we do this we find that much of the physics of the solar wind interaction depends on the relative size of the curvature of the obstacle and the particle kinetic motion.

4.　　HYBRID SIMULATIONS

To treat kinetic processes in the global interaction the hybrid model follows the ion motion and treats the electrons as a massless fluid. This hybrid approach reveals much about the interplay of the global interaction and kinetic processes. In fact we find that the kinetic processes depend on the scale size of the global interaction. In the simulations presented herein we use simulations with a two-dimensional box and three-dimensional fields. The Y direction is parallel or antiparallel to the IMF and X is parallel to the flow. The obstacle is a two-dimensional dipole (Ogino, 1993) as would be created by two antiparallel currents flowing in the Z direction. Resistivity has been added uniformly to produce a steady state. The bodies have no ionosphere, no atmosphere and no rotation. The scale size of the obstacle, D_p, is taken to be the distance at which the magnetic pressure of the dipole is equal to the dynamic pressure of the solar wind flow. This scale size is then compared to the ion inertial length that is equal to the ion gyroradius for a beta equal one plasma. We have examined interactions with D_p ranging from 0.05 to 63 ion inertial lengths. For reference Mercury would typically have a D_p of 85, Earth 640 and Jupiter 5800 ion inertial lengths.

Figure 9 gives a quick summary of the results of this investigation of the effect of scale size on the interaction (Omidi et al., 2002; Blanco-Cano et al., 2003; Omidi et al., 2003). The two columns of panels are colored with the B_x component on the left and the density on the right. In all cases shown in Figure 9 the interplanetary magnetic field is northward. The B_x component indicates the bending of the magnetic field around the obstacle. When D_p is very small (top) a whistler wave is created. The field is bent but there is no density perturbation. When D_p is larger but still less than an ion inertial length both a whistler mode wake and a magnetosonic mode wake are formed. This shows up in both the magnetic field and the density.

When D_p is close to unity the interaction begins to produce some magnetospheric characteristics (Omidi et al., 2003). There is a fast mode wave that is the precursor to the magnetosheath and bow shock and a slow mode wave in the center of the tail that is the precursor to the plasma sheet.

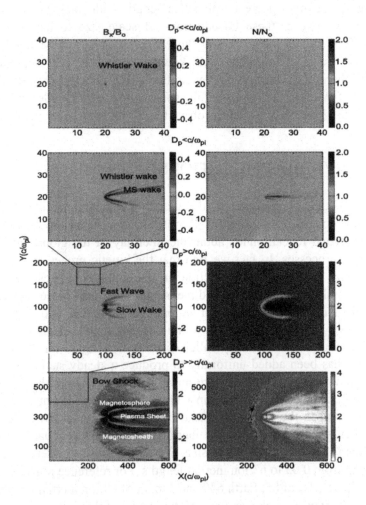

Figure 9. Magnetic field component along the flow and the plasma density for four different hybrid runs of increasing obstacle size illustrating the cross scale coupling of kinetic and global scales. The simulations use a two-dimensional dipole field equivalent to that produced by two current carrying wires into the page. The box size is longer in the bottom two cases as indicated. Interplanetary magnetic field is northward.

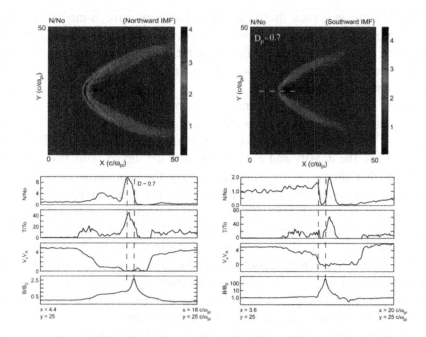

Figure 10. The hybrid solution for the interaction when the scale size of the obstacle D_p is approximately equal to the ion inertial length. Cuts through the subsolar points for both northward and southward IMF are shown. Vertical dashed lines are separated by D_p. Panels with the density color coded in the background show the effect of the magnetic field direction in excavating a cavity when the IMF is southward. Note that the color bars are different for the two cases.

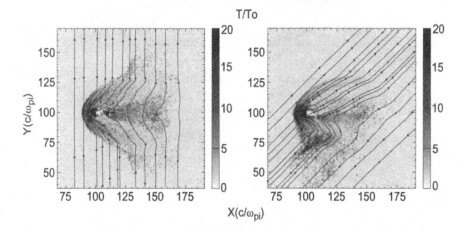

Figure 11. The temperature obtained in the hybrid simulation for D_p greater than the ion inertial length for two magnetic field orientations showing the formation of reflected ions.

When D_p is much greater than unity, there is a bow wave, a magnetosheath, a magnetopause, a magnetosphere and a plasma sheet. Moreover, periodic reconnection occurs producing flux transfer events and plasmoids. We can think of these transitions as phase changes in the interaction. Some of these states are worth further attention.

Figure 10 shows the precursor to reconnection that appears for D_p approximately one ion inertial length. In the lefthand panels where the IMF is parallel to the magnetic field on the leading edge of the dipole there is a bow wave in the density and a pile up around the dipole. This is evident in the panel that shows the measurements along the subsolar streamline. The righthand panels show the measurements along the same cut, and the density contours for southward IMF. There is now a density void at the subsolar point. We emphasize that in reconnection we expect that the electron kinetics will also play a role, especially in the initiation of reconnection but it is interesting that ion dynamics alone gives a reconnection like signature. Since near the polar cusp the radius of curvature of the magnetopause can be relatively small, this apparent facilitation of reconnection for small radius of curvature may be important in the Earth's magnetosphere.

Upstream ions are produced at this relative scale size as well. This is best seen in the temperature shown in Figure 11. The bow wave and tail are quite evident but so too are the ions reflected by the bow wave and flowing

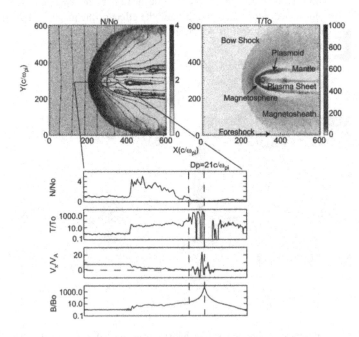

Figure 12. The density and temperature in the D_p greater than 20 simulations. The cut is along the nose.

upstream along the magnetic field. The righthand panel tilts the northward interplanetary magnetic field to show that upstream of the quasi-parallel shock ions are moving back against the flow.

When the scale size exceeds 20 times the ion inertial length the interaction creates many phenomena familiar to magnetospheric physicists. As shown in Figure 12 reconnection takes place periodically leading to flux transfer events and plasmoid formation in the tail. There is a bow shock, a magnetosheath, a magnetopause, a polar cusp, and a tail with an embedded plasma sheet. Radiation belts also appear. Figure 13 shows these radiation belts as evident in the plots of temperature and density. Because of the two dimensional nature of the dipole magnetic field these trapped ions stay principally on the upstream side and do not have the opportunity to complete a ring current around the Earth.

At this writing, three-dimensional hybrid simulations corresponding to these examples have been run to steady-state conditions up to a D_p slightly greater than unity validating many of the results of the two-dimensional code. Of course some phenomena, such as the formation of a trapped radiation belt, have significant differences in two and three dimensions. Table 1 summarizes the variations in upstream plasma changes, waves and features within the magnetosphere as the scale size is changed. The hybrid simulations produce a very realistic magnetosphere with kinetic processes clearly present and important. The hierarchy of processes that arise at different relative scales illustrates the sensitivity of the different kinetic processes to the coupling between scales. Hybrid simulations have also been successfully applied to mass-loading obstacles such as unmagnetized planets (Brecht, 1990) and comets (Lipatov et al., 2002).

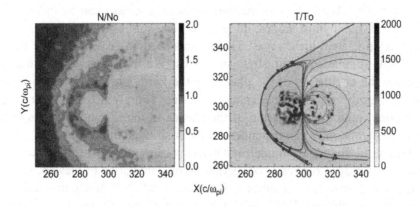

Figure 13. The density and temperature for D_p greater than 20 showing the formation of the radiation belts in the hybrid code.

Table 1. Summary of features seen in hybrid simulations as a function of scale size of the obstacle.

D_p	Upstream Plasma Changes	Waves	Magnetospheric features
$\ll c/\omega_{pi}$	None	Whistler	None
$< c/\omega_{pi}$	Some flow deflection, n increases, and v decreases at $r > D_p$	Whistler wake, fast and slow magnetosonic waves at wake edges	Precursor of a plasma tail
$> c/\omega_{pi}$	Pileup at $r \sim D_p$ Flow deflection, n, T, B increase, v decreases Reflected ions	Fast mode bow wave upstream Slow mode wake	Particle acceleration at dipole (Particle trapping at belts) Tail with hot plasma Reconnection precursor
$\gg c/\omega_{pi}$	Flow modified and deflected at bow shock at $r \gg D_p$ n, T, B increase, v decreases Magnetosheath	Bow shock	Magnetopause, cusp Tail with plasma sheet Radiation belts Reconnection, leading to ion acceleration and magnetic island formation

5.   THE EFFECT OF THE SOLAR WIND INTERACTION INSIDE THE MAGNETOSPHERE

The processes occurring inside the magnetosphere are controlled at the Earth by the mass, momentum and energy that enter the magnetosphere from the solar wind. While approximations to these processes can be included in MHD and the hybrid codes, in general we still use models based on observation to treat these areas. Thus in this section we abandon the use of numerical simulation to illustrate processes. We use observational data to compare the circulation of the plasma in the Earth's magnetosphere, which can to zeroth order be understood as due entirely to the solar wind interaction, with the circulation of the plasma in the jovian magnetosphere, which to zeroth order is independent of the solar wind interaction.

Figure 14 shows the model of Dungey (1961; 1963) for how the solar wind drives magnetospheric convection, or the circulation of plasma, for two diametrically-opposite conditions: due northward IMF and due southward IMF. The northward IMF case has reconnection of oppositely-directed magnetic fields behind the cusp. If only magnetic tension were important, the flow would be across the polar cap toward the sun and magnetic flux and plasma would return to the nightside at lower latitudes. Song and Russell (1992) have pointed out that this process also captures any solar wind momentum originally on the reconnected flux tube. This capture also assists in the circulation of plasma and could create the low latitude boundary layer.

When the interplanetary magnetic field is southward, reconnection occurs on the closed dayside field lines at low latitudes. This process leads to flow over the polar cap away from the sun. Reconnection of oppositely-directed field lines in the magnetotail completes the circulation by returning magnetic flux from the tail to the night side magnetosphere. This process provides an excellent mechanism for extracting momentum from the magnetosheath. Behind the polar cusp the curvature in the magnetic field is such as to slow down the flow. This extracts momentum flux from the solar wind and adds it to the tail in the form of magnetic energy. Magnetic energy can be tapped relatively rapidly for substorms from this tail reservoir.

We note that circulation in the magnetosphere is quite unsteady. This is generally related to the variability of the interplanetary magnetic field (Russell and McPherron, 1973; McPherron, 1991; Raeder et al., 2001).

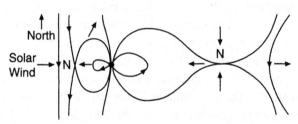

Figure 14. Dungey's (1961; 1963) mechanism for coupling solar wind momentum to the Earth's magnetosphere. (Top) Interplanetary field is northward, reconnection takes place behind the cusp and flow is sunward over the poles. (Bottom) Interplanetary magnetic field is southward, re-connection takes place in the subsolar region and drives flow tailward over the polar caps.

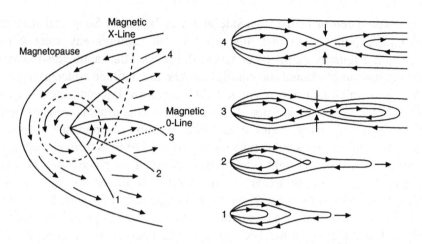

Figure 15. Vasyliunas' (1983) model for circulation in the jovian magnetosphere. The magnetosphere is driven into corotation by the ionosphere that corotates with the planet. Mass added to the magnetosphere by Io stretches field lines until reconnection occurs in the nighttime hemisphere. Shown on the right is the temporal evolution of these magnetic field lines. The net result of this reconnection is the formation of a magnetic island containing ions that are removed thereby from the magnetosphere. Emptied flux tubes can then return to the interior of the magnetosphere to eventually be mass-loaded once again.

When steady conditions endure for many hours, the activity will become quiet if the field is northward. If the field is strong, steady, and southward a magnetic storm will ensue.

In all of our discussion to date we have ignored the upper atmosphere and ionosphere but we know that ionospheric plasma can be found in the terrestrial magnetosphere. To see how the ionosphere can control a magnetosphere we turn to the jovian system.

For the jovian magnetosphere the solar wind mainly sets the size of the magnetosphere but appears not to be responsible for energization. The ionosphere corotates with the planet as does the Earth's ionosphere, but on Jupiter the ionosphere is able to enforce corotation of the magnetospheric plasma over a wide radial extent. This enforced corotation is especially important in the vicinity of the moon, Io, that supplies about a ton of ions per second to the magnetosphere. This material is accelerated to corotational energy and begins a slow and then more rapid outward radial drift in order to maintain a steady state ion content. This circulation pattern was first proposed by Vasyliunas (1983) and is shown in Figure 15. The plasma added to the field lines stretches the field lines as well as pulls them outward. Eventually the field lines stretch to the breaking point. The time sequence of magnetic meridians on the right shows the evolution of reconnection and the formation of a magnetized island of ions. These ions leave the system down the tail. While the picture drawn here is for a steady state, the process is episodic like terrestrial substorms (Russell et al., 1998b). Analogous disturbance to the magnetic storm has not been identified at Jupiter but one could imagine that a strong period of iogenic volcanism could produce a strong ring current in the jovian magnetosphere.

To complete the convection pattern magnetic flux must find its way back to Io to be mass-loaded again. The part of the flux tube inside the reconnection point is mainly devoid of plasma. These empty flux tubes are buoyant and move inward to be repopulated with plasma by Io. Hence at Jupiter the ionosphere and Io work together to power the circulation of plasma and provide the energy of magnetospheric process. The solar wind acts mainly to determine the size of the magnetosphere. In contrast the Earth's magnetosphere is very much controlled by the solar wind.

6.   SUMMARY AND CONCLUSIONS

At the beginning of the space age only simple models such as the gasdynamic codes were available to guide our understanding of magnetospheric processes. These models could produce a bow shock and magnetosheath but did not produce the magnetic cavity. Over the next couple of decades MHD models were perfected to produce a more realistic magnetosheath and a magnetic cavity. Reconnection is present in these models but the physical processes by which it occurs are different from the kinetic processes that we believe are at work.

The gasdynamic models and MHD models (to zeroth order) are self-similar and do not produce different physics as their scale size changes. Hybrid models now exist that include ion motion and modern computers are large enough and fast enough that global simulations including ion effects can be run cost effectively. These models show that scale size does matter to the physics of the system. Hybrid simulations produce whistler-mode waves, magnetosonic waves, the bow shock, the magnetosheath, upstream waves, the magnetopause, radiation belts, the plasma sheet and reconnection. It is possible that the approximation used in these simulations has produced physics that differs from the natural ones.

Our comparison of aspects of the terrestrial and jovian magnetospheres highlights the importance of the ionosphere in producing corotation. It also can provide a source of plasma, as it does on Earth. In contrast in the jovian magnetosphere Io provides most of the plasma and energizes the system by tapping the rotational energy of the planet. In short, while the solar wind interaction controls much of the physics of the terrestrial magnetosphere, there are other possible energy sources and the jovian magnetosphere is much different inside than that of the Earth.

REFERENCES

Blanco-Cano, X., Omidi, N., and Russell, C. T., 2003, Hybrid simulations of solar wind interaction with magnetized asteroids: Comparison with Galileo observations near Gaspra and Ida, *J. Geophys. Res.* **108**:1216, doi: 10.1029/2002JA009618.

Brecht, S. H., 1990, Magnetic asymmetries of unmagnetized planets, *Geophys. Res. Lett.* **17**:1243–1246.

Dungey, J. W., 1961, Interplanetary magnetic field and the auroral zones, *Phys. Rev. Lett.* **6**:47–48.

Dungey, J. W., 1963, The structure of the exosphere or adventures in velocity space, in: *Geophysics: The Earth's Environment*, C. Dewitt, J. Hieblot and A. Lebeau, eds., New York, Gordon and Breach, pp. 505–550.

Fedder, J. A., Lyon, J. G., and Giuliani, J. L. Jr., 1986, Numerical simulations of comets: Predictions for Comet Giacobini-Zinner, *EOS Trans. AGU* **67**:17–18.

Fedder, J. A., Slinker, S. P., Lyon, J. G., Russell, C. T., Fenrich, F. R., and Luhmann, J. G. , 1997, A first comparison of POLAR magnetic field measurements and magnetohydro-dynamic simulation results for field-aligned currents, *Geophys. Res. Lett.* **24**:2491–2494.

Lipatov, A. S., Motschman, U., Bagdonat, T., 2002, 3D hybrid simulations of the interaction of the solar wind with a weak comet, *Planet. Space Sci.* **50**(4):403–411.

Luhmann, J. G., Walker, R. J., Russell, C. T., Crooker, N. U., Spreiter, J. R., and Stahara, S. S., 1984a, Patterns of potential magnetic field merging sites on the dayside magnetopause, *J. Geophys. Res.* **89**:1739–1742.

Luhmann, J. G., Walker, R. J., Russell, C. T., Spreiter, J. R., Stahara, S. S., and Williams, D. J., 1984b, Mapping the magnetosheath field between the magnetopause and bow shock: Implications for magnetospheric particle leakage, *J. Geophys. Res.* **89**:6829–6834.

Lyon, J. G., 1994, MHD simulations of the magnetosheath, *Adv. Space Res.* **14**(7):21–28.

McPherron, R. L., 1991, Physical processes producing magnetospheric substorms and magnetic storms, in: *Geomagnetism*, J. Jacob, ed., Academic Press, p.593.

Omidi, N., Blanco-Cano, X., Russell, C. T., Karimabadi, H., and Acuna, M., 2002, Hybrid simulations of solar wind interaction with magnetized asteroids: General characteristics, *J. Geophys. Res.* **107**:1487, doi: 10.1029/2002JA009441.

Omidi, N., Blanco-Cano, X., Russell, C. T., and Karimabadi, H., 2003, Dipolar magneto-spheres and their characterization as a function of magnetic moment, *Adv. Space Res.*, in press.

Ogino, T., 1993, Two dimensional MHD code, in: *Computer Space Plasma Physics: Simulations and Software*, H. Matsumoto and Y. Omura, ed., 161, Terra, Tokyo.

Raeder, J., McPherron, R. L., Frank, L. A., Peterson, W. R., Sigwarth, J. B., Lu, G., Kokubun, S., Mukai, T., and Slavin, J. A., 2001, Global simulation of the geospace environment modeling substorm challenge event, *J. Geophys. Res.* **106**:381–396.

Raeder, J., 2003, Global geospace modeling: Tutorial and review, in: *Space Plasma Simulation*, J. Buchner, C. T. Dum, and M. Scholer, eds., 615, Springer Verlag, Heidelberg.

Russell, C. T., and McPherron, R. L., 1973, The magnetotail and substorms, *Space Sci. Rev.* **15**:205–266.

Russell, C. T., Fedder, J. A., Slinker, S. P., Zhou, X-W., Le, G., Luhmann, J. G., Fenrich, F. R., Chandler, M. O., Moore, T. E., and Fuselier, S. A., 1998a, Entry of the POLAR spacecraft into the polar cusp under northward IMF conditions, *Geophys. Res. Lett.* **25**:3015–3018.

Russell, C. T., Khurana, K. K., Huddleston, D. E., and Kivelson, M. G., 1998b, Localized reconnection in the near Jovian magnetotail, *Science* **280**:1061–1064.

Russell, C. T., Wang, Y. L., and Raeder, J., 2003, Possible dipole tilt dependence of dayside magnetopause reconnection, *Geophys. Res. Lett.* **30**:1037, doi: 10.1029/2003GL017725.

Siscoe, G. L., Crooker, N. U., Erickson, G. M., Sonnerup, B. U. O., Maynard, N. C., Schoendorf, J. A., Siebert, K. D., Weimer, D. R., White, W. W., and Wilson, G. R., 2002, MHD properties of magnetosheath flow, *Planet Space Sci.* **50**(5-6):461–471.

Song, P., and Russell, C. T, 1992, Model of the formation of the low-latitude boundary layer for strongly northward interplanetary magnetic field, *J. Geophys. Res.* **97**:1411–1420.

Song, P., Russell, C. T., Gosling, J. T., Thomsen, M., and Elphic, R. C. , 1990, Observations of the density profile in the magnetosheath near the stagnation streamline, *Geophys. Res. Lett.* **17**:2035–2038.

Song, P., Russell, C. T., and Thomsen, M. F., 1992, Slow mode transition in the frontside magnetosheath, *J. Geophys. Res.* **97**:8295–8305.

Spreiter, J. R., Summers, A. L., and Alksne, A. Y., 1966, Hydromagnetic flow around the magnetosphere, *Planet. Space Sci.* **14**:223–253.

Vasyliunas, V. M., 1983, Plasma distribution and flow, in: *Physics of the Jovian Magnetosphere*, A. J. Dessler, ed., London, Cambridge University Press, pp. 395–453.

Wang, Y. L., Raeder, J., Russell, C. T., Phan, T. D., and Manapat, M., 2003a, Plasma depletion layer: Event studies with a global model, *J. Geophys. Res.* **108**:1010, doi: 10.1029/2002JA009281.

Wang, Y. L., Raeder, J., and Russell, C. T., 2003b, Plasma depletion layer: Magnetosheath flow structure and forces, *Annales Geophysicae*, in press.

Wang, Y. L., Raeder, J., and Russell, C. T., 2003c, Plasma depletion layer: The role of the slow mode waves, *Annales Geophysicae*, submitted.

Winske, D., and Omidi, N., 1996, A nonspecialist's guide to kinetic simulations of space plasmas, *J. Geophys. Res.* **101**:17,287–17,304.

Wu, C. C., 1992, MHD flow past an obstacle: Large scale flow in the magnetosheath, *Geophys. Res. Lett.* **19**:87.

AN OVERVIEW OF NEW CONCEPTS DEDUCED FROM INTERBALL SOLAR WIND INVESTIGATIONS

G. N. Zastenker

Space Research Institute, RAS, Profsouznaya Str. 84/32, 117997, Moscow, Russia

Abstract: Several new features of the solar wind were found in the Interball project by multipoint observations and using high-resolution plasma measurements onboard Interball-1/Magion-4 satellites. These results allow us to suggest some new concepts of solar wind propagation and its interaction with the magnetosphere, namely:

- dimensions and a persistent time of the middle-scale structures,
- large and very sharp plasma density changes on borders of small-scale structures, significant inclinations of many sharp plasma fronts,
- geoeffectivity of sharp changes of the plasma dynamic pressure,
- magnetic field and plasma in phase fast variations in the foreshock,
- large amplitude, low and high-frequency plasma and magnetic field variations in the magnetosheath; their origin, dependence on the IMF direction and comparison with MHD models.

Key words: solar wind; solar-terrestrial relations; interplanetary magnetic field; foreshock; magnetosheath.

1. INTRODUCTION

An important part of the International Solar-Terrestrial Program was the Interball project. In frame of this project, the Interball-1 satellite and its subsatellite Magion-4 were launched in August 3, 1995 into the elliptic orbit with the apogee of 30 Re (Earth radii) and perigee about 0.1-3 Re and during the 1995-2000 years performed the measurements in the solar wind (SW), magnetosheath (MSH), and magnetotail (Galeev et al., 1996). For these measurements we used the instruments: VDP on Interball-1 and VDP-S on

J. –A. Sauvaud and Z. Němeček (eds.),
Multiscale Processes in the Earth's Magnetosphere:From Interball to Cluster, 37-56.
© 2004 *Kluwer Academic Publishers. Printed in the Netherlands.*

Magion-4 (Safrankova et al., 1997; Nemecek et al., 1997b; Zastenker et al., 2000a).

Based on these measurements, many previously unknown and important results that allow us to suggest some new concepts of the solar wind and magnetosheath were obtained. The reasons why these investigations are so successful are:

– systematic measurements both magnetic field and plasma fluxes with a high (one second or better) time resolution onboard the Interball-1 and Magion-4 satellites;
– using of the multipoint comparison of simultaneous plasma and magnetic field measurements onboard widely separated (up to $1.5*10^6$ km) spacecraft ACE, WIND, IMP 8, Geotail, and Interball-1 and closely (about 1000 km) separated spacecraft pair of Interball-1/Magion-4.

In this paper, we present several new results of the solar wind investigations based on above mentioned advantages, namely, determinations of the middle-scale structures, their persistence time, and dimensions (Section 2), descriptions of the features of large and sharp plasma disturbances (Section 3), observations of fast solar wind variations in the foreshock (Section 4), and evaluation of characteristics of low and high-frequency plasma and magnetic field variations in the magnetosheath (Section 5).

2.   DIMENSIONS AND PERSISTENT TIME OF THE MIDDLE-SCALE SOLAR WIND STRUCTURES

This investigation was done by calculation of the cross-correlations of plasma measurements from several wide-separated spacecraft – Interball-1, IMP 8, WIND. It is interesting that the correlation of the interplanetary magnetic field (IMF) were studied in the 80's by ISEE-1, 3 and IMP 8 data (Crooker et al., 1982) but the systematic investigations of solar wind plasma correlations began only since comparison of the Interball-1 data with other spacecraft (Nemecek et al., 1997a; Paularena et al., 1997; Zastenker et al., 1998; Paularena et al., 1998). In Figure 1, we present the typical example of a good correlation between the solar wind ion flux and velocity measurements of the Interball-1 and WIND spacecraft separated by about 65 Re along Xgse and about 10 Re along Ygse. WIND data were time-shifted by advection to the Interball-1 position. It can be seen that in this case the correlation level is very high - up to 0.96-0.97 for the ion flux and 0.88-0.90 for the solar wind velocity. The correlations are near the same as for Interball-1/WIND as for Interball-1/IMP 8 pairs in spite of very different positions.

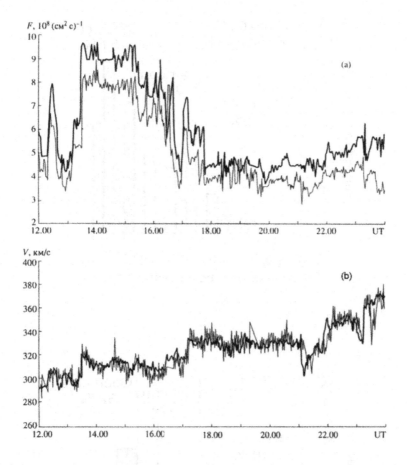

Figure 1. An example of good solar wind ion flux (a) and bulk velocity (b) correlations of Interball-1 (thick lines) and WIND (thin lines) data on 08.04.96.

However, in the real solar wind and IMF conditions, the correlation level changes in a broad range. Sufficiently large statistics calculated from the set of about five hundreds of 6-hour segments of measurements with 1-minute resolution for the ion flux and magnetic field are shown in Figure 2.

It can be seen that a good correlation (>0.8) takes place only for 30-50% of segments, and a very poor correlation (<0.5) is observed for 10-20% of segments. The average correlation coefficients are equal to about 0.75 for the ion flux and 0.71 for the magnetic field amplitude (Zastenker et al., 1998). Moreover, it is important that sometimes the plasma and magnetic field correlations are significantly different for the same segment.

Investigations of the factors controlling the correlation level have shown that the most important parameter is the amplitude of its variations. We

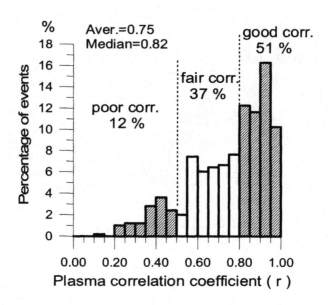

(a)

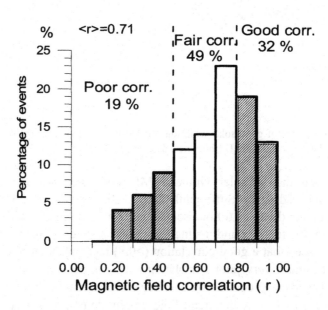

(b)

Figure 2. Statistics of solar wind ion flux (a) and IMF magnitude (b) correlations of Interball-1, IMP 8, and WIND data.

found that correlations became large enough if the relative standard deviations (i.e., the ratio of standard deviations to the average value of a parameter) of the ion flux or plasma density were higher than 0.2-0.3 (Paularena et al, 1998). It means that the large variations are associated with more global solar wind structures and small variations are local. Thus, the Space Weather predictions using observations onboard distant monitors are reliable enough only for large variations of the solar wind and IMF parameters.

We used the multifactor analysis to find the dependence of correlations on the spacecraft separation. A plot of the correlation coefficients versus dXgse (the spacecraft distance along the Sun-Earth line) allows us to estimate the persistence time of middle-scale structures. It was obtained that, in average, such structures are stable on the way from the L1 point to Earth (about $1.5*10^6$ km). An influence of dYZgse (the spacecraft distance in the plane perpendicular to the Sun-Earth line) allows us to estimate the dimensions of homogeneity of the solar wind middle-scale structures in this plane. It was shown that, in average, the correlations decrease only slightly as the separation in the YZgse plane increases up to 110 Re (Zastenker et al., 2000b).

Based on these investigations, we obtained following conclusions for the topics of the Section 2:
- the average level for both ion flux and IMF magnitude correlation coefficients between the well-separated spacecraft is about 0.70-0.75;
- the level of the correlations increase for large amplitudes of plasma and magnetic field variations;
- the persistence time of the middle-scale solar wind structures usually exceeds 1 hour;
- the correlation length of solar wind structures in the plane perpendicular to the Sun-Earth line is, in average, longer than 200-500 Re (i.e., significantly longer than the cross-section of the magnetosphere).

3.   SHARP AND LARGE PLASMA CHANGES AS THE BORDERS OF THE SOLAR WIND SMALL-SCALE STRUCTURES

An interesting feature of the solar wind is the sharp (from several seconds to several minutes of duration) and large (from tens to hundreds of percents) increase or decrease of the solar wind ion flux or density. This phenomenon was not studied before mainly because of a poor time resolution of plasma measurements. We made the investigations of such features by selection of the sharp (shorter than 10 minutes) and large (more than 20%) changes of

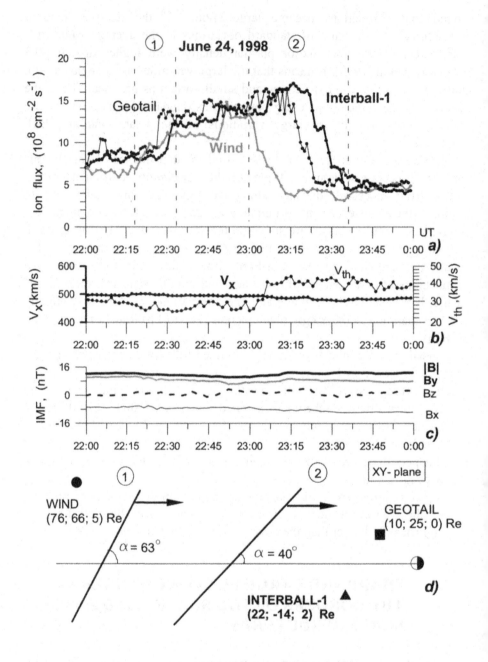

Figure 3. An example of the sharp and large solar wind ion flux pulse recorded by Interball-1, Geotail, and WIND (a); time behavior of the bulk Vx and thermal Vth velocities (b); time behavior of the IMF strength and components (c); positions of the spacecraft in the XYgse plane and calculated inclinations of both fronts (d).

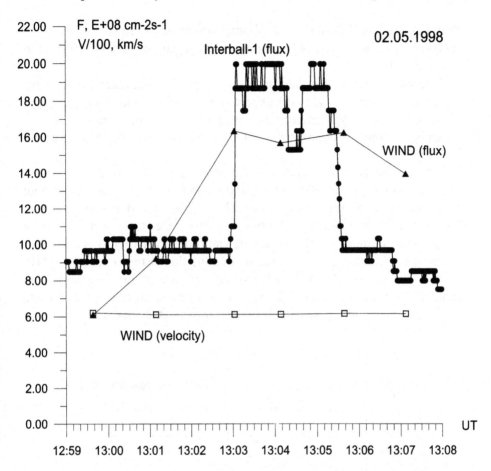

Figure 4. An example of very sharp solar wind ion flux pulse recorded by Interball-1 with the 1-s time resolution and by WIND with the 90-s resolution (the behavior of the velocity is shown).

the ion flux from systematic Interball-1 measurements with high-time resolution and by comparison them with other spacecraft (ACE, IMP 8, WIND, Geotail) data (Dalin et al., 2002a). A typical example of the large and sharp solar wind ion flux pulse is presented in Figure 3. This pulse was observed "simultaneously" by Interball-1, Geotail, and WIND spacecraft (all data are shifted by propagation time to the Interball-1 position). Spacecraft positions are shown in Figure 3d. It is clear that the same ion flux pulse is observed onboard each spacecraft but its fronts (especially the trailing ones) are significantly shifted.

Another example of the very sharp ion flux pulse is presented in Figure 4. A very high-time resolution of Interball-1 measurements (1s) allows us to show that in this event the leading front of the pulse is very short – increase of the flux by a factor of two during 2 s only. It means that the width of such narrow boundary between two plasma structures is equal to about of 1000 km, i.e., for observed conditions it is approximately 10 proton gyroradii only.

It should be noted that the ion flux pulses under investigations are very often an increase/decrease in the plasma density only, while other solar wind parameters (velocity, temperature, magnetic field magnitude, and its components) remain almost constant (see for example, Figures 3b,c and Figure 4).

Using an analysis of systematic solar wind observations of the Interball-1 satellite during 1996-1999, we have found a large number (about 20 thousands) of events related to this subject. Their statistics is presented in Figure 5 as the frequency of observations of the sharp ion flux front dependence on their relative amplitude (ratio of larger (F_2) to smaller (F_1) fluxes at the front) under the condition that the flux change exceeds $0.5*10^8$ cm^{-2}s^{-1}. One can see the smooth but strong decrease of this frequency from about 30 "weak" (with $F_2/F_1 > 1.2$) events per day up to one "strong" (with $F_2/F_1 > 3.0$) event during about 2 days (Riazantseva et al., 2003a).

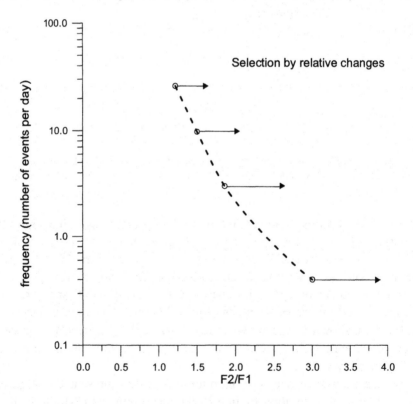

Figure 5. Statistics of the sharp solar wind ion flux increases or decreases; F2 is the large value, F1 is the small value of the flux at the change.

It is well known that the interplanetary magnetic field is, in average, directed along the Parker's spiral. However, the orientation of the solar wind plasma structures is not enough adequately investigated. Some results of the orientation of middle-scale structures were obtained from the correlation study (Richardson et al., 1998; Coplan et al., 2001; Dalin et al., 2002b) but for the sharp plasma fronts there is necessary to use another method. We determined their orientation by an analysis of the time delay between observations of corresponding plasma fronts at two or more well-separated spacecraft in the solar wind (Riazantseva et al., 2003b). The key hypothesis that such fronts are planar on the scales of several tens of Re was checked by comparison between several pairs of measurements. An example of such study is present in Figure 3. It is seen that using the values of observed time delays and spacecraft positions, we estimate the inclinations as 63° and 40° with respect to the Sun-Earth line for leading and trailing fronts of this pulse, respectively.

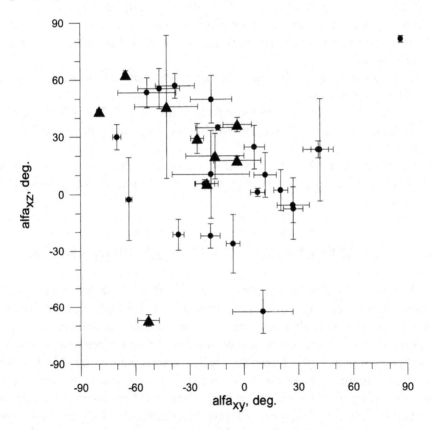

Figure 6. Inclinations of the sharp solar wind ion flux fronts to XYgse (alfaXY) and XZgse (alfaXZ) planes; triangles show the most reliable estimates; error bars in both angles are shown by lines.

Using these delays for three spacecraft data, we can calculate the inclinations of the fronts in a 3D space. Such result is shown in Figure 6. Here the angle alfa$_{XY}$ means the angle between the normal to the front and the XYgse plane. The similar angle, alfa$_{XZ,}$ is the angle to the XZgse plane. It is seen that there are a large scattering of the front inclinations and in many cases the value of inclination is large - up to 60-80°. Usually, the large tilt to one plane coincides with a large one to another plane.

This study shows us that, in spite of the common point of view, in many cases the planes of the sharp and large plasma fronts are not perpendicular to the Sun-Earth line but they are inclined to it at a significant angle.

The special topic of our study was the determination of the influence of the sharp and large solar wind ion flux (i.e., dynamic pressure) changes on the magnetosphere. As it was shown in several our papers, they cause a very strong compression or decompression of the magnetosphere and the large disturbances of several magnetospheric current systems (Sibeck et al., 1996; Riazantseva et al., 2003a). Based on these investigations, we obtained following conclusions for the topics of the Section 3:

– large and sharp ion flux (density) changes are observed rather often in the solar wind;
– the edges of small-scale solar wind structures can be thin as several tens of proton gyroradii;
– in about 50% of the analyzed cases, fronts of ion flux pulses are not perpendicular to the Sun-Earth line but the inclination angle can be less than 60°;
– the sharp ion flux (or dynamic pressure) changes can be very geoeffective and can create significant disturbances of the geomagnetic field.

4.   PLASMA VARIATIONS IN THE FORESHOCK

The bow shock upstream region - the foreshock - ahead of the quasi-parallel bow shock is one of the most interesting but complicated phenomena in a space. It was widely explored in many theoretical and experimental papers (see for example the review of Fuselier (1994) and references therein) but many features of the foreshock are not known well so far. Most of the previous studies concentrated to waves of the magnetic field and backstreaming particles but there is no data based on solar wind plasma fluctuations in the foreshock. Plasma measurements with the high-time resolution onboard Interball-1 and Magion-4 satellites allow us to observe the solar wind fast changes (modification) in the foreshock region. Figure7

INTERBALL-1 May 25, 1997

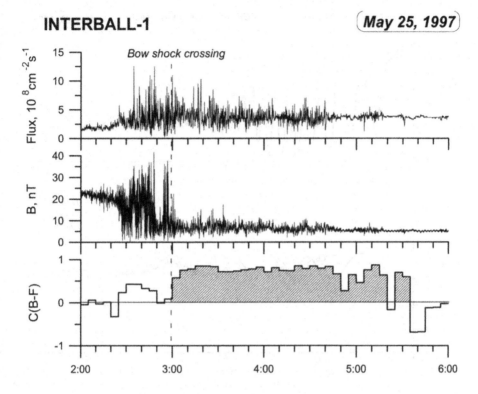

Figure 7. An example of solar wind ion flux and IMF magnitude variations in the foreshock; the windowed correlation between these parameters is shown at the bottom panel.

presents an example of such event of the INTERBALL-1 observations during the period of two hours after the bow shock crossing:
- the solar wind ion flux with 1-s time resolution;
- the magnitude of IMF with the same resolution;
- windowed (on 5-minute intervals) correlation coefficient between the solar wind ion flux and IMF values.

It is seen from Figure 7 that after a crossing of the bow shock by Interball-1 (from the magnetosheath to solar wind) at about 0259 UT, the large variations in the solar wind ion flux and IMF magnitude are observed until 0520 UT with gradually decreasing amplitude. It allows us to identify this interval as a foreshock – a region with large amplitude (tens of percents) and fast (1–100 mHz) fluctuations of the magnetic field and plasma. This high level of the fast solar wind ion flux variations (never published previously) means that the foreshock does modify the solar wind before it gets to the bow shock.

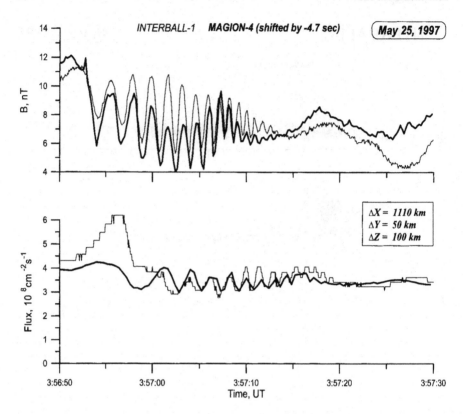

Figure 8. An example of quasi-harmonic oscillations of the solar wind ion flux and IMF magnitude in the foreshock recorded by Interball-1 (thin line) and Magion-4 (thick line); a separation between both satellites is shown in the bottom panel.

The new topic is a study of correlations between field and plasma fluctuations in the foreshock. As can be seen from Figure 7, high positive correlation (about 0.7-0.8) does exist during the whole interval of foreshock observations. Such a phenomenon seems to be unique for the foreshock and may be interpreted as an evidence of a fast mode of the magnetosonic waves generated in this region (Lacombe et al., 1995). It is worth to note that the correlation between the ion flux and magnetic field strongly decreases both in the magnetosheath and in the undisturbed solar wind.

Another remarkable foreshock feature found by high-time resolution measurements in the Interball project is the existence of quasi-harmonic fast plasma oscillations (Zelenyi et al., 2000; Eiges et al., 2002) Several papers described such structures in IMF (see for example Russell, 1994) but plasma data were never presented till our publications. An example of such measurements is shown in Figure 8. Simultaneous data of the solar wind ion flux and IMF from Interball-1 and Magion-4 are compared. It is seen that both values - plasma and magnetic field - show almost harmonic oscillations with a period of about 2 s and a duration of about 15 s. A good coincidence between the plasma and IMF variations suggests a compressional nature of waves.

We can note a good correlation between measurements (shifted by 4.7 s) of Interball-1 and Magion-4 with a spacecraft separation of about 1200 km. The time delay between these observations can be used as an estimate of the lower limit of the "life time" of these features.

Based on aforementioned investigations, we can conclude for the topics of the Section 4:

- both magnetic field magnitude and solar wind ion flux exhibit the large (about tens of percents) and fast (with frequencies of 1-100 mHz) variations in the foreshock;

- a unique foreshock feature is the large positive cross-correlation (with correlation coefficient of about 0.7-0.8) between plasma and field variations; it may be interpreted as generation of the fast magnetosonic waves;

- rather short (about 10-15 s) bursts of quasi-harmonic ion flux oscillations (in the phase of the magnetic field) with the period of about 2 s can be observed sometimes in the foreshock; these fluctuations are in a good correlation in measurements of two closely separated (<1 Re) spacecraft.

5.   PLASMA AND MAGNETIC FIELD VARIATIONS IN THE MAGNETOSHEATH

The magnetosheath (MSH) as an interface between the solar wind and the magnetosphere plays an important role in the transfer of the solar wind plasma and IMF to the magnetopause. However, the MSH properties have been studied worse than those of any other region of the near-Earth space. Only recently, this region received more attention (see for example Song et al., 1999a; 1999b; Zastenker et al., 1999). It is common meaning that the variability of all parameters of plasma and magnetic field in MSH is dramatically larger than that in the undisturbed SW but any quantitative comparison of its characteristics was not previously done in detail.

From simultaneous observations of the plasma and magnetic field onboard the WIND spacecraft (used as a SW monitor) and the Interball-1 satellite in the MSH we studied the plasma and field variations (Zastenker et al., 1999; Nemecek et al., 2001; Nemecek et al., 2002; Zastenker et al., 2002). Figure 9 shows an example of the Interball-1 ion flux and magnetic field magnitude for a typical MSH crossing. Upper panels present about 5 hours of measurements from the magnetopause (MP) to the bow shock (BS) with the 1-minute time resolution. Lower panels present about 20 minute of data recorded with the 1-s time resolution for the intervals shown on the upper time axes. The top panels also show the solar wind ion flux and IMF from WIND with ~1.5-minute resolution; the time-shifted by the plasma propagation speed from the WIND to Interball-1 position.

In accordance with the general picture of the plasma flow around the magnetosphere (Spreiter et al., 1966), both ion flux and magnetic field magnitude significantly exceed values in an undisturbed SW. However, the main MSH feature demonstrated in Figure 9 is that the ion flux and magnetic field magnitude values exhibit variations with the amplitude much larger than those in the corresponding solar wind on both time scales – minutes and seconds.

To estimate a level of variations, we have used values of standard deviations (SD) and the relative standard deviation (RSD). A comparison

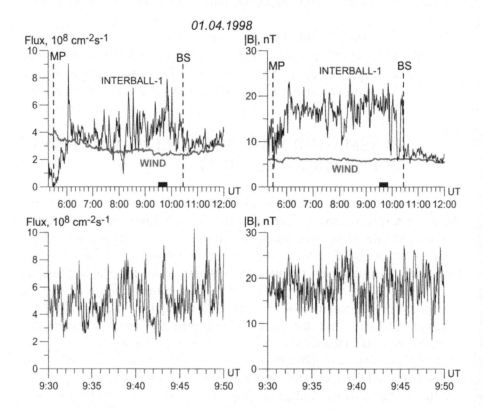

Figure 9. An example of the low-frequency (upper panels) and high-frequency (bottom panels) ion flux and magnetic field variations in the magnetosheath recorded by the Interball-1 data; simultaneous solar wind data from WIND (shifted by propagation time) are shown. Solid bars on the time axes of upper panels show the intervals of high frequency data.

between the ion flux and magnetic field magnitude fluctuations in MSH and SW for high-frequency variations is presented in Figure 10 that shows histograms of the percentage of RSD values in each RSD bin. It is clearly seen that relative variations in both ion flux and magnetic field in MSH are, in average, 2-2.5 times higher than those in the undisturbed SW.

The radial profile of parameters in the MSH can help to reveal a source of plasma and magnetic field variations. In papers of Nemecek et al. (2000; 2002), it was shown that this profile (from MP to BS) could be investigated in terms of the Spreiter model (Spreiter et al., 1966). To account for simultaneous changes in SW, we calculated the normalized ion flux compression coefficient, FCC = Flux(MSH)/Flux(SW) (Zastenker et al., 1999). The average radial profiles of FCC and its SD for high-frequency variations are presented in Figure 11 (Shevyrev et al., 2003).

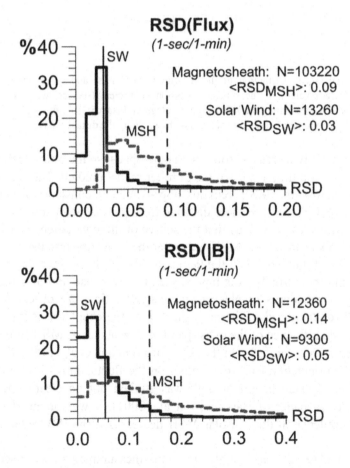

Figure 10. Statistics of relative high-frequency ion flux and magnetic field magnitude variations in the solar wind and magnetosheath from the Interball-1 data.

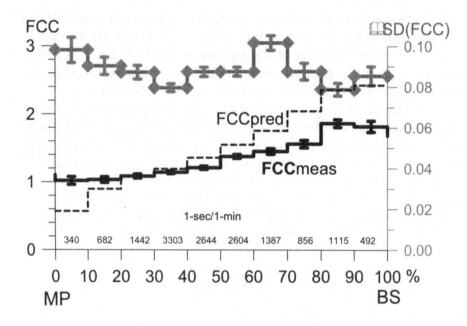

Figure 11. Averaged distribution of the measured (FCCmeas) and predicted by Spreiter model (FCCpred) values of the flux compression coefficient (see the text) and standard deviations of FCCmeas along the MSH (in percentage of distance) from the magnetopause to bow shock. The amount of cases for each bin is shown along the distance axis.

FCC gradually increases from the magnetopause to the bow shock – from 1 to about 1.8 – a little less than the Spreiter model predicts. But the level of variations across the MSH remains approximately constant: SD(FCC) is about of 0.08-0.10 for the whole distance from the magnetopause to the bow shock. Thus, we can conclude that the source of these variations is located at the bow shock or inside the MSH but not at the magnetopause itself.

As it is well known, when the θ_{Bn} angle (the angle between the vector of the IMF and the normal to the bow shock) is lower than approximately $45°$, very large fluctuations of plasma and magnetic field are observed in the foreshock region ahead of the bow shock (see the chapter 4 of this paper). However, the influence of the quasi-parallel bow shock on MSH fluctuations was not examined carefully. In the paper of Luhmann et al. (1986), the effect of the IMF orientation (the cone angle) on the fluctuations of the magnetic field in the dayside magnetosheath was studied but no clear conclusions were made. In paper of Barkhatov et al. (2001), it was obtained that the plasma turbulence in the subsolar MSH decreases if the IMF cone angle is near $90°$.

To calculate the θ_{Bn} angle for our measurements, we used flow streamlines from the Spreiter model to connect a normalized satellite position to a point at the model BS. At this BS point, we can determine the normal to the bow shock surface and the θ_{Bn} angle using time-shifted IMF data from WIND. The statistical study of the θ_{Bn} influence is presented in Figure 12 (Shevyrev et al., 2003). It is seen that normalized values of the ion flux, FCC, and magnetic field amplitude, BCC (defined by the same way as FCC) in the MSH do not depend on the IMF orientation but the level of plasma and magnetic field fluctuations in the MSH strongly decrease when the θ_{Bn} angle increases. In average, behind the quasi-parallel bow shock these fluctuations are approximately two times larger than behind the quasi-perpendicular bow shock.

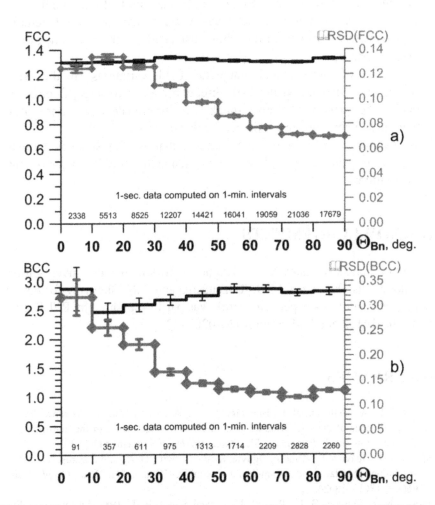

Figure 12. The averaged dependence of compression coefficients of the ion flux, FCC (a) and magnetic field, BCC (b) and their relative standard deviations RSD(FCC) and RSD(BCC) of high-frequency variations on the IMF orientations (the θ_{Bn} angle is in degrees). The amount of cases for each bin is shown along the angle axes.

These results bring further evidence that the bow shock is a source of the plasma and magnetic field magnetosheath fluctuations because their amplitudes strongly depend on the bow shock features. Further study is needed to understand the mechanism of this influence in detail.

Based on these investigations, we obtained such conclusions for the topics of the Section 5:

- large plasma and magnetic field variations are observed in the MSH in a wide range of frequencies (1-100 mHz);
- relative variations of both ion flux and magnetic field in the MSH are in average 2-2.5 times larger than those in the undisturbed solar wind;
- the level of plasma and field variations across the MSH remains approximately constant for the whole distance from the bow shock to the magnetopause;
- the level of plasma and magnetic field variations in the MSH significantly depend on the IMF direction and, in average, fluctuations behind the quasi-parallel bow shock are approximately two times larger than those behind the quasi-perpendicular shock;
- all features described allow us to suppose that the bow shock is the main source of the large plasma and magnetic field magnetosheath fluctuations.

ACKNOWLEDGEMENTS

Author thanks to many his colleagues with whom these investigations were done. Also he thanks M.O. Riazantseva and N.N. Shevyrev for help in the preparation of this paper. This work was partly supported by RFBR grant No. 04-02-16152 and NSF grant ATM-0203723.

REFERENCES

Barkhatov, N. A., Bellustin, N. S., Bougeret, J.-L., Sakharov, S. Yu., and Tokarev, Yu. V., 2001, Influence of the solar wind magnetic field on the turbulence of the transition zone behind the bow shock, *Izvestiya VUZov, Radiophysics* (in Russian) **44**(12):993.

Coplan, M. A., Ipavich, F., King, J., Ogilvie, K. W., Roberts, D. A., and Lazarus, A. J., 2001, Correlation of solar wind parameters between SOHO and WIND, *J. Geophys. Res.* **106**(A9):18615–18624.

Crooker, N. U., Siscoe, G. L., Russell, C. T., and Smith, E. J., 1982, Factors controlling degree of correlation between ISEE 1 and ISEE 3 interplanetary magnetic field, *J. Geophys. Res.* **87**:2224–2230.

Dalin, P. A., Zastenker, G. N., Paularena, K. I., and Richardson, J. D., 2002a, A Survey of large, rapid solar wind dynamic pressure changes observed by INTERBALL-1 and IMP-8, *Ann. Geophys.* **20**(3):293–299.

Dalin, P. A., Zastenker, G. N., and Richardson, J. D., 2002b, Orientation of the solar wind plasma middle-scale structures, *Kosmich. Issled.* (in Russian) 40(4):343–348.

Eiges, P., Zastenker, G. N., Nozdrachev, M., Rybyeva, N., Safrankova, J., and Nemecek, Z., 2002, Small scale solar wind ion flux and IMF quasi-harmonical structures in the Earth's foreshock: INTERBALL-1 and MAGION-4 observations, *Adv. Space Res.* 30(12):2725–2729.

Fusilier, S. A., 1994, Suprathermal ions upstream and downstream from the Earth's bow shock, in: *Solar wind sources of magnetospheric ultra-low-frequency waves*, M. J. Engebretson, ed., *AGU Geophysical monograph, 81*, pp.107–119.

Galeev, A. A., Galperin, Yu. I., and Zelenyi, L. M., 1996, The INTERBALL Project to study solar-terrestrial physics, *Cosmic Res.* 34(4): 313–333.

Lacombe, C., Belmont, G., Hubert, D., Harvey, C. C., Mangeney, A., Russell, C. T., Gosling, J. T., And Fuselier, S. A., 1995, Density and magnetic field fluctuations observed by ISEE 1-2 in the quiet magnetosheath, *Ann. Geophys.* 13:343–357.

Luhmann, J. G., Russell, C. T., and Elphic, R. C., 1986, Spatial distributions of magnetic field fluctuations in the dayside magnetosheath, *J. Geophys. Res.* 91:1711–1715.

Nemecek, Z., Safrankova, J., Zastenker, G. N., and Triska, P., 1997a, Multipoint study of the solar wind: INTERBALL contribution to the topic, *Adv. Space Res.* 20(4/5):659–670.

Nemecek, Z., Fedorov, A., Safrankova, J., and Zastenker, G. N., 1997b, Structure of the low-latitude magnetopause: MAGION-4 observations, *Ann. Geophys.* 15(5):553–561.

Nemecek, Z., Safrankova, J., Zastenker, G. N., Pisoft, P., Paularena, K. I., and Richardson, J. D., 2000, Observations of the radial magnetosheath profile and a comparison with gasdynamic model predictions, *Geophys. Res. Lett.* 27(17):2801–2804.

Nemecek, Z., Safrankova, J., Zastenker, G. N., Pisoft, P., 2001, Statistical study of ion flux fluctuations in the magnetosheath, *Czech. J. Phys.* 51(8):853–862.

Nemecek, Z., Safrankova, J., Zastenker, G. N., Pisoft, P., and Jelinek, K., 2002, Low frequency variations of the ion flux in the magnetosheath, *Planet. Space Sci.* 50(5/6):567–575.

Paularena, K. I., Richardson, J. D., Lazarus, A. J., Zastenker, G. N., and Dalin, P. A., 1997, IMP 8, WIND, and INTERBALL observations of the solar wind, *Phys. Chem. Earth* 22(7/8):629–637.

Paularena, K. I., Zastenker, G. N., Lazarus, A. J., and Dalin, P. A., 1998, Solar wind plasma correlations between IMP 8, INTERBALL-1, and WIND, *J. Geophys. Res.* 103(A7):14601–14617.

Riazantseva, M. O., Dalin, P. A., Zastenker, G. N., Parhomov, V. A., Eselevich, V. G., Eselevich, M. V., and Richardson, J., 2003a, Features of the sharp and large changes of the solar wind ion flux (density), *Kosmich. Issled.* (in Russian) 41(4):405–416.

Riazantseva, M. O., Dalin, P. A., Zastenker, G. N., and Richardson, J. D., 2003b, Orientation of the sharp solar wind plasma fronts, *Kosmich. Issled.* (in Russian) 41(4):395–404.

Richardson, J. D, and Paularena, K. I., 1998, The orientation of the plasma structure in the solar wind, *Geophys. Res. Lett.* 25(12):2097–2100.

Russell, C. T., 1994, Magnetospheric and SW studies with co-orbiting s/c, in: *Solar System Plasma in Space and Time*, J. L. Burch and J. H. White, eds., *AGU Geophys. Monograph 84*, pp.85.

Safrankova, J., Zastenker, G. N., Nemecek, Z., Fedorov, A., Simersky, M., and Prech, L., 1997, Small scale observations of magnetopause motion: Preliminary results of the INTERBALL project, *Ann. Geophys.* 15(5):562–569.

Sibeck, D. G., Borodkova, N. L., and Zastenker, G. N., 1996, Solar wind pareameter variations as a source of the short disturbances of the dayside magnetospere magnetic field, *Kosmich. Issled.* (in Russian) 34(3):248–263.

Shevyrev, N. N., Zastenker, G. N., Nozdrachev, M. N., Nemecek, Z., Safrankova, J., Richardson, J. D., 2003, High and low frequency large amplitude variations of plasma and magnetic field in the magnetosheath: radial profile and some features, *Adv. Space Res.* **31**(5):1389–1394.

Song, P., Russell, C. T., Gombosi, T. I., Spreiter, J. R., Stahara, S. S., Zhang, X. X., 1999a, On the processes in the terrestrial magnetosheath, 1. Scheme development, *J. Geophys. Res.* **104**(A10):22345–22355.

Song, P., Russell, C. T., Zhang, X. X., Stahara, S. S., Spreiter, J. R., Gombosi, T. I., 1999b, On the processes in the terrestrial magnetosheath 2. Case study, *J. Geophys. Res.* **104**(A10):22357–22373.

Spreiter, J. R., Summers, A. L., and Alksne, A. Y., 1966, Hydromagnetic flow around the magnetosphere, *Planet. Space. Sci.* **14**:223–253.

Zastenker, G. N., Dalin, P. A., Lazarus, A. J., and Paularena, K. I., 1998, Comparison of the solar wind parameters measured simultaneously aboard several spacecraft, *Cosmic Research* **36**(3):214–226.

Zastenker, G. N., Safrankova, J., Nemecek, Z., Paularena, K. I., Fedorov, A. O., Kirpichev, I. P., and Borodkova, N. L., 1999, Fast and large variations of magnetosheath parameters: 1. Variations of ion flux and other plasma characteristics, *Kosmich. Issled.* (in Russian) **37**(6):605–615.

Zastenker, G. N., Fedorov, A. O., Sharko, Yu. V., et al., 2000a, Some features of integral Faraday cup using onboard Interball-1 satellite, *Kosmich. Issled.* (in Russian) **38**(1):248–263.

Zastenker, G. N., Dalin, P. A., Paularena, K. I., Richardson, J. D., and Dashevskiy, F., 2000b, Solar wind correlation features obtained from a multi-spacecraft study, *Adv. Space Res.* **26**(1):71–76.

Zastenker, G. N., Nozdrachev, M. N., Nemecek, Z., Safrankova, J., Paularena, K. I., Richardson, J. D., Lepping, R. P., and Mukai, T., 2002, Multispacecraft measurements of plasma and magnetic field variations in the magnetosheath: comparison with Spreiter models and motion of the structures, *Planet. Space Sci.* **50**:601–612.

Zelenyi, L., Zastenker, G. N., Dalin, P. A., Eiges, P., Nikolaeva, N., Safrankova, J., Nemecek, Z., Triska, P., Paularena, K., Richardson, J., 2000, Variability and structures in the solar wind-magnetosheath-magnetopause by multiscale multipoint observations, Proceedings of the ESA Workshop in London, *ESA SP-449*, 29–38.

INTERPLANETARY DISCONTINUITIES AND SHOCKS IN THE EARTH'S MAGNETOSHEATH

Adam Szabo

NASA Goddard Space Flight Center, Greenbelt, Maryland, 20771, USA

Abstract   The study of the propagation of interplanetary disturbances, shocks and discontinuities, through the magnetosheath is critical to improve our understanding of the Sun-Earth connected system. In this paper, the current status of both theoretical and observational studies in this crucial area is reviewed separately for interplanetary shocks and discontinuities. It is suggested that tangential and rotational discontinuities suffer significant geometrical distortions traveling between the Earth's bow shock and magnetopause. On the other hand, the pressure fronts of the transmitted interplanetary shocks most likely remain unaltered promising the possibility of improved space weather forecasting accuracies.

Key words: interplanetary magnetic field; interplanetary shock; magnetosheath; interplanetary discontinuity.

1.   INTRODUCTION

The magnetosheath is the site where the solar wind and the magnetosphere couples, therefore considerable observational and modeling attention has been paid to this region. To first order, space measurements agree very well with the gasdynamic model of [Spreiter and Alksne, 1969] that calculates the thickness of the magnetosheath and the properties of the solar wind flow around the magnetosphere. Our current knowledge of the inner boundary of the magnetosheath, the magnetopause is sufficiently high for static states that its dependence on upstream conditions can be relatively accurately predicted [Safrankova et al., 2002a]. The same cannot be said of the outer boundary, the bow shock [see for recent reviews Song, 2000, and Merka et al., 2003]. Within the magnetosheath, there are still observed structures that are not well understood, moreover it is not clear if they have exogenous or endogenous origins. Even less understood

J. −A. Sauvaud and Z. Němeček (eds.),
Multiscale Processes in the Earth's Magnetosphere:From Interball to Cluster, 57-71.

are dynamic features such as the propagation of interplanetary (IP) shocks and discontinuities from the bow shock to the magnetopause.

A strong correlation of IP shocks impinging on the magnetosphere and geomagnetic disturbances have been reported by many observers [e.g., Tsurutani et al., 1995, Gonzalez et al., 1999, and Tsurutani and Gonzalez, 1997]. IP shocks, as all solar wind pressure events, tend to disrupt the magnetopause surface leading to magnetopause transient events [e.g., Sibeck and Newell, 1995] that, in turn, can initiate reconnection resulting in substorm onset [Wu, 2000]. IP shocks have also been connected to sudden commencements and auroral brightening [Tsurutani et al., 2001]. Solar wind discontinuities interacting with the magnetosphere have been linked to hot flow anomalies (HFA) [e.g., Schwartz et al., 2000; Sibeck et al., 1999], magnetic impulse events in the high latitude ionosphere [e.g., Lanzerotti et al., 1987; Chen et al., 2000], and even the formation of unusual substorm activity [Sergeev et al., 1998]. Therefore, the study of the transition of solar wind disturbances through the magnetosheath is vital for the understanding of a wide range of geomagnetic activities.

2.   THE SHAPE OF SHOCKS
        AND DISCONTINUITIES
        IN INTERPLANETARY SPACE

Before reviewing the characteristics of interplanetary disturbances in the magnetosheath, it is necessary to establish their topological properties in the undisturbed solar wind. Since multi-spacecraft observations are necessary for the determination of the three-dimensional geometry or shape of shocks [e.g., Thomsen, 1988], very few studies have attempted to address this question. Generally it is assumed that the incoming IP shocks are planar on the scale-size of the magnetosphere. Indeed, in a recent study Russell et al., 2000 analyzed a single IP shock with four solar wind satellites and found that three of them were consistent with the planarity assumption. However, deviation from planarity has been reported before [Russell et al., 1983; Safrankova et al., 1998; Szabo et al., 1999]. Specifically, Szabo et al., 2001 has found that IP shocks driven by small magnetic clouds have highly corrugated surface geometries on magnetospheric scale lengths. Hence care should be taken to establish the interplanetary shape of shocks before their geometry in the magnetosheath can be addressed.

Even less systematic work has been carried out with regards to the geometry of rotational and tangential discontinuities, RDs and TDs. Even though Lepping et al., 2003 found that the radius of curvature of large angle IP discontinuities are between 290 to 380 Earth radii, that is they can be considered planar on magnetospheric scale lengths, they investigated only the thickest of discontinuities that took many minutes to cross and are not the most common

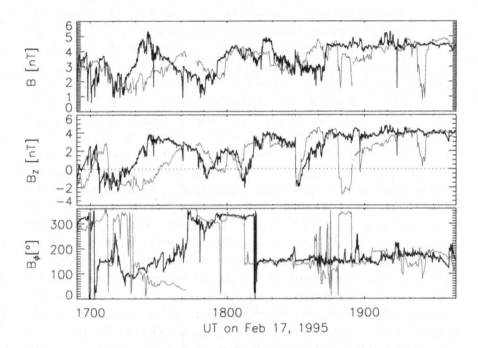

Figure 1. WIND (black) and IMP 8 (red) interplanetary magnetic field magnitude, North-South component and azimuth angle (from top to bottom). The large, near 180° rotations are current sheet crossing. Note that IMP 8 encounters at least one extra pair of current sheet crossings that implies a local discontinuity curvature on the same order of magnitude as the interspacecraft separation of a few tens of Earth radii.

type. Moreover, Szabo and Lepping, 1996b studied discontinuities associated with the heliospheric current sheet and found that some multi-spacecraft observations required highly curved discontinuity surfaces with radii of curvatures in the few tens of Earth radii (see Figure 1). However, this preliminary study was based on only two-point observations by WIND and IMP 8 that placed only an upper limit on the three-dimensional geometry of directional discontinuities. Therefore, for both IP shocks and discontinuities their interplanetary shape has to be established before their magnetosheath deformation can be studied.

3.   TRANSITION OF INTERPLANETARY DISCONTINUITIES THROUGH THE MAGNETOSHEATH

The interaction of TDs and RDs with the Earth's bow shock and their subsequent propagation through the magnetosheath has been studied from very early on especially after Burlaga and Ogilvie, 1969 showed that most geomagnetic sudden impulses in the second half of 1967 were caused by TDs with density jumps. The one-dimensional calculations of Volk and Auer, 1974 suggested that if the plasma density increased through the incoming TD then a shock would be generated in the magnetosheath, while if the density dropped through the TD then a shock and a rarefaction wave, bounded by two weak discontinuities, would occur. Neubauer, 1976 suggested that as a result of TD/bow shock interaction slow shocks would form in the magnetosheath. In a more recent one-dimensional MHD simulation, Wu et al., 1993 found that in the case of a TD, with an enhanced dynamic pressure behind it, the discontinuity is transmitted through the bow shock with a reduced dynamic pressure jump across it. Also a fast compressional shock is excited ahead of the transmitted TD and propagates toward the Earth's magnetosphere. For the case in which the dynamic pressure is reduced behind the interplanetary TD, the pressure jump across the transmitted discontinuity is substantially weakened and a rarefaction wave that propagates downstream is excited. Most recently, much attention has been given to the formation of hot flow anomalies, that occasionally result in massive disruption of the magnetopause [Sibeck et al., 1998], as a result of TDs interacting with the bow shock [Paschmann et al., 1988; Schwartz et al., 1988, 2000; Lin, 1997; Safrankova et al., 2000, 2002b; Sibeck et al., 2000].

For the case of RDs impinging on the bow shock, the one-dimensional MHD model of Yan and Lee, 1996 predicts that in the magnetosheath at least two slow shocks and an intermediate shock should be observable. In the hybrid simulation of Lin et al., 1996a a RD replaces the intermediate shock. A similar result is reported by Lin, 1997 coupled with a noticeable sunward deflection of the magnetosheath flow. In the most recent two-dimensional hybrid simulation of Lin et al., 1996b and three-dimensional MHD simulations of Cable and Lin, 1998 it was found that the intermediate and slow shock structures coalesced into a single pressure pulse followed by a second pressure pulse associated with reflected ions at the bow shock and the foreshock especially when the IMF changed its direction. This second pressure pulse also convects through the bow shock and travels through the magnetosheath. The amplitude of the downstream pressure pulses can be up to 100% of the background magnetosheath value, hence readily observable. However, so far very little observational evidence exists of these newly excited wave modes. This was pointed out earlier by Hassam, 1978 and more recently by Szabo and Lepping, 1996a, with the possible exception of the study of Sibeck et al., 1997 where reverse Alfvenic fluctuations were measured in the magnetosheath, and the observations of transient density events in the sheath following the impinging of a RD [Hubert and Harvey, 2000].

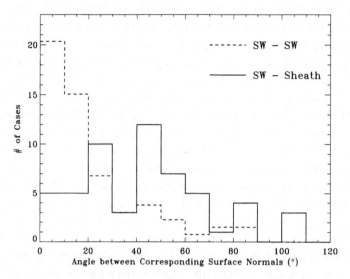

Figure 2. Distribution of discontinuity surface normal angular deviations measured between two spacecraft. WIND is the solar wind monitor and IMP 8 is either in the solar wind (dashed line) or in the sheath (solid line). See the text for details.

4. THE GEOMETRY OF DISCONTINUITIES IN THE MAGNETOSHEATH

Even though most interplanetary discontinuities represent very small pressure jumps that are impossible to separate out from the general fluctuations of the magnetosheath, those with larger magnetic field rotations are readily identifiable even in the deep sheath. While the ambient solar wind magnetic field strength does increase in the magnetosheath, due to compression at the bow shock, the orientation of the magnetic vectors remains largely unimpacted and the sharp rotations of discontinuities are readily apparent. The unmodified field rotations allow not only the identification of the time and position of the discontinuity crossing, but also the determination of the local surface normal directions (see the Appendix for a quick review of minimum variance techniques).

Since it is very rare to have simultaneous multi-point solar wind and magnetosheath observations when a discontinuity with a large field rotation passes through the Earth system, we have to rely on a statistical approach when studying the deformation of discontinuities in the magnetosheath. Figure 2 shows 50 cases when IMP 8 observed a discontinuity with very large (near 180°) field rotations while simultaneous solar wind data was available from WIND. The local surface normal is determined for both the solar wind and magnetosheath observations of the same discontinuity and the difference between these directions, binned in $10°$ increments, is histogramed with the solid outline. Since the solar wind and magnetosheath monitors are never on the same streamline, deviations in the local surface normal directions are expected just from the interplanetary curvature of these surfaces, as discussed above. In order to make a meaningful comparison, observation pairs of the same discontinuities, when both monitors were in the undisturbed solar wind, were also compiled. Over 100 pairs of observations are used when WIND and IMP 8 had a similar orbital separation. The thus obtained histogram, normalized by total number to the sheath study of 50 cases, is plotted with a dashed outline on the same plot. It is apparent that while significant deviations can be seen even for the solar wind – solar wind cases, the solar wind – magnetosheath cases result in much more pronounced differences suggesting that the discontinuity surfaces undergo a systematic and significant deformation in the magnetosheath. This is not an unexpected result as once the directional discontinuities (RDs and TDs) enter the magnetosheath, their propagation velocities – especially for the strictly advecting TDs – become significantly different from their ambient solar wind values [Szabo and Lepping, 1996a].

5.   TRANSITION OF INTERPLANETARY SHOCKS THROUGH THE MAGNETOSHEATH

The interaction of interplanetary shocks with the bow shock and its transmission through the magnetosheath to the boundary of the magnetosphere, the magnetopause, has been studied mainly by gas dynamic modeling [e.g., Shen and Dryer, 1972; Dryer, 1973; Stahara and Spreiter, 1992]. These models, by construction, allow the generation and propagation of only fast mode waves in the magnetosheath. Therefore, it is not surprising that they find only a single, fast mode pressure pulse (or fast shock in the supersonic flanks) propagating through the magnetosheath. Interestingly the predicted disturbance shape in the magnetosheath remains nearly planar (see figure 3) in line with the IP shock [Stahara and Spreiter, 1992]. This prediction seems to be supported by observations reported by Szabo et al., 2000.

Attempts have been made to include the effect of the magnetic field on the interaction by using various MHD and hybrid formulations. The one-dimensional model of Whang, 1991 was very successful at describing outer heliospheric observations of IP shocks. It allowed the merger of two IP shocks if they are both forward or both reverse, and predicted the transmission of the two interacting shocks if one is forward and the other reverse, with a TD forming between them. This model, however, is limited to the treatment of perpendicular shocks. Cargill, 1990 relaxed this requirement with the use of a one-dimensional hybrid

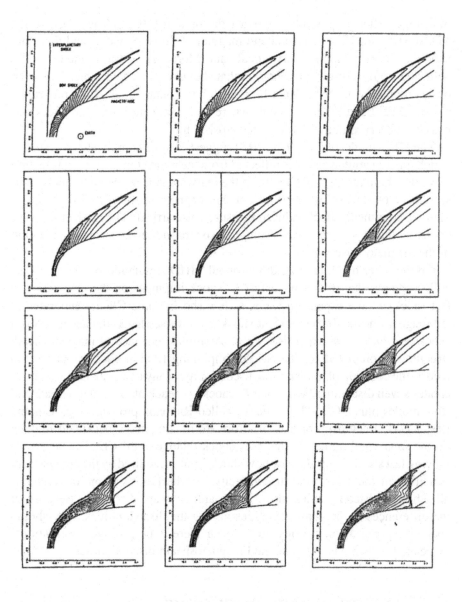

Figure 3.   Plasma pressure contours based on gas dynamic model of an interplanetary shock passing through a planetary magnetosheath [*Stahara and Spreiter*, 1993].

code. He showed that while the collision between two perpendicular collision-less shocks gives rise to a TD located between the two transmitted shocks, as predicted by one-dimensional MHD theory, the collision between two oblique shocks produces a much more extensive and turbulent region between the two transmitted shocks, possibly a contact discontinuity (CD). The CD, like the TD, shows jumps in the plasma and magnetic field components and therefore should be identifiable in observational data. Moreover the transmitted shock is deflected from its original orientation that should also be detectable. However, observational evidence is rather limited so far. Zhuang et al., 1981 reported a case of ISEE 1 and 3 measurements where such a sequence of disturbances was possibly observed in the Earth's magnetosheath.

The MHD model of Grib et al., 1979 suggests a more complicated scenario in which the transmitted shock is reflected from the magnetopause as a rarefaction wave which, in turn, is reflected from the rearward side of the bow shock. This secondary rarefaction wave arrives at the magnetosphere after a time of 3-5 minutes after the IP shock induced pressure pulse arrival. The rarefaction wave decreases the flow pressure on the magnetosphere and causes an outward motion of the magnetopause.

On the other hand, the one-dimensional MHD simulation of Yan and Lee, 1996 suggests that if a forward fast IP shock impinges on the bow shock, a fast shock, a slow expansion wave, a slow shock, and a CD are generated in the magnetosheath. If the incident shock is a reverse shock, the generated fast shock becomes a fast expansion wave. A similar sequence of magnetosheath structures is predicted for the case of impinging IP slow shocks. In fact, in theory, the collision of a solar wind discontinuity of any type with the bow shock creates seven discontinuities: a pair of shocks for each of fast, intermediate and slow modes plus a CD. This is the so-called Riemann problem [e.g. Lin and Lee, 1994]. Each pair consists of a forward moving and a backward moving (in the solar wind frame of reference) shock relative to the CD. The backward moving fast shock forms the new bow shock front as required by the downstream condition of the solar wind discontinuity. In the Earth's frame of reference all other six discontinuities move in the anti-sunward direction. Because of the differences in the phase velocities among the discontinuities, they should spatially spread as they propagate. So far there is no positive observational evidence for such a list of generated magnetosheath discontinuities.

6.   THE GEOMETRY OF SHOCKS IN THE MAGNETOSHEATH

Assuming a planar geometry for the incoming IP shocks, the gasdynamic model of Stahara and Spreiter, 1992 predicts that the transmitted magnetosheath pressure pulse will maintain planarity (see Figure 3). This result is supported by the preliminary results of Szabo et al., 2000. In particular, during the time period of 1998–1999 twelve IP shocks have been identified that were observed by at least 2 solar wind monitors (WIND and ACE) to place some limit on their

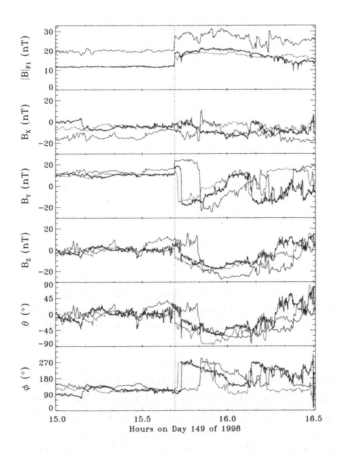

Figure 4.   Magnetic field magnitude, GSE Cartesian components, elevation and azimuth angles (from top to bottom) measured on day 149, 1998 by WIND (black), ACE (red) and IMP 8 (blue). The Wind and ACE data has been time-shifted to line up the IP shock marked by the dashed line. Note that the shock is readily identifiable in the magnetosheath observations of IMP 8.

interplanetary curvatures, while IMP 8 provided magnetosheath observations [Szabo et al., 2003]. Some tentative Geotail sheath events have also been identified. For some of these IP shocks a corresponding pressure pulse is clearly identifiable in the sheath observations. Figure 4 shows 90 minutes of observations of the magnetic field and its components by WIND, ACE and IMP 8 on May 29, 1998. The WIND and ACE data has been time shifted by 29 and 37 minutes, respectively to line up the IP shock with the sheath pressure pulse event (dashed vertical line). The sheath pressure pulse is very clear in both magnetic field and plasma (not shown) observations. Also it should be noted that the nearby large field rotation corresponding to a TD has a markedly different advection time delay. This is consistent with the pressure

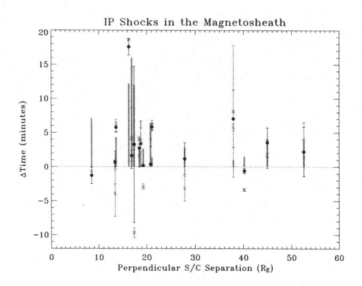

Figure 5.   Difference in the predicted – based on planar geometry – and observed arrival times of IP shocks in the magnetosheath as a function of the perpendicular separation of the solar wind monitor (WIND or ACE) and sheath spacecraft (IMP 8 or Geotail). See the text for details.

pulse corresponding to the IP shock that travels faster than the strictly advecting TD. However, such a clear sheath signature is not always apparent. Weaker and reverse IP shocks produce significantly broader sheath pressure ramps, some reaching over 10 minutes. On the other hand, even for the clearest cases no other discontinuity types nearby could be identified. This does not prove that the secondary discontinuities, predicted by MHD and hybrid codes, do not exist as the general magnetosheath background fluctuations could easily mask small variations. However, it does point out that for magnetospheric energy and momentum input the leading fast-mode pressure jump or shock is the most significant.

In order to make some assessment of the magnetosheath geometry of the transmitted disturbance, the IP shock surface normal directions and speeds, fitted in the solar wind data, are used to estimate predicted arrival times at the magnetosheath monitor. This predicted arrival time is compared to the actual observed time. The thus obtained difference is plotted in Figure 5. A positive difference time refers to the actual magnetosheath observation being later than predicted based on the upstream shock fit results. This would be consistent with the pressure front decelerating in the sheath as is the case for TDs. Negative difference times correspond to earlier than expected arrival times. This could be due to an unlikely acceleration of the front or, more likely, to the intrinsic curvature of the IP shock. All time differences are calculated with respect to the beginning of the sheath pressure pulse. The length of the pressure ramp is indicated by the blue bars. Then a time difference within the blue bars could be still consistent with an unaltered shock disturbance front. The same procedure is repeated for both solar wind monitors of WIND and ACE. The results corresponding to the solar wind monitor closest to the sheath monitor (IMP 8 or Geotail) perpendicular to the Sun-Earth line is plotted as a solid circle (the other result is plotted as a cross) as a function of this cross-wind separation. As is apparent, there is no clear dependence on the spacecraft separation from the Sun-Earth line indicating no systematic deformation in the pressure front surfaces.

While the data set presented is very limited and the question of the transition of IP shocks through the magnetosheath complicated, at least a few preliminary assessments can be made. A clear magnetosheath fast-mode shock or pressure pulse with a wider ramp could be identified for most solar wind observed IP shocks. However, the model predicted secondary discontinuities were not apparent in the highly fluctuating sheath background indicating that for energetics the leading pressure pulse is the most relevant. Even though the intrinsic curvature or corrugation of IP shocks complicates the determination of the geometrical effects of the magnetosheath on the transmitted shocks, the data presented suggests that there is no systematic deceleration of the pressure front. That is, the uncertainty of the arrival time of an IP shock at the magnetopause based on upstream solar wind observations (a value that could be near 10 minutes) is not due to the effects of the magnetosheath but most likely to the unknown interplanetary geometry of the shock surface fronts.

APPENDIX: DETERMINATION OF DISCONTINUITY NORMALS

Traditionally the minimum variance technique of Sonnerup and Cahill, 1967 is used to determine the individual discontinuity normal orientations. However, the statistical errors appearing in this method are difficult to estimate accurately because of the complicated form of the eigenvalue decomposition. In most studies, the determination of the discontinuity normal directions alone, without a true assessment of the uncertainty associated with them, is not sufficient to determine the three-dimensional shape of these structures with any degree of certainty. The equations of Sonnerup, 1971 give upper bounds for the errors with the underlying assumption that all terms contributing to the errors are independent and have a Gaussian distribution. However, it is known that not all of the terms are independent and there is an ongoing debate as to how Gaussian the IMF distributions are [See Marsch and Tu, 1994 and Padhye et al., 2001 and references within]. Lepping and Behannon, 1980 attempted to overcome this problem by establishing an empirical formula for the errors based on the eigenvalue ratios. While their numerical experiments appear valid, the resulting formula is limited by their experimental database and lacks the certainty of an analytical approach. Recently, Kawano and Higuchi, 1995 demonstrated the superiority of the bootstrap method [Efron, 1979] for the error estimation of the minimum variance analysis.

Though their method requires significant computational power, it does not assume a Gaussian distribution for the data or the linear independence of the contributing variables.

REFERENCES

Burlaga, L. F., and Ogilvie, K. W., 1969, Causes of sudden commencements and sudden impulses, *J. Geophys. Res.* **74**:2815.

Cable, S., and Lin, Y., 1998, Three-dimensional MHD simulations of interplanetary rotational discontinuities impacting the Earth's bow shock and magnetosheath, *J. Geophys. Res.* **103**(A12):29551–29567.

Cargill, P. J., 1990, The formation of discontinuities as a result of shock collisions, *J. Geophys. Res.* **95**(A12):20731–20741.

Chen, G. X., Lin, Y., and Cable, S., 2000, Generation of traveling convection vortices and field-aligned currents in the magnetosphere by response to an interplanetary tangential discontinuity, *Geophys. Res. Lett.* **27**(21):3583–3586.

Dryer, M., 1973, Bow shock and its interaction with interplanetary shocks, *Radio Sci.* **8**:893.

Efron, B., 1979, Bootstrap methods: Another look at the jackknife, *Annals of Statistics* **7**:1.

Gonzalez, W. D., Tsurutani, B. T., and De Gonzales, A. L., 1999, Interplanetary origin of geomagnetic storms, *Space Sci. Rev.* **88**(3–4):529–562.

Grib, S. A., Brunelli, B. E., Dryer, M., and Shen, W. W., 1979, Interaction of interplanetary shock waves with the bow shock – magnetopause system, *J. Geophys. Res.* **84**:5907.

Hassam, A. B., 1978, Transmission of Alfvenic waves through the Earth's bow shock: Theory and observation, *J. Geophys. Res.* **83**:643.

Hubert, D., and Harvey, C. C., 2000, Interplanetary rotational discontinuities: From the solar wind to the magnetosphere through the magnetosheath, *Geophys. Res. Lett.* **27**:3149–3152.

Kawano, H., and Higuchi, T., 1995, The bootstrap method in space physics: Error estimation for the minimum variance analysis, *Geophys. Res. Lett.* **22**(2):307–310.

Lanzerotti, L. J., Hunsucker, R. D., Rice, D., Lee, L. C., Wolfe, A., Maclennan, C. G., and Medford, L. V., 1987, Ionosphere and ground-base response to field-aligned currents near the magnetospheric cusp regions, *J. Geophys. Res.* **92**(A7):7739–7743.

Lepping, R. P., and Behannon, K. W., 1980, Magnetic field directional discontinuities: I. Minimum variance errors, *J. Geophys. Res.* **85**(NA9):4695–4703.

Lepping, R. P., Wu, C.-C., and McClernan, K., 2003, The two dimensional curvature of large angle interplanetary MHD discontinuity surfaces: IMP 8 and WIND observations, *J. Geophys. Res.* **108**:1279, doi: 10.1029/2002JA009640.

Lin, Y., and Lee, L. C., 1994, Structure of reconnection layers in the magnetosphere, *Space Sci. Rev.* **65**(1-2):59–179.

Lin, Y., Lee, L. C., and Yan, M., 1996a, Generation of dynamic pressure pulses downstream of the bow shock by variations in the interplanetary magnetic field orientation, *J. Geophys. Res.* **101**(A1):479–493.

Lin, Y., Swift, D. W., and Lee, L. C., 1996b, Simulation of pressure pulses in the bow shock and magnetosheath driven by variations in interplanetary magnetic field direction, *J. Geophys. Res.* **101**(A12):27,251–27269.

Lin, Y., 1997, Generation of anomalous flows near the bow shock by its interaction with interplanetary discontinuities, *J. Geophys. Res.* **102**:24,265–24,281.

Marsch, E., and Tu, C. Y., 1994, Non-Gaussian probability distributions of solar wind fluctuations, *Ann. Geophys.* **12**:1127–1138.

Merka, J., Szabo, A., Narock, T. W., King, J. H., Paularena, K. I., and Richardson, J. D., 2003, A comparison of IMP 8 observed bow shock positions with model predictions, *J. Geophys. Res.* **108**(A2):1077, doi: 10.1029/2002JA009384.

Neubauer, F. M., 1976, Nonlinear interaction of discontinuities in the solar wind and the origin of slow shocks, *J. Geophys. Res.* **81**:2248.

Padhye, N. S., Smith, C. W., and Matthaeus, W. H., 2001, Distribution of magnetic field components in the solar wind plasma, *J. Geophys. Res.* **106**:18635–18650.

Paschmann, G., Haerendel, G., Sckopke, N., Möbius, E., Luhr, H., and Carlson, C. W., 1988, Three-dimensional plasma structures with anomalous flow directions near the Earth's bow shock, *J. Geophys. Res.* **93**:11279–11294.

Russell, C. T., Gosling, J. T., Zwickl, R. D., and Smith, E. J., 1983, Multiple spacecraft observations of interplanetary shocks: ISEE three-dimensional plasma measurements, *J. Geophys. Res.* **88**(NA12):9941–9947.

Russell, C. T., Wang, Y. L., Reader, J., Tokar, R. L., Smith, C. W., Ogilvie, K. W., Lazarus, A. J., Lepping, R. P., Szabo, A., Kawano, H., Mukai, T., Savin, S., Yermolaev, Y. I., Zhou, X.-Y., and Tsurutani, B. T., 2000, The interplanetary shock of September 24, 1998: Arrival at Earth, *J. Geophys. Res.* **105**(A11):25143–25154.

Safrankova, J., Nemecek, Z., Prech, L., Zastenker, G., Paularena, K. I., Nikolaeva, N., Nozdrachev, M., Skalsky, A., and Mukai, T., 1998, The January 10-11, 1997 magnetic cloud: Multipoint measurements, *Geophys. Res. Lett.* **25**:2549–2552.

Safrankova, J., Prech, L., Nemecek, Z., Sibeck, D. G., and Mukai, T., 2000, Magnetosheath response to the interplanetary magnetic field tangential discontinuity, *J. Geophys. Res.* **105**:25,113–25,121.

Safrankova, J., Nemecek, Z., Dusik, S., Prech, L., Sibeck, D. G., and Borodkova, N. N., 2002a, The magnetopause shape and location: A comparison of the Interball and Geotail observations with models, *Ann. Geophys.* **20**:301–309.

Safrankova, J., Prech, L., Nemecek, Z., and Sibeck, D. G., 2002b, The structure of hot flow anomalies in the magnetosheath, *Adv. Space Res.* **30**(12):2737–2744.

Schwartz, S. J., Kessel, R. L., Brown, C. C., Woolliscroft, L. J. C., Dunlop, M. W., Farrugia, C. J., and Hall, D. S., 1998, Active current sheets near the Earth's bow shock, *J. Geophys. Res.* **93**:11,295–11,310.

Schwartz, S. J., Paschmann, G., Sckopke, N., Bauer, T. M., Dunlop, M., Fazakerley, A. N., and Thomsen, M. F., 2000, Conditions for the formation of hot flow anomalies, *J. Geophys. Res.* **105**:12,639–12,650.

Shen, W. W., and Dryer, M., 1972, Magnetohydrodynamic theory for the interaction of an interplanetary double–shock ensemble with the Earth's bow shock, *J. Geophys. Res.* **77**:4627.

Sergeev, V. A., Kamide, Y., Kokubun, S., Nakamura, R., Deehr, C. S., Hughes, T. J., Lepping, R. P., Mukai, T., Petrukovich, A. A., Shue, J.-H., Shiokawa, K., Troshichev, O. A., and Yumoto, K., 1998, Short–duration convection bays and localized interplanetary magnetic field structures on November 28, 1995, *J. Geophys. Res.* **103**:23,593–23,609.

Sibeck, D. G., and Newell, P. T., 1995, Pressure–pulse driven surface waves at the magnetopause: A rebuttal, *J. Geophys. Res.* **100**:21,773–21,778.

Sibeck, D. G., Takahashi, K., Kokubun, S., Mukai, T., Ogilvie, K. W., and Szabo, A., 1997, A case study of oppositely propagating Alfvenic fluctuations in the solar wind and magnetosheath, *Geophys. Res. Lett.* **24**(24):3133–3136.

Sibeck, D. G., Borodkova, N. L., Zastenker, G. N., Romanov, S. A., and Sauvaud, J.-A., 1998, Gross deformation of the dayside magnetopause, *Geophys. Res. Lett.* **25**:453–456.

Sibeck, D. G., Borodkova, N. L., Schwartz, S. J., Owen, C. J., Kessel, R., Kokubun, S., Lepping, R. P., Lin, R. P., Liou, K., Luhr, H., McEntire, R. W., Meng, C.-I., Mukai, T., Nemecek, Z., Parks, G., Phan, T. D., Romanov, S. A., Safrankova, J., Sauvaud, J.-A., Singer, H. J., Solovyev, S. I., Szabo A., Takahashi, K., Williams, D. J., Yumoto, K., and Zastenker, G. N., 1999, Comprehensive study of the magnetospheric response to a hot flow anomaly, *J. Geophys. Res.* **104**:4577–4593.

Sibeck, D. G., Kudela, K., Lepping, R. P., Lin, R. P., Nemecek, Z., Nozdrachev, M. N., Phan, T. D., Prech, L., Safrankova J., Singer, H., and Yermolaev, Y., 2000, Magnetopause motion driven by interplanetary magnetic field variations, *J. Geophys. Res.* **105**:25,155–25,169.

Song, P., 2000, Forecasting Earth's magnetopause, magnetosheath, and bow shock, *IEEE Trans on Plasma Sci.* **28**:1966–1975.

Sonnerup, B. U. O., and Cahill, L. J., 1967, Magnetopause structure and attitude from Explorer 12 observations, *J. Geophys. Res.* **72**:171.

Sonnerup, B. U. O., 1971, Magnetopause structure during the magnetic storm of September 24, 1961, *J. Geophys. Res.* **76**:6717.

Spreiter, J. R., and Alksne, A., 1969, Plasma flow around the magnetosphere, *Rev. Geophys.* **7**:11.

Stahara, S. S., and Spreiter, J. R., 1992, Computer modeling of solar wind interaction with Venus and Mars, in: *Venus and Mars: Atmospheres, Ionospheres and Solar Wind Interactions*, J. G. Luhmann, M. Tatrallyay, and R. O. Pepin, eds., AGU Monograph **66**, Washington, D.C., pp. 345–383.

Szabo, A., and Lepping, R. P., 1996a, The interaction of the heliospheric current sheet with Earth's bow shock, *EOS Supp.* **77**:F582.

Szabo, A., and Lepping, R. P., 1996b, The heliospheric current sheet: A Sun–solar wind–magnetosphere connection, *EOS Supp.* **77**:S221.

Szabo, A., Smith, C. W., Tokar, R. L., and Skoug, R. M., 1999, Multi–spacecraft observations of interplanetary shocks, *EOS Supp.* **80**:S265.

Szabo, A., Smith, C. W., and Skoug, R. M., 2000, The transition of interplanetary shocks through the magnetosheath, *EOS Supp.* **81**:F966.

Szabo, A., Lepping, R. P., Merka, J., Smith, C. W., and Skoug, R. M., 2001, The evolution of interplanetary shocks driven by magnetic clouds, in: *Proceedings of Solar Encounter: The First Solar Orbiter Workshop*, Tenerife, Spain.

Szabo, A., Smith, C. W., and Skoug, R. M., 2003, The transition of interplanetary shocks through the magnetosheath, in: *Solar Wind Ten*, M. Velli, R. Bruno, and F. Malara, eds., AIP Conference Proceedings **679**, Melville, New York, pp.782–785.

Thomsen, M. F., 1988, Multi–spacecraft observations of collisionless shocks, *Adv. Space Res.* **8**(9):157.

Tsurutani, B. T., and Gonzalez, W. D., 1997, The interplanetary causes of magnetic storms: A Review, in: *Magnetic Storms*, B. T. Tsurutani, W. D. Gonzales, Y. Kamide, and J. K. Arballo, eds., AGU Monograph **98**, Washington, D.C., pp. 77–90.

Tsurutani, B. T., Gonzalez, W. D., Gonzalez, A. L. C., Tang, F., Arballo, J. K., and Okada, M., 1995, Interplanetary origin of geomagnetic activity in the declining phase of the solar cycle, *J. Geophys. Res.* **100**:21717-21733.

Tsurutani, B. T., Zhou, X.-Y., Arballo, J. K., Gonzalez, W. D., Lakhina, G. S., Vasyliunas, V. M., Pickett, J. S., Araki, T., Yang, H., Rostoker, G., Hughes, T. J., Lepping, R. P., and

Berdichevsky, D., 2001, Auroral zone dayside precipitation during magnetic storm initial phases, *J. Atmos Sol.-Terr. Phys.* **63**:513–522.

Volk, H. J., and Auer, R.-D., 1974, Motions of the bow shock induced by interplanetary disturbances, *J. Geophys. Res.* **79**:40.

Whang, Y. C., 1991, Shock interactions in the outer heliosphere, *Space Sci. Rev.* **57**(3-4):339–388.

Wu, B. H., Mandt, M. E., Lee, L. C., and Chao, K. J., 1993, Magnetospheric response to solar wind dynamic pressure variations: Interaction of interplanetary tangential discontinuities with the bow shock, *J. Geophys. Res.* **98**(A12):21297–21311.

Wu, C. C., 2000, Shock wave interaction with the magnetopause, *J. Geophys. Res.* **105**(A4):7533–7543.

Yan, M., and Lee, C. L., 1996, Interaction of interplanetary shocks and rotational discontinuities with the Earth's bow shock, *J. Geophys. Res.* **101**(A3):4835–4848.

Zhuang, H. C., Russell, C. T., Smith, E. J., and Gosling, J. T., 1981, Three–dimensional interaction of interplanetary shock waves with the bow shock and magnetopause: A comparison of theory with ISEE observations, *J. Geophys. Res.* **86**(A7):5590–5600.

MAGNETOSHEATH INVESTIGATIONS: INTERBALL CONTRIBUTION TO THE TOPIC

Jana Šafránková, Mykhaylo Hayosh, Zdeněk Němeček, and Lubomír Přech
Faculty of Mathematics and Physics, Charles University,
V Holešovičkách 2, 180 00 Prague 8, Czech Republic

Abstract   We review the statistical processing of four years of INTERBALL-1 observations in the nightside magnetosheath and discuss peculiarities of the magnetosheath ion flux and magnetic field radial profiles. Our investigations reveal that the magnetosheath ion flux profile is similar to but flatter than that predicted by the gasdynamic and MHD models. The most pronounced difference seen at the bow shock region is attributed to kinetic processes not involved in these models. On the other hand, the magnetic field magnitude profile is nearly constant. It indicates that magnetic forces contribute significantly to the formation of the magnetosheath flow and frozen-in approximation should be used with a care. According to our investigations, the rise of the ion flux from the magnetopause toward the bow shock is much steeper during intervals of a radial IMF orientation.

Statistical processing has shown (1) the limitations of gasdynamic and MHD models, (2) the conditions favorable for the creation of a plasma depletion layer adjacent to the flank magnetopause, (3) a strong dawn-dusk asymmetry of the ion fluxes, (4) that the presence of high-energy particles influences the total ion flux only weakly, and (5) that the coupling between high-energy particles and the ion flux and/or magnetic field fluctuation level is strong.

Key words: magnetosheath; energetic particles; ion flow; magnetic field; radial profile; bow shock; magnetopause.

1.   INTRODUCTION

Over the past 40 years, the observational knowledge and physical understanding of a region of compressed and heated plasma — the magnetosheath — have greatly increased; from theoretical speculations to experimental ex-

J. –A. Sauvaud and Z. Němeček (eds.),
Multiscale Processes in the Earth's Magnetosphere:From Interball to Cluster, 73-94.
© 2004 *Kluwer Academic Publishers. Printed in the Netherlands.*

aminations and from gasdynamic theories to theories describing processes that occur in the actual magnetosheath. Changes in the solar wind plasma and interplanetary magnetic field (IMF) influence the processes in the magnetosheath and are important sources of many dynamic features observed in the magnetosphere (Elphic and Southwood, 1987; Song et al., 1992; Russell et al., 1992).

Since upstream variations in the solar wind plasma can be significantly modified upon traversing the magnetosheath from the bow shock to the magnetopause (Yan and Lee, 1994), the magnetosheath region itself is difficult to simulate. Nevertheless, several theoretical models have been developed to understand the evolution of the plasma and magnetic field properties in this transition region. The first magnetosheath model was developed by Spreiter et al. (1966). In this model, the plasma flow around the magnetosphere was considered in a gasdynamic approximation with $B = 0$, and then the magnetic field was calculated from frozen-in conditions in a kinematic approximation. In the model, the solar wind flows along the Sun-Earth line, strikes the subsolar magnetopause and then is diverted radially from this point. The model further predicts that velocities decrease from the bow shock to the magnetopause, whereas the density and temperature increase in the vicinity of the stagnation streamline. Farther from the subsolar region, the density and the velocity decrease but the temperature increases through radial profiles from the bow shock to the dayside magnetopause. Along the flanks of the near-Earth magnetotail, minimum velocities and maximum temperatures occur in the middle magnetosheath. The plasma flows radially away from the stagnation streamline. This flow accelerates up to the solar wind speed and becomes increasingly like solar wind toward the flanks, where the bow shock is weaker.

Zwan and Wolf (1976) used the results of the Spreiter model at the bow shock and the magnetopause and provided a formulation and numerical estimation of the magnetosheath flow using a magnetohydrodynamics (MHD) approach. Their model describes a magnetic flux tube moving from the bow shock to the magnetopause and predicted a new effect: an increase of the magnetic field strength which is coupled with a plasma depletion. The Zwan and Wolf theory predicts a monotonic density decrease from the bow shock to the magnetopause. Observations confirm their theory, showing the presence of a depletion layer near the subsolar magnetopause where the density drops and the magnetic field strength increases.

Southwood and Kivelson (1992, 1995) revisited the Zwan and Wolf model and proposed a solution for a few inconsistencies by adding a compressional front between the two depletion regions of the Zwan and Wolf model. This front compresses the plasma, while rarefying the magnetic field and diverts the plasma flow from the Earth-Sun line.

Wu (1992) made numerical simulations of the magnetosheath profile using a 3-D MHD calculation, taking into account the formation of a plasma depletion layer. In this model, the magnetosphere is a solid impermeable obstacle. The density increases first and then decreases from the bow shock toward the magnetopause along the Sun-Earth line. In the inner magnetosheath, the decrease with distance from the magnetopause is more abrupt than the increase in density within the outer magnetosheath.

Spreiter and Stahara (1980) formulated a gasdynamic convected magnetic field (GDCF) model, and calculated magnetic field by convecting the field lines along with the fluid. The results of simulations depend on the shape of obstacle, the Mach number of the flow, and the polytropic index. The flow was considered to be cylindrically symmetric around the Sun-Earth line and convects the three-dimensional magnetic field through the gasdynamic flow. In this model, all mass flux crossing the bow shock must flow around the obstacle. The flux has a minimum value near the subsolar point and increases toward the magnetotail. The plasma density is largest at the stagnation point, whereas the plasma velocity is small. Along the magnetosheath flanks, the velocity increases more abruptly than the density decreases and therefore, the flux increases. The model does well at predicting the observed magnetic field and dayside magnetopause position, somewhat less well at predicting the magnetosheath thickness, and least well at predicting plasma parameters.

Song et al. (1999a) used observed, time-varying solar wind data to produce time-dependent predictions of the magnetosheath flux in the GDCF model and compared these predictions with observations. Their results showed that a nearly constant speed component toward the magnetopause is almost always present in the real magnetosheath. In the GDCF model, the predicted magnetosheath temperature is always too low and the predicted magnetosheath is too thin. Song et al. (1999b) applied an artificially elevated solar wind temperature and received improved magnetosheath temperature and thickness predictions.

Siscoe et al. (2002) discussed some aspects of the magnetosheath flow if magnetic forces are included in the framework of ordinary gasdynamics (Spreiter et al., 1966). The authors suggested four such aspects and illustrated them with computations using a numerical MHD code that simulates the global magnetosphere and its magnetosheath. The four inherently MHD aspects of a magnetosheath flow that the authors considered were the depletion layer, the magnetospheric sash, MHD flow deflections, and the magnetosheath's slow-mode expansion into the magnetotail. They introduced a few new details of these aspects or illustrated known details in a new way. These include the following: the dependence of the depletion layer on interplanetary magnetic field (IMF) clock angle; the agreement between the locations of the antiparallel regions (Luhmann et al., 1984) and the magnetospheric sash (e.g., Maynard et al., 2002) and isolated flow deflections corresponding to a stagnation line and magnetic reconnection.

Simultaneously with the evolution of theoretical models, the magnetosheath has been investigated actively based on *in situ* spacecraft measurements. However, only a few studies have used plasma data from the magnetosheath flanks. Howe and Binsack (1972) studied the magnetosheath $20 - 60$ R_E (Earth radii) downstream from the Earth. They found that the flow pattern agrees well with hydrodynamic theory. Kaymaz et al. (1992) used the IMP-8 magnetic field data to investigate the magnetic field configuration in Earth's magnetosheath. They confirmed the draping of the magnetic field predicted by magnetohydrodynamic as well as gasdynamic models and showed that the rotation of the draping patterns caused by reconnection varies with the IMF clock angle. Petrinec et al. (1997) looked at several GEOTAIL passes 25-45 R_E downstream of the Earth and found that magnetosheath speeds near the magnetopause are larger when the magnetic field is perpendicular to the flow vector. Sibeck et al. (2000) reported a survey of MHD waves in the magnetosheath and presented that the fluctuations almost invariably propagate antisunward in the magnetosheath, independent of their propagation direction relative to IMF. In terms of data from the IMP 8, WIND, ISEE-1, and ISEE-3 spacecraft, Paularena et al. (2001) showed a significant dawn-dusk asymmetry of the plasma density in the magnetosheath near solar maximum, with larger density values on the dawn than dusk side. The observed asymmetry does not depend on the IMF orientation, ruling out both foreshock effects and different compression by parallel and perpendicular shocks as causes. They compared observations with MHD and gas dynamic models. These comparisons showed that the MHD simulation corresponded to the measurements better than the gasdynamic predictions but neither of these calculations are able to explain the observed phenomena.

Němeček et al. (2000a) used measurements of the WIND, INTERBALL-1, and GEOTAIL spacecraft to investigate magnetosheath ion fluxes. Their results showed that the gasdynamic model can predict the magnetosheath ion flux profile with an accuracy $\pm 40\%$. Němeček et al. (2000b) showed that a difference between the averaged ion flux radial profile and its gasdynamic prediction decreases with increasing ion plasma beta and/or Alfvénic Mach number. This result can be easily understood because the influence of magnetic forces decreases in both cases.

The present paper summarizes and comments on a large number of statistical studies of the ion flux and magnetic field profiles in the nightside magnetosheath. The review is based on INTERBALL-1 observations in the magnetosheath and simultaneous solar wind and IMF monitoring carried out by the WIND spacecraft. We discuss mutual relations between the plasma flow and the magnetic field as well as the dependence of both quantities on upstream parameters. The results are compared with gasdynamic and global MHD models.

2. DATA PROCESSING AND PRESENTATION

Upstream solar wind parameters influence the magnetosheath flow and magnetic field several ways. The locations of magnetosheath boundaries are a function of solar wind dynamic pressure and IMF. The magnetopause moves inward with increasing solar wind dynamic pressure and changes its shape with the IMF B_Z component (see e.g., Shue et al., 1997; Šafránková et al., 2002). The bow shock also moves in response to changes of the solar wind dynamic pressure, and its location is sensitive to the upstream Mach number, especially in the low-Mach number range (e.g., Formisano et al., 1979; Fairfield and Feldman, 1975). Moreover, Němeček and Šafránková (1991) suggested that the bow shock moves outward with increasing IMF magnitude. Even if the locations of boundaries were constant, the magnetosheath density, velocity, and magnetic field depend on the corresponding upstream parameters which vary over large ranges.

In order to present the magnetosheath data in a consistent way and to carry out a statistical study, several normalizations are required. Since these normalizations are essential for understanding the presented results, we will describe them in detail.

2.1 INFLUENCE OF THE EARTH'S ORBITAL MOTION

The whole magnetosphere and, consequently, the magnetosheath is not aligned with the Sun-Earth line but rather with the solar wind direction. The declination of the solar wind from the X_{GSE} axis is caused by two effects: the Earth's orbital motion and perpendicular components of the solar wind velocity. However, Šafránková et al. (2002) showed that the effect of non-radial solar wind flow is not statistically significant.

To account for the Earth's motion, we have recalculated the locations of all measuring points in the magnetosheath, rotating them about the Z_{GSE} axis by the amount of the aberration angle.

2.2 MAGNETOSHEATH COORDINATES

We assume rotational symmetry of the magnetosheath with the inner boundary given by the Petrinec and Russell (1996) magnetopause model and the outer boundary (bow shock) given by the equation (derived by fitting the INTERBALL-1 bow shock crossings):

$$R_{BS} = 10.414 \times (nv^2)^{\frac{-1}{6}} (530 - 0.43X^2 - 47X)^{\frac{1}{2}} \times \frac{0.66M_A^2 + 2}{2.66(M_A^2 - 1)}.$$

The estimated magnetosheath thickness in the direction perpendicular to the aberrated X coordinate is thus given as $R_{BS} - R_{MP}$. The distance of a par-

ticular point in the magnetosheath from the magnetopause can be expressed
as $R - R_{MP}$ where $R = (Y^2 + Z^2)^{1/2}$ is the distance of the point from
aberrated X coordinate. For our study of magnetosheath radial profiles, we
are using the normalized distance, D, which can be written as a ratio $D = ((R - R_{MP})/(R_{BS} - R_{MP})).100\%$.

2.3   NORMALIZED MAGNETOSHEATH FLUX

A reliable determination of parameters in the highly turbulent magnetosheath plasma is a difficult task. To avoid the problem of intercalibration
of different devices, we are using the data provided by the INTERBALL-1
spacecraft which was launched into a highly elongated polar orbit and passed
through the magnetosheath twice per four-day orbit. The trajectories cover a
magnetosheath region from $X_{GSE} = 0$ to $X_{GSE} \sim -18\ R_E$. The data set
was obtained during August - October (dawn flank) and January - March (dusk
flank) in the years 1995 through 1999 years.

In the Maxwellian approximation, the plasma flow can be described by the
density, velocity vector, and temperature. Separation of these quantities requires measurements of the velocity distribution by scanning analyzers. The
resulting time resolution is thus low (usually about 1 minute) and thus the
reliable determination of all quantities is impossible. For this reason, we
use the ion flux data computed from the INTERBALL-1 Faraday cup (VDP)
(Šafránková et al., 1997). These data were measured 16 times per second and
either the full resolution or 1 s averages were transmitted to Earth. The source
of magnetic field data is the MIF-M magnetometer (Klimov et al., 1997) with
a similar time resolution. To avoid the influence of high-frequency magnetosheath waves, we calculated 5-minute averages of all parameters. This averaging allows us to better compare our results with those presented by Paularena
et al. (2001). After averaging, we obtained more than ~ 4000 magnetosheath
measurements. As a solar wind monitor, we used WIND plasma and magnetic
data (lagged by the propagation time between WIND and INTERBALL-1 positions).

According to Spreiter et al. (1966), the magnetosheath plasma parameters
can be normalized with respect to corresponding upstream parameters. The
results of our analysis are thus presented as plots of $FCCm$ (measured flux
compression coefficient, defined as the ion mass density times the bulk flow
speed downstream of the bow shock divided by the same parameters upstream
of the bow shock) and $BCCm$ (measured magnetic field compression coefficient defined in the same way as $FCCm$) versus the normalized distance from
the magnetopause, D. For a particular task, the data in a bin were sorted in
accordance with other parameter(s). It should be noted that all data were measured in the magnetosheath but, due to a well known inaccuracy of the bow

shock and magnetopause models, a part of our measurements lay either below the predicted magnetopause or upstream of the predicted bow shock. We have processed these data in the same way as the rest of the data set.

The normalized magnetosheath ion flux $FCCm$ typically rises by a factor of ~ 2 from the magnetopause to the bow shock To remove this effect, we computed av_{FCCm}, the mean magnetosheath flux as a function of D. In several presentations, we normalize $FCCm$ using this averaged profile.

2.4 INFLUENCE OF ALPHA PARTICLES

The INTERBALL-1 Faraday cup provides the total ion current which is composed mainly of proton and helium contributions but WIND regularly supplies only the proton parameters. However, Aellig et al. (2001) investigated the helium abundance and its variations in the solar wind on a time scale of years. Based on data from the WIND/SWE experiment (Ogilvie et al., 1995) between the end of 1994 and early 2000, the authors found a clear dependency of the He/H ratio in the solar wind on the solar cycle and solar wind velocity. We used their results to remove He from INTERBALL-1 data and thus we can calculate $FCCm$, the proton flux compression coefficient. This $FCCm$ is then used throughout this paper to allow a comparison with previous results.

2.5 MODELS USED

The measured $FCCm$ was compared with the value ($FCCpr$) predicted by the Spreiter et al. (1966) gasdynamic model and by two MHD models:

- BATS-R-US (http://csem.engin.umich.edu/docs/), the Block-Adaptive-Tree-Solarwind-Roe-Upwind-Scheme which was developed by the Computational Magnetohydrodynamics (MHD) Group at the University of Michigan

- GGCM (http://www-ggcm2.igpp.ucla.edu/index.html) originally developed by J. Raeder as a magnetohydrodynamic model of the Earth's magnetosphere at UCLA in the early 1990's

The value of $FCCpr$ in the case of the gasdynamic model is determined by the following procedure: averaged (1 min or 1 hour) solar wind and IMF data were introduced into the model of the magnetopause position and the subsolar distance R_o was determined. The satellite coordinates were expressed in R_o units and these "normalized" coordinates were used for determination of the position in the magnetosheath described by the Spreiter et al. (1966) model. For the sake of simplicity, $M_A = 8$ and $\gamma = 5/3$ were used as parameters of the model.

MHD models provide the magnetopause and bow shock locations a self-consistent way. However, the model determined locations of these boundaries

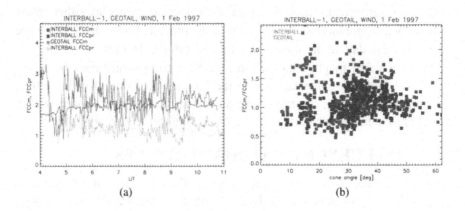

Figure 1. Profiles of measured and predicted FCC at two different sites in the magnetosheath (INTERBALL-1 and Geotail) - (a); normalized FCC (i.e., a ratio of $FCCm/FCCpr$) as a function of the IMF cone angle - (b).

do not correspond exactly to those determined experimentally. For this reason, we determined the locations of the boundaries in the model data and then processed the model data in same way as for the experimental data.

3.   SIMULTANEOUS MEASUREMENTS OF TWO SPACECRAFT IN THE MAGNETOSHEATH

As a first step, we compared simultaneous magnetosheath observations of two spacecraft located in the same flank. During the time interval from 4 to 11 UT on February 1, 1997, the solar wind parameters were nearly constant and simultaneous magnetosheath observations from INTERBALL-1 and GEOTAIL were available. Both spacecraft moved in the dusk magnetosheath; GEOTAIL at GSE $= (-2, 18.5, -2)R_E$ and INTERBALL-1 at GSE $= (-7, 20, 14)R_E$. The $FCCm$ computed from their measurements are plotted in Figure 1a together with gasdynamic predictions, $FCCpr$. The values of $FCCpr$ for the GEOTAIL position are lower than that for INTERBALL-1 because GEOTAIL is located nearer to the magnetopause. The measured $FCCm$ for both spacecraft exhibits sharp and fast enhancements and measured fluxes are higher than the predicted ones most of the time. Some of the distinct spikes seem to be observed by both spacecraft but the time delay is changes from 3 to 7 minutes which suggests either a complicated spatial structure of the ion density or highly turbulent velocity of the magnetosheath plasma.

Magnetosheath fluctuations seem to be controlled by the IMF direction as can be seen from Figure 1b where the ratio of $FCCm/FCCpr$, the normalized flux, is plotted as a function of the IMF cone angle. The scatterplot involves the measurements of both spacecraft during the time interval depicted in Figure 1a. The highest values of normalized FCC are observed for the cone angle 15° and they gradually decrease as the cone angle increases. On the other hand, the lowest fluxes increase and thus the spread of the normalized FCCs, which is ± 70% for a cone angle of 15°, decreases to ± 20% when the cone angle increases to 50°. A similar trend of decreasing magnetosheath fluctuations with the increasing cone angle was observed during all analyzed passes of INTERBALL-1 or GEOTAIL through the magnetosheath. It is interesting to note that in all cases the magnetosheath fluxes peaked for angles between 15° ÷ 30°, not for the minimum of the cone angle range. (A complex analysis of magnetosheath fluctuations is published in Němeček et al. (2002)).

This preliminary study shows that the gasdynamic description of the magnetosheath flow is not sufficient and thus we compare our observations with gasdynamic and MHD simulations, as well as with results of long-time statistics in next sections.

4. COMPARISON OF ONE MAGNETOSHEATH PASS WITH MODELS

For a comparison of the measured magnetosheath properties with predictions of gasdynamic and MHD models, we have selected one transition through the magnetosheath registered by INTERBALL-1 on August 8 and 9, 1997 as an example. The IMF and solar wind plasma flux detected by WIND and magnetosheath magnetic field and plasma flux registered by INTERBALL-1 are plotted in Figure 2. The sign of IMF B_X and B_Y components changed during this period. Coincident with the IMF rotation, the solar wind flux slightly increased. As a consequence of these changes, the magnetosheath ion flux underwent a more rapid increase. The IMF strength was nearly constant throughout the event but the magnetosheath magnetic field magnitude gradually decreases. During the interval under study, INTERBALL-1 crossed the whole magnetosheath from the magnetopause (2344 UT) to the bow shock (0459 UT). Since the solar wind dynamic pressure was nearly constant and IMF B_Z changed only slightly, we expect that the locations of both boundaries did not change significantly. This allows us to re-plot the data in normalized magnetosheath co-ordinates and compare them with predictions of MHD and gasdynamic models as shown in Figure 3.

As we noted above, the problem with this comparison is the accuracy of the determination of both boundaries in the MHD results. Whereas the magnetopause can be determined from the IMF rotation, the MHD bow shock exhibits very smooth changes of all parameters. For this reason, we have chosen the lo-

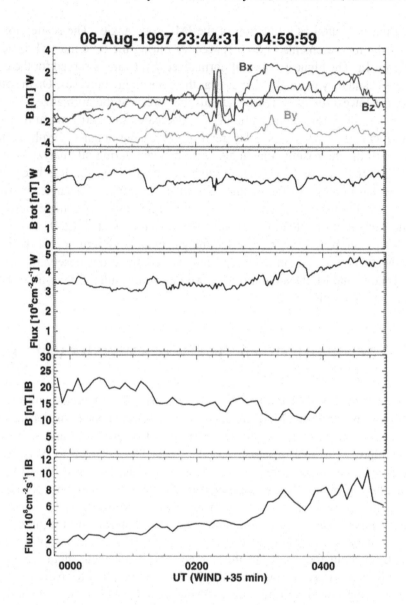

Figure 2.   The IMF and solar wind conditions from WIND during the analyzed event (from top to bottom: IMF components; the IMF strength; solar wind flux) and an overview of INTERBALL-1 measurements (the magnetic field strength and magnetosheath tailward ion flux). Note that the WIND data are shifted on estimated propagation time, ~ 35 minutes.

cation of the bow shock in the MHD simulations as the point of maximum density. This choice provides the best agreement with observations. However, this method of bow shock determination may put the bow shock closer to the Earth

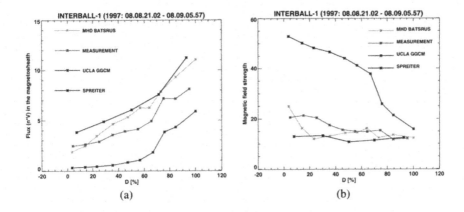

Figure 3. A comparison of observed magnetosheath flux (a) and magnetic field (b) profiles with gasdynamic and both 3-D MHD model predictions.

and thus provide a thinner magnetosheath than observed. This effect would result in a lower measured magnetosheath flux. MHD models qualitatively reproduce the increase of the observed flux in the middle of the magnetosheath (from 5 to 8 at D of about 70% as seen in Figure 3). This increase can be connected either with the increasing latitude of INTERBALL-1 during the observations or with above mentioned change of the IMF orientation. In order to distinguish these possibilities, we will analyze these effects separately.

However, if we compare the model results with our experimental data for this particular pass (Figure 3), we can conclude that the BATS-R-US model predicts the magnetosheath ion profile better than the UCLA GGCM and the gasdynamic Spreiter et al. (1966) models. For this reason, we will use the BATS-R-US model for further comparison of simulation results with observations.

5. STATISTICAL COMPARISON OF MHD MODELS AND OBSERVATIONS

A qualitative agreement of MHD simulations with observations encouraged us to carry out a detailed analysis. Paularena et al. (2001) and Němeček et al. (2003) have shown a mysterious dawn-dusk magnetosheath asymmetry. MHD modeling reveals that this asymmetry can be a result of the latitudinal distribution of the ion flux in combination with influences of the IMF direction and or/and the tilt angle of the Earth's dipole. For this reason, we have studied the influence of the tilt angle and IMF direction on the dawn-dusk asymmetry of the ion flux for high and low latitudes separately. For tilt angles close to

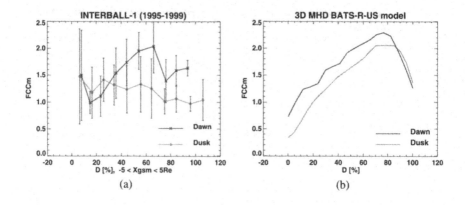

Figure 4.   The observed dawn-dusk asymmetry (a) and its prediction by the BATS-R-US model (b) for high latitudes and tilt angles close to zero.

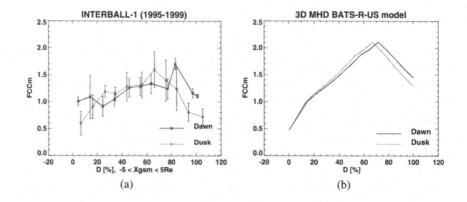

Figure 5.   The observed dawn-dusk asymmetry (a) and its prediction by the BATS-R-US model (b) for low latitudes and tilt angles close to zero.

zero, the spread of measurements remains large for high latitudes. We showed this spread as error bars for both flanks in the left panels of Figures 4 and 5. As can be seen, they are very wide but the trends of the profiles are clear enough. At low latitudes, the measurement coverage is compact and both flank profiles are within the same error bar. This suggests that our limitations did not exclude some factors, which influence the magnetosheath plasma at high latitudes. The model result corresponds to observations rather well. However, if we compare measured and simulated profiles, the significant difference is on the dusk side at high latitudes. We assume that this difference can be caused possibly by the plasma entry to the cusp region in combination with the sign of the IMF B_Y component. At low latitudes, flank differences do not develop in both the model and the observations.

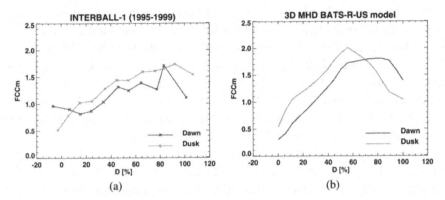

Figure 6. The observed dawn-dusk asymmetry (a) and its prediction by the BATS-R-US model (b) for low latitudes and negative tilt angles.

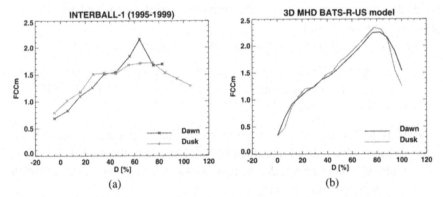

Figure 7. The observed dawn-dusk asymmetry (a) and its prediction by the BATS-R-US model (b) for high latitudes and negative tilt angles.

For negative tilt angles, the predictions are again similar to the measurements. At low latitudes (Figure 6), both simulation and observation show a clear dawn-dusk asymmetry with larger values of $FCCm$ on the dusk side (Němeček et al., 2003). On the other hand, at high latitudes, the dawn-dusk asymmetry disappears (Figure 7).

As a conclusion, we can note that we have compared observations of the magnetosheath ion flux with predictions of the BATS-R-US MHD model. We have found a qualitative agreement of the model predictions with observations at low latitudes but the spread of experimental points in high latitudes prevents us from drawing concrete conclusions. We suggest that this spread may be caused by plasma inflow into the cusp. The cusp is able to influence the behavior of the magnetosheath flow. Němeček et al. (2000c) have shown that the high-latitude cusp location dependents significantly on the dipole tilt angle. They demonstrated that for negative tilt angles the cusp moves to lower latitudes. On the other hand, a direction of the cusp-region plasma flow strongly depends on the IMF orientation and variations of the solar wind dynamic pressure (Weiss et al., 1995). For our case, where only the negative IMF B_Y component was chosen (IMF B_Z is close to zero), the cusp precipitation at low latitudes shifts toward dawn in the northern hemisphere. The combinations of these two factors mean that the plasma of the low-latitude magnetosheath penetrates into the cusp more on dawn than dusk flanks. Therefore, the plasma flow on the dusk side is larger than that on the dawn side.

For tilt angles close to zero, the cusp moves to higher latitudes and its influence on the plasma flow decreases. The cusp plasma flow is the same from both flanks. Due to the choice of the IMF B_Y orientation, reconnection takes place on the dawn magnetopause at high latitudes. The energy that is released during the reconnection process accelerates the protons in the tailward direction, resulting in plasma flow increases and a proton flux greater than that on the dusk side. At low latitudes, the influences of the cusp and reconnection positions may cancel out.

The results of this particular study can be summarized as follows:

- The BATS-R-US MHD model simulates the magnetosheath properties better than the gasdynamic (Spreiter et al., 1966) and UCLA-GGC MHD models.

- Both MHD simulations and experimental data show a significant change of the magnetosheath parameters with latitude in the considered X_{GSM} interval. The magnetosheath flux does not exhibit the radial symmetry which is generally expected.

- The IMF direction in combination with the orientation of Earth's dipole play an important role in the formation of the magnetosheath ion flow.

- The tilt angle influences the dawn-dusk asymmetry more for the high-latitude than for the low-latitude magnetosheath.

However, we note that the BATS-R-US model describes magnetosheath parameter changes qualitatively but not quantitatively, and thus further experimental investigations are needed.

6.   INFLUENCE OF THE IMF ORIENTATION ON THE FLUX AND MAGNETIC FIELD COMPRESSION

Since previous results suggest that the observed dawn-dusk asymmetry can be connected with the IMF direction, we have investigated two limiting cases - radial and perpendicular IMF orientations. The selection criterion was the value of the upstream cone angle. Figure 8 showes that the averaged $FCCm$

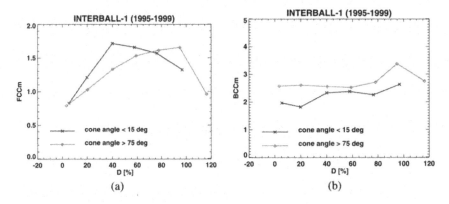

Figure 8. Normalized ion flux (a) and magnetic field (b) profiles for radial (cone angle $< 15°$) and perpendicular IMF (cone angle $> 75°$).

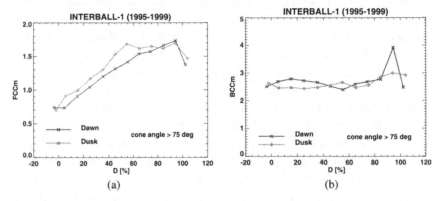

Figure 9. The dawn-dusk difference between ion flux (a) and magnetic field (b) profiles for a large cone angle.

profile peaks at the middle of the magnetosheath for radial IMF (cone angle $< 15°$) but that it has a maximum near the bow shock when the IMF is perpendicular to the solar wind flow. However, $BCCm$ profiles remained nearly constant for both IMF orientations but the magnetic field is less compressed when the IMF is radial.

Both investigated IMF orientations would not change the proportion of dawn and dusk ion and magnetic fluxes. Nevertheless, Figure 9 demonstrates that the dusk ion flux is larger during intervals of perpendicular IMF, whereas $BCCm$ does not exhibit any dawn-dusk asymmetry. On the other hand, the situation is opposite during intervals of radial IMF, $FCCm$ is, within statistical errors, the same on both flanks but dawn $BCCm$ is larger as can be seen from Figure 10.

Our investigations have shown that a connection between the magnetosheath proton flux and magnetic field is rather weak. The magnetic field magnitude neither follows the plasma compression nor compensates for the total pressure

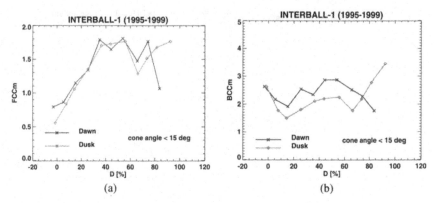

Figure 10.   FCCm (a) and *BCCm* (b) profiles for a small cone angle.

but is nearly constant across the magnetosheath. On the other hand, the magnetosheath proton flux profile is strongly affected by the IMF orientation. A maximum of the plasma compression shifts from the bow shock region toward the magnetosheath center when IMF becomes more radial. It means that expected weaker plasma compression at the quasiparallel shock is compensated by a further compression in the magnetosheath. This fact can explain why the bow shock location does not depend on the angle between IMF and bow shock normal (Šafránková et al., 2003) but the source of this additional compression is unknown. Němeček et al. (2000a,b) attributed the difference between gasdynamic model and observations at the bow shock region to kinetic effects forming the magnetosheath flow just behind the bow shock. These effects would be more pronounced during intervals of radial IMF because nearly entire bow shock is quasiparallel and thus they can be responsible for the ion flux profile shown in Figure 8.

7.   RELATION BETWEEN THE ION FLUX AND HIGH-ENERGY PARTICLES

According to previous investigations (e.g., Kudela et al., 2000), the θ_{Bn} angle (the angle between the magnetic field and the shock normal) is a good parameter controlling the high-energy fluxes in near upstream region. Thus, assuming a dependence of the magnetosheath energetic population on the bow shock type, we calculated θ_{Bn} at two locations: (1) the point where the magnetosheath fluid parcel crossed the bow shock, and (2) the point on the bow shock to which the magnetic field currently connects the fluid parcel. These two procedures include:

(1) For each measuring point (i.e., each INTERBALL-1 position), we determined an orientation of the IMF vector. Then, we calculated intersections of

this vector with the model bow shock surface. After that, we defined the $M\theta_{Bn}$ angle between the IMF vector and the normal vector to the model bow shock in intersection points and estimated corresponding distances of each measuring point to the bow shock, D_{BS}. The model bow shock surface is an ellipsoid of revolution and thus two intersection points would result from calculations. We are using that providing shorter D_{BS} distance.

(2) We applied the gasdynamic model (Spreiter et al., 1966) to map the plasma flow along the streamline to the dayside region and determined the $P\theta_{Bn}$ angle at the cross-section of the streamline with the bow shock. The detailed description of these procedures is in Hayosh et al. (2004).

High-energy ion fluxes were registered by the DOK-2 instrument (Lutsenko et al., 1995; Kudela et al., 1995). This instrument consisted of narrow surface-barrier silicon detectors measuring the flux of ions in 3 different ranges of energy. We use the energy range of $22 - 29$ keV in our study. DOK-2 was equipped with two ion detectors that covered different spatial angles. The first detector, $Fp1$, was oriented along the rotation axis of the spacecraft and observed particles flowing toward the Sun, while the second detector, $Fp2$, was inclined by $62°$ with respect to the rotation axis and spun with a 2-minute period. To remove the spin modulation in this detector, the 10-minute averages were computed and a sum, $Fp1 + Fp2$, was used as a measure of the presence of energetic particles.

Since the main task of this section is a correlation of energetic particles and plasma fluxes, we present these quantities as a function of both the $M\theta_{Bn}$ and $P\theta_{Bn}$ angles in Figure 11. These angles are not fully independent because they are defined as angles between the IMF vector and normals to the model bow shock in two different locations. For this reason, the measurements are concentrated along the line determined by an equality of both angles with a spread of $\sim \pm 30°$. Figure 11a shows the behavior of the energetic particles. This plot demonstrates that the highest fluxes of energetic particles are observed when the $P\theta_{Bn}$ angle is lower than $30°$, and these fluxes do not depend on the $M\theta_{Bn}$ angle. We suggest that the observed particles are generated at the quasiparallel bow shock, trapped in local magnetic field inhomogeneities and carried downstream with the magnetosheath flow. The fraction of the particles streaming to the measuring point along magnetic field lines is measurable but much lower. These particles are observed when $P\theta_{Bn}$ is large and $M\theta_{Bn}$ is low.

Figure 11b illustrates a very surprising result; larger plasma fluxes are observed when the $P\theta_{Bn}$ angle is low, i.e., when streamlines connect magnetosheath points to the quasiparallel bow shock. On the other hand, when $P\theta_{Bn}$ is high (top part of the panel), we can see the opposite trend - the plasma flux increases with increasing $M\theta_{Bn}$. Comparing Figures 11a and 11b, one can note that the plasma and energy particle fluxes are roughly anticorrelated when $P\theta_{Bn} > 40°$ but they are nearly in correlation for lower values of $P\theta_{Bn}$. How-

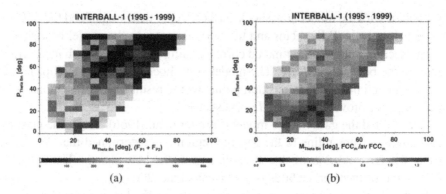

Figure 11. A connection between $M\theta_{Bn}$ and $P\theta_{Bn}$ for two parameters: (a) - high-energy particle flux $(Fp1 + Fp2)$; (b) - normalized ion flux $(FCCm/avFCCm)$.

ever, sorting the data according to θ_{Bn} angles does not reflect the position of investigated points in the magnetosheath. We have described this position by the distance from the magnetopause in units of the magnetosheath thickness, D. Since we cannot exclude the possibility that the relation of plasma and energetic particle fluxes would be different in the inner and outer magnetosheath, we have plotted the energetic particle flux as a function of this distance in Figures 12 and 13. $M\theta_{Bn}$ is plotted on horizontal axes of both panels in Figure 12 and the D distance is on vertical axes. Figure 12a clearly reveals that the largest fluxes of energetic particles can be observed in the region of the quasiparallel $(M\theta_{Bn} < 30°)$ bow shock and this flux decreases toward the magnetopause. On the other hand, the plot in Figure 12b shows that there is no connection between the plasma flux and $M\theta_{Bn}$ because the squares with a higher plasma flux are spread randomly. It means that a value of the plasma flux is given by conditions on the entry and that a further evolution of a radial profile due to energetic particles coming along magnetic field lines is negligible.

The same data sorted according to $P\theta_{Bn}$ are shown in Figure 13. The behavior of energetic particles in this figure (Figure 13a) is similar to the previous plot. This means that energetic particles can come to a particular magnetosheath point either along the magnetic field line or that the plasma can carry these particles embedded in local magnetic inhomogenities and the both sources are probably equally important.

The other possible source - leakage of particles from the magnetosphere - can be probably excluded because Figures 12a and 13a show a clear minimum of energetic particles near the magnetopause $(D = 0)$ regardless of the θ_{Bn} angle.

The magnetopause as a source of energetic particles was suggested many times but we are analyzing the particles measured by the DOK-2 device and, as we noted above, this device has a limited view angle and particles leaking

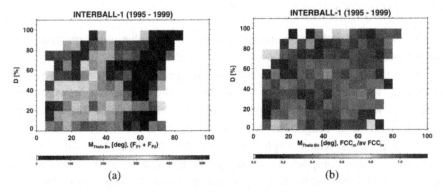

Figure 12. Distributions of high-energy particle fluxes (a) and normalized ion fluxes (b) as a function of normalized magnetosheath distance and parameter, $M\theta_{Bn}$.

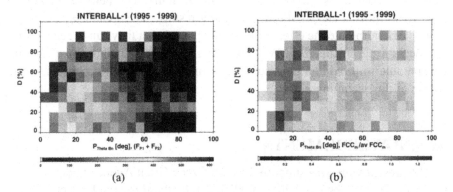

Figure 13. Distributions of high-energy particle fluxes (a) and normalized ion fluxes (b) as a function of normalized magnetosheath distance and parameter, $P\theta_{Bn}$.

from the magnetopause would occupy a narrow range of pitch angles and thus they would be frequently missed in DOK-2 measurements.

Strong magnetic field fluctuations typical for the magnetosheath flow would lead to the pitch-angle scattering of the particles. We assume that the streaming particles can be often out of the DOK-2 view angle and thus they are observed only when their distribution becomes more isotropic. Consequently, although being generated at the magnetopause, the particles are observed in the distance from the magnetopause is sufficient for isotropization to have occurred.

Figure 13b confirms our finding that the plasma flux is larger at the magnetosheath points connected via streamlines to a nearly parallel bow shock. However, this flux excess is observed only for $D < 70\%$, whereas the flux is depleted closer to the bow shock.

This complicated dependence of the plasma flux on two parameters can explain why the previous attempts to find a clear dependence of the plasma flux on one parameter only were not too successful.

8. CONCLUSION

The magnetosheath as an interface between the solar wind and magneto-sphere mediates all manifestations of solar wind activity. However, the present paper demonstrates that our knowledge of magnetosheath processes is still insufficient. The partial success of an analysis of the connection between energetic particles and plasma flow presented in the last figure of this paper suggests that further investigations of the magnetosheath should be based on multifactorial analysis. Each magnetosheath parameter depends on three spatial coordinates, three components of the upstream magnetic field, the solar wind density, the velocity, and the temperature and, maybe to a lesser extent, on many others. We think that even all the magnetosheath measurements carried out over the last 40 years are not sufficient for a reliable study. The increasing capacity of computers and development of new numerical methods leading to new MHD models are providing promising results. These models cannot, in principle, incorporate the effect of energetic particle but they can suggest a way to process experimental data. We hope that a joint effort of all the investigators will be able to provide a consistent view of magnetosheath processes in the course of the oncoming years.

ACKNOWLEDGMENTS

The present work was supported by the Czech Grant Agency under Contracts 205/02/0947 and 205/03/0953, and by the Research Project MSM 113200004.

REFERENCES

Aellig, M. R., Lazarus, A. J., and Steinberg, J. T., 2001, The solar wind helium abundance: Variation with wind speed and the solar cycle, *Geophys. Res. Lett.* **28**(14):2767–2770.

Elphic, R. C., and Southwood, D. J., 1987, Simultaneous measurements of the magnetopause and flux-transfer events at widely separated sites by AMPTE UKS and ISEE-1 and ISEE-2, *J. Geophys. Res.* **92**:13666–13672.

Fairfield, D. H., and Feldman, W. C., 1975, Standing waves at low Mach number laminar bow shock, *J. Geophys. Res.* **80**:515–522.

Formisano, V., 1979, Orientation and shape of the Earth's bow shock in three dimensions, *Planet. Space Sci.* **27**:1151–1161.

Hayosh, M., Šafránková, J., Němeček, Z., Přech, L., Kudela, K., and Zastenker, G. N., 2004, Relationship Between High-energy Particles and Ion Flux in the Magnetosheath, *Planet. Space Sci.*, accepted.

Howe Jr., H. C., and Binsack, J. H., 1972, Explorer 33 and 35 plasma observations of magnetosheath flow, *J. Geophys. Res.* **77**:3334–3344.

Kaymaz, Z, Siscoe, G. L., and Luhmann, J. G., 1992, IMF draping around the GEOTAIL - IMP 8 observations, *Geophys. Res. Lett.* **19**:829–832.

Klimov, S., et al., 1997, ASPI experiment: Measurements of fields and waves onboard the Interball-1 spacecraft, *Ann. Geophys.* **15**(5):514–527.

Kudela, K., Slivka, M., Rojko, J., and Lutsenko, V. N., 1995, The apparatus DOK-2 (Project Interball): Output data structure and models of operation, in: *preprint UEF - 01 -95*, IEP, Kosice, pp. 35.

Kudela, K., Slivka, M., Sibeck, D. G., Lutsenko, V. N., Sarris, E. T., Němeček, Z., Šafránková, J., Kiraly, P., and Kecskemety, K., 2000, Medium energy proton fluxes outside the magnetopause: INTERBALL-1 data, *Adv. Space Res.* **25**:1517–1522.

Luhmann, J. G., Walker, R. J., Russell, C.-T., Crooker, N. U., Spreiter, J. R., and Stahara, S. S., 1984, Patterns of magnetic field merging sites on the magnetopause, *J. Geophys. Res.* **89**:1739–1742.

Lutsenko, V. N., Rojko, J., Kudela, K., Gretchko, T. V., Balaz, J., et al., 1995, Energetic particle experiment DOK-2, in: *INTERBALL Mission and Payload*, Y. Galperin, T. Muliarchik, and J-P. Thouvenin, eds., CNES-IKI-RSA, Paris-Moscow, pp. 249–255.

Maynard, N. C, Sonnerup, B. U. O., Siscoe, G. L., Weimer, D. R., Siebert, K. D., Erickson, G. M., White, W. W., Schoendorf, J. A., Ober, D. M., Wilson, G. R., and Heinemann, M. A., 2002, Predictions of magnetosheath merging between IMF field lines of opposite polarity, *J. Geophys. Res.* **107**(A12):1456.

Němeček, Z., and Šafránková, J., 1991, The Earth's bow shock and magnetopause position as a result of solar wind — magnetosphere interaction, *J. Atm. Terr. Phys.* **53**(11-12):1049–1054.

Němeček, Z., Šafránková, J., Přech, L., Zastenker, G. N., Paularena, K. I., and Kokubun, S., 2000a, Magnetosheath study: INTERBALL observations, *Adv. Space Res.* **25**(7/8):1511–1516.

Němeček, Z., Šafránková, J., Zastenker, G. N., Pišoft, P., Paularena, K. I. I., and Richardson, J. D., 2000b, Observations of the radial magnetosheath profile and a comparison with gasdynamic model predictions, *Geophys. Res. Lett.* **27**(17):2801–2804.

Němeček, Z., Měrka, J., and Šafránková, J., 2000c, The tilt angle control of the outer cusp position, *Geophys. Res. Lett.* **27**(1):77–80.

Němeček, Z., Šafránková, J., Zastenker, G., Pišoft, P., Jelínek, K., 2002, Low-frequency variations of the ion flux in the magnetosheath, *Planet. Space Sci.* **50**(5/6):567–575.

Němeček, Z., Hayosh, M., Šafránková, J., Zastenker, G., and Richardson, J., 2003, The dawndusk asymmetry of the magnetosheath: INTERBALL-1 observations, *Adv. Space Res.* **31**(5): 1333–1340.

Ogilvie, K. W., Chornay, D. J., Fritzenreiter, R. J., Hunsaker, F., Keller, J., Lobell, J., Miller, G., Scudder, J. D., Sittler, E. C., Torbert, R. B., Bodet, D., Needell, G., Lazarus, A. J., Steinberg, J. T., And Tappan, J. H., 1995, SWE, A comprehensive plasma instrument for the Wind spacecraft, *Space Science Reviews* **71**(1-4):55–77.

Paularena, K. I., Richardson, J. D., Kolpak, M. A., Jackson, C. R., and Siscoe, G. L., 2001, A dawn-dusk density asymmetry in Earth's magnetosheath, *J. Geophys. Res.* **106**:25377–25394.

Petrinec, S. M., and Russell, C. T., 1996, Near-Earth magnetotail shape and size as determined from the magnetopause flaring angle, *J. Geophys. Res.* **101**(A1):137–152.

Petrinec, S. M., Mukai, T., Nishida, A., Yamamoto, T., Nakamura, T. K., and Kokubun, S., 1997, Geotail observations of magnetosheath flow near the magnetopause, using Wind as a solar wind monitor, *J. Geophys. Res.* **102**:26943–26959.

Russell, C. T., Ginskey, M., Petrinec, S., and Le, G., 1992, The effect of solar wind dynamic pressure changes on low and midlatitude magnetic records, *Geophys. Res. Lett.* **19**:1227–1230.

Šafránková, J., Zastenker, G., Němeček, Z., Fedorov, A., Simerský, M., and Přech, L., 1997, Small scale observation of the magnetopause motion: Preliminary results of the INTER-BALL project, *Ann. Geophys.* **15**(5):562–569.

Šafránková, J., Němeček, Z., Dušík, Š., Přech, L., Sibeck, D. G., and Borodkova, N. N., 2002, The magnetopause shape and location: a comparison of the Interball and Geotail observations with models, *Ann. Geophys.* **20**(3):301–309.

Šafránková, J., Jelínek, K., and Němeček, Z., 2003, The bow shock velocity from two-point measurements in frame of the INTERBALL project, *Adv. Space Res.* **31**(5):1377–1382.

Shue, J.-H., Chao, J. K., Fu, H. C., Khurana K. K., Russell, C. T., Singer, H. J., and Song, P., 1997, A new functional form to study the solar wind control of the magnetopause size and shape, *J. Geophys. Res.* **102**(A5):9497–9511.

Sibeck, D. G., Phan, T.-D., Lin, R. P., Lepping, R. P., Mukai, T., and Kokubun, S., 2000, A survey of MHD waves in the magnetosheath: International Solar Terrestrial Program observations, *J. Geophys. Res.* **105**:129–137.

Siscoe, G. L., Crooker, N. U., Erickson, G. M., Sonnerup, B. U. O., Maynard, N. C., Schoendorf, J. A., Siebert, K. D., Weimer, D. R., White, W. W., and Wilson, G. R., 2002, MHD properties of magnetosheath flow, *Planet. Space Sci.* **50**(5/6):461–471.

Song, P., Russell, C. T., and Thomsen, M. F., 1992, Slow mode transition in the frontside magnetosheath, *J. Geophys. Res.* **97**:8295–8305.

Song, P., Russell, C. T., Gombosi, T. I., Spreiter, J. R., Stahara, S. S., Zhang, X. X., 1999a, On the processes in the terrestrial magnetosheath 1. Scheme development, *J. Geophys. Res.* **104**(A10):22345–22355.

Song, P., Russell, C. T., Zhang, X. X., Stahara, S. S., Spreiter, J. R., Gombosi, T. I., 1999b, On the processes in the terrestrial magnetosheath 2. Case study, *J. Geophys. Res.* **104**(A10): 22357–22373.

Southwood, D. J., and Kivelson, M. G., 1992, On the form of the flow in the magnetosheath, *J. Geophys. Res.* **97**(A3):2873–2879.

Southwood, D. J., and Kivelson, M. G., 1995, Magnetosheath flow near the magnetopause: Zwan-Wolf and Southwood-Kivelson theories reconciled, *Geophys. Res. Lett.* **22**(23):3275–3278.

Spreiter, J. R., Summers, A. L., and Alksne, A. Y., 1966, Hydromagnetic flow around the magnetosphere, *Planet. Space Sci.* **14**:223–253.

Spreiter, J. R., Stahara, S. S., 1980, A new predictive model for determining solar wind-terrestrial planet interactions, *J. Geophys. Res.* **85**(NA12):6769–6777.

Weiss, L. A., Reiff, P. H., Weber, E. J., Carlson, H. C., Lockwood, M., Peterson, W. K., 1995, Flow-aligned jets in the magnetospheric cusp: Results from the Geospace Environment Modeling Pilot Program, *J. Geophys. Res.* **100**(A5):7649–7659.

Wu, C. C., 1992, MHD flow past and obstacle: Large-scale flow in the magnetosheath, *Geophys. Res. Lett.* **19**(2):87–90.

Yan, M., and Lee, L. C., 1994, Generation of slow mode waves in the front of the dayside magnetopause, *Geophys. Res. Lett.* **21**:629–632.

Zwan, B. J., and Wolf, R. A., 1976, Depletion of the solar wind plasma near a planetary boundary, *J. Geophys. Res.* **81**:1636–1648.

PRESSURE PULSES AND CAVITY MODE RESONANCES

David G. Sibeck
NASA Goddard Space Flight Center, Greenbelt, MD 20771

Abstract: Theory predicts that abrupt variations in the solar wind dynamic pressure trigger widespread compressional cavity mode resonances within the magnetosphere. We inspect solar wind and magnetospheric observations at the times of previously reported events seen in ground magnetograms. We find evidence for abrupt solar wind pressure variations in the form of direct observations of solar wind dynamic pressure, motion of the bow shock, or fluctuations in the location of the foreshock. We also find evidence for widespread compressions of the magnetospheric magnetic field in observations by geosynchronous spacecraft. However, in contrast to the predictions of the model for an abrupt increase in wave activity followed by a gradual decay, we find that the periodicity seen in previously reported events occurs primarily in response to repeated impulsive excitations. If cavity mode resonances are present, they dissipate very rapidly within two cycles.

Key words: foreshock; cavity mode resonances; pulsations.

1.   INTRODUCTION

Theory predicts the resonant oscillation of geomagnetic field lines in both compressional and transverse modes at frequencies dependent upon the plasma density, magnetic field strength, and field line length. For a simple box model of the magnetosphere, the oscillation frequency can be expressed as $f = nB(4l^2\mu_o\rho)^{-1/2}$, where ρ is the mass density, B the magnetic field strength, l the length of the magnetic field line, n any positive integer, and μ_o the permeability of free space (Kivelson and Russell, 1995). A combination of high plasmaspheric densities within the inner magnetosphere, a radial decrease in magnetic field strength, and a radial increase in the length of magnetic field lines leads to predictions that oscillation frequencies will peak

95

J. –A. Sauvaud and Z. Němeček (eds.),
Multiscale Processes in the Earth's Magnetosphere:From Interball to Cluster, 95-110.
© 2004 *Kluwer Academic Publishers. Printed in the Netherlands.*

near 100 mHz just outside the plasmapause. Frequencies should diminish to values of 25 mHz or less in the outer magnetosphere, to values of 10 or less in the outer plasmasphere (e. g., Poulter et al., 1988), but rise again to values of 40 to 70 mHz on magnetic field lines equatorward of the plasmasphere.

Oscillations with these periods are common on dayside magnetospheric magnetic field lines and at their footprints in the ionosphere. Oscillation periods within the outer magnetosphere invariably exhibit the expected decrease in frequency with radial distance from Earth (Anderson and Engebretson, 1994). While a similar decrease can occasionally be observed with increasing latitude on the ground (corresponding to increasing radial distance), it is far more common for the pulsations observed by ground stations over a wide range of latitude to exhibit a common frequency (Siebert, 1964; Voelker, 1968). Nevertheless, that common frequency tends to be less for pulsations events observed at higher latitudes than for those observed at lower latitudes (Samson and Rostoker, 1972). The sense of polarization within bands of pulsations with common frequencies reverses across local noon and the latitude at which the pulsation amplitude peaks (Samson et al., 1971).

These reversals, and the observations of common frequencies over a range in latitudes, led to the development of new models for pulsations within the magnetosphere (Chen and Hasegawa, 1974; Southwood, 1974). In these models, monochromatic compressional waves generated by the Kelvin-Helmholtz instability at the magnetopause propagate into the magnetosphere. The compressional waves excite natural azimuthal resonances deep within the magnetosphere. The amplitude of the azimuthal resonances diminishes with both increasing and decreasing radial distance away from resonant shells where the frequencies of the compressional and azimuthal waves match. The oscillations should be confined to a very narrow range of radial distances when there are sharp gradients in the magnetospheric density. The sense of polarization reverses across both the resonant shell and local noon.

Despite this success, Kivelson and Southwood (1985) questioned the existence of narrow band compressional wave mode sources at the magnetopause or the existence of the sharp plasma density gradients needed to confine the pulsations to a small range of radial distances within the magnetosphere. Instead, they proposed an alternative 'cavity-mode' model for both compressional and transverse pulsations within the magnetosphere. In this model, a single abrupt compression launches a fast mode wave into the magnetosphere that is then reflected from a sharp density gradient (e.g. the plasmapause). Resonant global compressional oscillations then appear within the outer magnetosphere at discrete frequencies. These standing compressional oscillations decay by exciting transverse mode oscillations that dissipate in the ionosphere. In the cavity mode model, compressional wave amplitudes and periods are similar at all radial positions outside the sharp gradient in density, whereas transverse mode frequencies and amplitudes depend strongly on radial distance (Allan et al., 1986). The frequencies of the oscillations depend upon the size of the cavity, but remain greater than 5 mHz for realistic magnetospheric dimensions.

When observations by ground radars identified oscillations at a number of clearly identifiable discrete frequencies, it seemed natural to invoke the cavity mode model (Samson et al., 1992). However, a number of problems were immediately apparent. First, the pulsations propagated antisunward, either at frequency-independent (Olson and Rostoker, 1978) or frequency-dependent (Mathie and Mann, 2000) phase velocities. These observations were reconciled with the model's prediction of standing waves by allowing the compressional oscillations to bounce back and forth between the magnetopause and inner density gradient as they simultaneously propagated antisunward. Opening the cavity to loss via the magnetotail requires waves amplitudes to decrease faster than they would have done in a closed system, as does allowing the magnetopause to move freely following the perturbation that initiates the oscillations (Freeman, 2000). However, inclusion of a realistic magnetotail waveguide does not change the expected frequencies of the waves, which remain 4 to 5 mHz (Allan and Wright, 2000).

A more serious threat to the cavity mode model emerged when the discrete frequencies were found to remain nearly constant from event to event, while the dimensions of the magnetosphere are known to vary greatly (Harrold and Samson, 1992). Despite the emphasis originally placed on these observations, it was subsequently argued that the periods do not in fact remain constant from event to event (Mathie et al., 1999). Observed frequencies on the order of 2 mHz forced Harrold and Samson (1992) to argue that confining cavity lies between an inner density gradient (e. g., the plasmapause) and the bow shock, rather than the magnetopause. To accommodate the lower frequencies, Mann et al. (1999) invoked free magnetopause motion and offered an alternative interpretation in which only quarter-wavelength modes were present within the magnetosphere.

Efforts to retain the cavity mode model were truly imperiled when Kepko et al. (2002) noted the frequent presence of compressional oscillations within the magnetosphere at frequencies below 1 mHz. They argued that no possible combination of boundaries and parameters would allow these oscillations to be explained in terms of the cavity mode model. Instead, Kepko et al. (2002) showed that they corresponded in a one-to-one manner to naturally-occurring variations in the solar wind density.

In fact, a similar correspondence between solar wind density variations and compressional oscillations of the dayside magnetospheric magnetic field

Table 1. "Pressure Pulse" Events

Frequency (mHz)	Solar Wind Monitor	M-sphere Response	Author
2.1	IRM	CCE/GOES	Sibeck et al. (1989)
2.4	IMP-8	GOES	Korotova et al. (1995)
3.3/5.6	ISEE-1/3	IMP-8	Sarafopoulos (1995)
2.8	Wind	IMP-8	Moldwin et al. (2001)
1.3/1.9/2.7	Wind	GOES	Kepko et al. (2002)
1.3/1.9/2.7/3.3	Wind	Ground	Stephenson/Walker (2002)
1.3/1.9/2.6/3.4	Wind	GOES	Kepko/Spence (2003)

had been noted some time before by Sibeck et al. (1989). They presented a case study that established a direct relationship between 2 mHz pulsations in the solar wind density with magnetopause motion, compressions of the geosynchronous magnetic field, and pulsations in high-latitude ground magnetograms. However, as clearly demonstrated by Fairfield et al. (1990), the driving solar wind density variations were not intrinsic solar wind features but rather pressure pulses generated by thermalized ions within the foreshock. A small fraction of solar wind ions incident upon the bow shock streams back into the solar wind. The backstreaming ions generate high frequency (10 to 50 mHz) waves on magnetic field lines connected to the bow shock (Fairfield, 1969). The waves scatter and thermalize the ions. Enhanced pressures associated with the ions generate cavities of depressed magnetic field strength and density on those magnetic field lines connected to the bow shock (Thomas and Brecht, 1985). The expanding cavities compress plasma on nearby magnetic field lines unconnected to the bow shock (Sibeck et al., 2002). Table 1 surveys past studies that attributed magnetospheric pulsations to both intrinsic solar wind and induced foreshock pressure pulse drivers.

Both intrinsic solar wind and foreshock-generated pressure pulses provide a convenient alternative to the monochromatic Kelvin-Helmholtz waves whose existence was questioned by Kivelson and Southwood (1985). Simulations indicate that the pulses are simply transmitted through the bow shock (Thomas et al., 1995). They drive antisunward-moving waves on the magnetopause and generate pulsations within the magnetosphere whose sense of polarization should reverse across local noon. The pressure pulses also launch fast mode waves into the magnetosphere that propagate antisunward and initiate resonant transverse oscillations at appropriate radial distances from Earth. These are precisely the characteristics of previously reported pulsations.

Consequently, the question now arises as to whether events previously interpreted in terms of cavity mode oscillations can be interpreted in terms of near-monochromatic solar wind pressure variations striking the magnetosphere. The purpose of this paper is to survey the solar wind observations corresponding to previously reported events in an effort to identify density variations capable of driving the events.

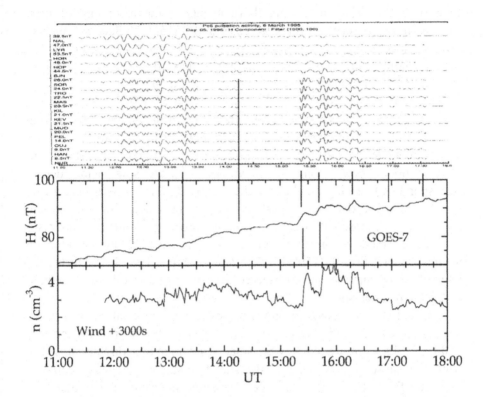

Figure 1. A comparison of IMAGE ground magnetograms with GOES-7 geosynchronous magnetometer and Wind solar wind plasma observations. From top to bottom, the figure shows the H components of magnetometers in the IMAGE array, the H component observed by NOAA/GOES-7 at dayside geosynchronous orbit, and the solar wind density observed by the GSFC SWE instrument on Wind from 1100 to 1800 UT on March 6, 1995. Wind observations have been lagged 3000s. Solid vertical lines connect corresponding features. A dashed vertical line indicates the absence of an impulsive event in the ground magnetograms at the time of one compression in the geosynchronous magnetic field.

2. CASE STUDY: MARCH 6, 1995

The top panel of Figure 1 presents the H (north-south) component of observations from auroral zone IMAGE ground magnetometers located at latitudes ranging from Ny Alesund (NAL, Corrected Geomagnetic Latitude 75.25°, Longitude 112.08°) to Nurmijärvi (NUR, CGL Latitude 56.89°, Longitude 102.89°) from 1100 to 1800 UT on March 6, 1995. Magnetic local time for Ny Alesund is UT + 3.2 hours. The observations have a time resolution of 10s and have been filtered between 1 and 10 mHz. Mathie and Mann (2000) identified two distinct pulsation packets from 1200 to 1330 and

1515-1630 UT. They noted impulsive events at the beginning of these packets and two further events at 1250 and 1315 UT. Spectral analysis revealed spectral peaks at 1.3, 1.9, 2.6, 3.3, 3.7, and 4.4 mHz in the first packet. Mathie and Mann (2000) interpreted these frequencies as evidence for waveguide harmonics. Peak amplitudes reached 50 nT.

The bottom panel of Figure 1 presents Wind SWE solar wind density observations for the same interval. The SWE observations have a time resolution of 83s and have been lagged by 3000s to account for the fact that they were made some 207 R_E upstream from Earth. As noted by Mathie and Mann (2000), there is an abrupt variation in the solar wind density corresponding to each ground event after 1500 UT. The variations recur each ~ 12 min. Clearly the largest amplitude pulsations in the ground magnetograms, those with frequencies on the order of 1.4 mHz, do not result from cavity mode resonances, but rather solar wind features striking the magnetosphere. Prior to 1500 UT, the relationship is less clear. During this interval, Wind may not have observed the solar wind features that actually struck the magnetosphere.

We will employ geosynchronous magnetic field observations to test this hypothesis and identify the cause of the ground events observed prior to 1500 UT. Past studies employing solar wind monitors directly upstream from the subsolar bow shock have shown that dayside geosynchronous magnetic field observations provide a sensitive indicator of the solar wind dynamic pressure applied to the magnetosphere (Sibeck et al., 1989; Fairfield et al., 1990). The middle panel in Figure 1 presents GOES-7 dayside magnetospheric magnetic field strength observations at 1 min time resolution. During the interval shown, GOES-7 was located at LT = UT - 9 Hours. Vertical dashed lines indicate times when the geosynchronous magnetic field strength increased abruptly. After 1500 UT, the relationship of these abrupt changes to variations in the solar wind density is clear. Prior to 1500 UT, the relationship between the geosynchronous magnetic field strength and the density variations observed by Wind is unclear. However, there is an almost one-to-one relationship between transient events in the auroral zone ground magnetograms and abrupt variations in the geosynchronous magnetic field strength.

The close relationship between abrupt changes in the geosynchronous magnetic field and transient events on the ground leads us to conclude that each burst of wave activity on the ground was triggered by a variation in the solar wind dynamic pressure. After 1500 UT, we were able to identify corresponding variations in the lagged Wind SWE solar wind density observations. Prior to 1500 UT, we were not able to identify corresponding variations in the SWE observations. We conclude that Wind did not observe the solar wind features that struck the magnetosphere prior to 1500 UT.

Finally, we note that no resonant oscillation in the ground magnetograms endured for more than two cycles following an impulsive excitation. Although we cannot be sure that the ground oscillations resulted from cavity mode (as opposed to resonant transverse) oscillations, the observations indicate very rapid ionospheric damping.

3.   CASE STUDY: MARCH 8, 1994

The top panel of Figure 2 presents 10s time resolution H component observations from auroral zone IMAGE ground magnetometers at latitudes ranging from Ny Alesund to Nurmija rvi. The observations have been filtered between 1 and 10 mHz. Spectral analysis indicates the presence of waves with frequencies of ~1.8, 2.7, 3.4, and 4.1 mHz from 0620 to 0800 UT (Mathie and Mann, 2000). In contrast to the predictions of the cavity mode model, wave amplitudes did not decay with time following a single impulsive excitation. Instead, the wave train exhibited several impulsive events with similar (<200 nT) amplitudes at 0624, 0648, 0709, 0724, and 0742 UT.

The repeated occurrence of impulsive events with similar amplitudes suggests an interpretation in which a sequence of solar wind dynamic pressure variations strike the magnetosphere. This hypothesis can be tested by examining simultaneous solar wind observations. The lower three panels of Figure 2 presents IMP-8 MIT Faraday cup plasma and GSFC magnetic field observations for the corresponding interval at 60 and 15.36s time resolution, respectively. IMP-8 moved from GSE (x, y, z) = (-8.9, 28.4, 18.4) R_E at 0500 UT to (-12.0, 27.3, 17.6) R_E at 1000 UT. The spacecraft was in the nominal vicinity of the dusk flank bow shock.

Ideally, we should be able to identify corresponding solar wind density and pressure variations in the plasma observations. However, these observations terminated with a prolonged data gap at 0700 UT. Furthermore, the observations prior to this time, exhibited a number of isolated data spikes. However, there are reasons to believe that these impulsive variations were instrumental artifacts. In the pristine solar wind, density and magnetic field strengths vary in antiphase (Burlaga, 1968). Within the foreshock, they vary in phase (Sibeck et al., 1989; Fairfield et al., 1990). However, there was no clear relationship between the density and magnetic field strength variations shown in the second and third panels of Figure 2.

Thus, we must rely upon the magnetic field observations alone to infer variations in the solar wind density and dynamic pressure applied to the magnetosphere. The magnetic field observations can be divided into three categories. From 0500 to 0615 and 0740 to 0815 UT, the magnetic field

strength remained nearly constant at 6 nT. These are intervals of pristine
solar wind observations. The absence of any bow shock crossings within
these intervals indicates that the solar wind dynamic pressure remained
relatively constant, while the large cone angles indicate that the foreshock

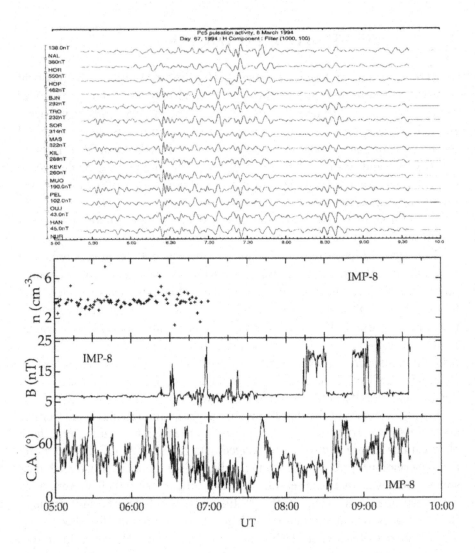

Figure 2. A comparison of IMAGE ground magnetograms with IMP-8 solar wind plasma and
magnetic field observations from 0500 to 1000 UT on March 8, 1994. From top to bottom, the
figure shows the H components observed by ground magnetometers in the IMAGE array, the
IMP-8 MIT Faraday cup solar wind density, the magnetic field strength observed by the IMP-
8 GSFC magnetometer, and the cone angle observed by the IMP-8 GSFC magnetometer. The
cone angle is the angle between the magnetic field and the Sun-Earth line.

did not lie directly upstream from the subsolar bow shock. Consistent with our inference that the pressure applied to the magnetosphere remained nearly constant, we note the absence of significant events in the ground magnetograms from 0500 to 0620 and from 0755 to 0825 UT.

From 0615 to 0740 UT, the cone angle diminished to values well below 30° and the magnetic field strength exhibited large transient variations. These observations typify the foreshock. During this interval, the magnetosphere should have been battered by a sequence of foreshock-generated pressure pulses. Consistent with this inference, we note a sequence of large-amplitude transient events in the ground magnetograms from 0620 to 0755 UT. Because IMP-8 was not located directly upstream from the subsolar bow shock, we do not expect (nor do we observe) a one-to-one correspondence between individual events in the IMP-8 and ground observations.

After 0815 UT, IMP-8 observed a sequence of bow shock crossings between regions of interplanetary (6 nT) and magnetosheath (21-25 nT) magnetic field strengths. Since the bow shock moves inward during periods of enhanced solar wind density and dynamic pressure and outward during periods of depressed density and dynamic pressure, these crossings constitute evidence for a sequence of intrinsic solar wind density and dynamic pressure variations. From the crossings themselves, we cannot estimate the amplitude of the bow shock motion or the pressure variations driving it. However, the fact that the ground pulsation with the largest amplitude occurred at 0830 UT suggests that the bow shock crossing at this time exhibited the largest amplitude motion.

In summary, we conclude that there was strong evidence for both foreshock-generated (0625 to 0750 UT) and intrinsic (after 0830 UT) solar wind density and dynamic pressure variations battering the magnetosphere during the intervals when ground pulsations were observed. By contrast, solar wind parameters remained steady during intervals when pulsation amplitudes diminished. The ground observations can best be interpreted in terms of a direct response to a sequence of pressure variations. Once again, we cannot be sure that these are compressional, rather than azimuthal, resonant oscillations. However, in either case they damp rapidly within two cycles.

4. CASE STUDY: APRIL 13, 1994

The top panel of Figure 3 presents the filtered H component observations of mid-latitude Australian ground magnetometers from 1830 to 1850 UT on April 13, 1994. During this interval, the ground stations were located

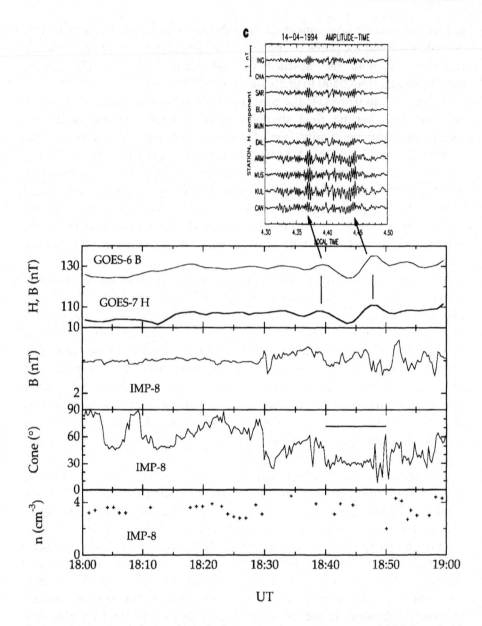

Figure 3. A comparison of Australian ground magnetograms, GOES-6 and -7 geosynchronous magnetic field observations, and IMP-8 solar wind magnetic field and plasma observations from 1800 to 1900 UT on April 13, 1994. From top to bottom, the figure shows H component observations from the Australian ground magnetometers, the GOES-6 B and GOES-7 H magnetic field observations, IMP-8 magnetic field strength and cone angle observations, and the IMP-8 solar wind density. A horizontal bar indicates an interval of low IMF cone angle. Vertical bars identify two prominent compressions of the magnetospheric magnetic field strength. Two arrows connect these compressions to intensifications of the wave activity in the ground magnetograms.

between 0430 and 0450 LT (on April 14, 1994). The latitudes of the stations ranged from Ingam (ING, CGL Latitude = -28.0°, Longitude = 219.6°) to Canberra (CAN, CGL Latitude = -45.6°, Longitude = 226.7°). The observations have been filtered between 10 and 100 mHz. Menk et al. (2000) noted the presence of spectral structuring with frequencies between 40 and 60 mHz from 1832 to 1848 UT. The wave activity does not exhibit the single impulsive excitation followed by an exponential decay predicted by the cavity mode model. Instead, arrows indicate at least two prominent enhancements in wave activity (both with amplitudes less than 1 nT) at 1836 and 1844 UT. Both enhancements decay rapidly, within 1-2 min.

Now consider the GOES-6 and -7 magnetometer observations shown in the second panel of Figure 3 at 60s time resolution. During the interval from 1800 to 1900 UT, the GOES-6 and -7 geosynchronous spacecraft moved through the dayside magnetosphere at LT = UT - 7 and LT = UT - 8, respectively. The variations in the geosynchronous magnetic field strength that both spacecraft recorded during this interval were presumably caused by fluctuations in the solar wind dynamic pressure applied to the magnetosphere.

To test this hypothesis, the lower panels in Figure 3 present IMP-8 MIT Faraday cup plasma and GSFC magnetometer observations. During the interval shown, IMP-8 moves from GSE (x, y, z) = (7.7, 31.3, 18.2) to (7.1, 31.5, 18.0) R_E. Although IMP-8 was not located directly upstream from the subsolar bow shock and there were large data gaps in the MIT Faraday cup plasma observations, we can infer conditions upstream from the subsolar bow shock from the IMP-8 magnetometer observations. Prior to 1830 UT, IMP-8 recorded a nearly constant magnetic field strength and cone angles exceeding 50°. By contrast, the spacecraft recorded significant variations in the magnetic field strength and much lower cone angles from 1830 UT onward. Transient decreases in the cone angle to 30° placed the foreshock directly upstream from the subsolar bow shock. During this interval, the magnetosphere should have been bombarded with the high frequency waves that are generated within the foreshock. These waves probably account for the weak (~1 nT), high frequency (40-60 mHz) oscillations seen in the ground magnetograms from 1832 to 1848 UT.

From 1840 to 1848 UT, the cone angle remained steady at low values near 30°. A prolonged period of low cone angles should result in the formation of a foreshock cavity with depressed densities bounded by regions of enhanced density. We suppose that one such cavity formed during the interval from 1840 to 1849 UT. Upon being swept downstream and impacting the magnetosphere, such a cavity should produce two magnetospheric compressions bounding a rarefaction, precisely the signatures observed by GOES-6 and -7 from 1840 to 1848 UT.

To summarize, a transient turning of the IMF towards a more radial orientation placed the foreshock directly upstream from the subsolar bow shock and resulted in the formation of a foreshock cavity. Pressure variations associated with both the foreshock cavity and high-frequency waves generated within the foreshock battered the magnetosphere. The former generated transient compressions in the geosynchronous magnetic field, while the latter provide a plausible source for the 40-60 mHz fluctuations observed in mid-latitude ground magnetograms. There is no need to invoke the cavity mode to explain the observations.

5.   SURVEY OF PREVIOUSLY REPORTED CASES

We examined simultaneous solar wind observations for a number of other previously reported case studies of geomagnetic pulsations that were interpreted in terms of cavity mode resonances. Table 2 summarizes our results. The previously reported case studies can be divided into three categories: those that considered oscillations with frequencies exceeding 40 mHz, those that considered oscillations with frequencies less than 10 mHz, and those that consider oscillations solely within the nightside magnetosphere and ionosphere. Events in the first category were invariably associated with low cone angles, indicating near-radial IMF orientations. Under these conditions, upstream waves with frequencies on the order of 30 mHz are common. Given the similarity of the frequencies seen within the foreshock and those observed on the ground during periods of radial IMF orientations, it seems natural to attribute the high frequency ground oscillations to upstream waves rather than cavity mode resonances.

By contrast, events exhibiting frequencies at and below 10 mHz could be associated with a wider variety of solar wind features, including pressure pulses, bow shock crossings, and radial IMF orientations. As we can infer variations in the solar wind dynamic pressure from bow shock crossings and expect significant variations to be generated within the foreshock during periods of radial IMF orientation, the observations again suggest solar wind pressure variations as direct drivers for the largest amplitude transient events seen in the ground magnetograms. They do not, however, rule out cavity mode resonances for the much weaker higher frequency waves seen in conjunction with these events.

The final category involves events seen in nightside ground magnetometer and optical observations, magnetospheric magnetic field measurements, as well as nightside ionospheric plasma flows (Sanchez et al., 1997; Lyons et al., 2002). Frequencies are similar in each location and range

Table 2. "Cavity Mode" Events

Date	UT	Frequency (mHz)	Solar Wind Feature	Author
11/24/77	1800–2400	2.1	ND	1
10/28/84	0900–1030	2.6	PP/R IMF	8
1/1/93	1230–1430	0–7	PP	2
3/8/94	0610–0810	10	BSX	3
3/17/94	1215–1345	1.4	PP	4
4/12/94	2110–2140	>40	R IMF	5
4/13/94	0130–0150	>40	R IMF	5
4/13/94	1830–1850	>40	R IMF	5
4/17/94	0239–0259	>40	ND	5
4/22/94	2305–2325	>40	R IMF (ND)	5
3/6/95	1145–1345	1–10	PP	3
3/9/96	1030–1230	1.9	PP	4
3/22/96	0530–0700	3.3	R IMF	4
4/23/96	1600–1630	7–8	R IMF	6
12/10/02	0000–0600	2.4–2.8	R IMF	7

Here R IMF indicates a radial IMF, PP indicates pressure pulses, ND indicates no data, and BSX indicates bow shock crossings. Authors: (1) Kivelson et al. (1984), (2) Ziesolleck and McDiarmid (1994), (3) Mathie and Mann (2000), (4) Mathie et al. (1999), (5) Menk et al. (2000), (6) Yeoman et al. (1997), (7) Mann et al. (2002), (8) Mann et al. (1998)

from 0.5 to 3.9 mHz. It seems unlikely that variations in the solar wind density (or magnetic field orientation) trigger individual intensifications, but rather more likely that internal instabilities associated with reconnection and/or current disruption trigger these 'waves'.

6. SUMMARY AND CONCLUSION

In this paper, we have reviewed the arguments in favor and against cavity mode resonances within the magnetospheric cavity and examined both solar wind and geosynchronous observations at the times of previously reported cavity mode events. Our survey of the literature reveals that resonance frequencies do not remain nearly constant from event to event (see Tables 1 and 2). There is no need for an explanation in terms of 'magic numbers' for frequencies that remain invariant despite varying solar wind conditions. As previously noted, frequencies on the order of 1 to 2 mHz are common. Such frequencies are too low to be supported by standard cavity mode models invoking compressional resonances between an inner plasmapause boundary and the outer dayside magnetopause boundary. They might be supported by models invoking quarter-wavelength modes within the magnetosphere.

In each case, the ground observations indicated nearly constant frequencies over a wide range of geomagnetic latitudes. Since the frequencies of toroidal events should vary with latitude, these observations imply the presence of compressional oscillations. To distinguish between the resonant global compressions predicted by the cavity mode models and oscillations directly driven by repetitive density variations striking the magnetosphere, we inspected simultaneous geosynchronous and solar wind observations. We found an abrupt change in the geosynchronous magnetic field strength corresponding to almost all intensifications in the wave activity at the ground. We were also able to observe or infer corresponding solar wind density variations for each event. Our survey of solar wind observations at the times of the variously reported events revealed numerous instances when they could be associated with solar wind density variations, bow shock crossings, and radial IMF orientations, all conditions from which we can infer strong density variations striking the magnetosphere. From this we conclude that the dominant oscillation mode, that with the lowest frequency and greatest amplitude, is not a cavity mode but rather directly driven by variations in the solar wind dynamic pressure. Without higher time resolution solar wind observations, we cannot comment on the origin of the much weaker higher frequency oscillations. They might be cavity mode resonances.

In each case presented, the large-amplitude, low-frequency ground oscillations died out within two cycles following impulsive excitation. If the oscillations are interpreted as evidence for cavity mode resonances, then the observations imply rapid damping. Possible loss modes for the energy within the oscillations include conversion to transverse modes and damping in the ionosphere, loss via propagation down the magnetotail, and loss into the magnetosheath via interaction with a freely-moving magnetopause.

ACKNOWLEDGMENTS

We thank A. Szabo for the IMP-8 GSFC magnetometer observations, A. J. Lazarus for the IMP-8 MIT plasma observations, K. W. Ogilvie for the Wind SWE plasma observations, and H. J. Singer for the GOES magnetometer observations. These data sets were supplied by the NASA/GSFC CDA Web data server.

REFERENCES

Allan, W., White, S. P., and Poulter, E. M., 1986, Impulse-excited hydromagnetic cavity resonances in the magnetosphere, *Planet. Space Sci.* **34**:371–385.

Allan, W., and Wright, A. N., 2000, Magnetotail waveguide: Fast and Alfvén waves in the plasma sheet boundary layer and lobe, *J Geophys Res.* **105**:317–328.

Anderson, B. J., and Engebretson, M. J., 1994, Relative intensity of toroidal and compressional Pc 3-4 wave power in the dayside outer magnetosphere, *J. Geophys. Res.* **100**:9591–9603.

Burlaga, L. F., 1968, Micro-scale structures in the interplanetary medium, *Solar Phys.* **4**:67–92.

Chen, L., and Hasegawa, A., 1974, A theory of long-period magnetic pulsations: 1. Steady state excitation of field line resonances, *J. Geophys. Res.* **79**:1024–1032.

Fairfield, D. H., 1969, Bow shock associated waves observed in the far upstream interplanetary medium, *J. Geophys. Res.* **74**:3541–3553.

Fairfield, D. H., Baumjohann, W., Paschmann, G., Lühr, H., and Sibeck, D. G., 1990, Upstream pressure variations associated with the bow shock and their effects on the magnetosphere, *J. Geophys. Res.* **95**:3773–3786.

Freeman, M. P., 2000, Effect of magnetopause leakage on the lifetime of magnetospheric cavity modes, *J. Geophys. Res.* **105**:5463–5470.

Harrold, B. G., and Samson, J. C., 1992, Standing ULF modes of the magnetosphere: A theory, *Geophys. Res. Lett.* **19**:1811–1814.

Kepko, L., Spence, H. E., and Singer, H. J., 2002, ULF waves in the solar wind as direct drivers of magnetospheric pulsations, *Geophys. Res. Lett.* **29**(8):1197, doi: 10.1029/2001GL014405.

Kepko, L., and Spence, H. E., 2003, Observations of discrete, global magnetospheric oscillations directly driven by solar wind density variations, *J. Geophys. Res.* **108**(A6):1257, doi: 10.1029/2002JA009676.

Kivelson, M. G., Etcheto, J., and Trotignon, J. G., 1984, Global compressional oscillations of the terrestrial magnetosphere: The evidence and a model, *J. Geophys. Res.* **89**:9851–9856.

Kivelson, M. G., and Southwood, D. J., 1985, Resonant ULF waves: A new interpretation, *Geophys. Res. Lett.* **12**:49–52.

Kivelson, M. G., and Russell, C. T., 1995, *Introduction to Space Physics*, Cambridge University Press, Cambridge.

Korotova, G. I., and Sibeck, D. G., 1995, A case study of transient event motion in the magnetosphere and in the ionosphere, *J. Geophys. Res.* **100**:35–46.

Lyons, L. R., Zesta, E., Xu, Y., Sanchez, E. R., Samson, J. C., Reeves, G. D., Ruohoniemi, J. M., and Sigwarth, J. B., 2002, Auroral poleward boundary intensifications and tail bursty flows: A manifestation of large-scale ULF oscillation? *J. Geophys. Res.* **107**(A11):1352, doi: 10.1029/2001JA000242.

Mann, I. R., Chisham, G., and Bale, S. D., 1998, Multi-satellite and ground-based observations of a tailward propagating Pc5 magnetospheric waveguide mode, *J. Geophys. Res.* **103**:4657–4669.

Mann, I. R., Wright, A. N., Mills, K. J., and Nakariakov, V. M., 1999, Excitation of magnetospheric waveguide modes by magnetosheath flows, *J. Geophys. Res.* **104**:333–353.

Mann, I. R., Voronkov, I., Dunlop, M., Donovan, E., Yeoman, T. K., Milling, D. K., Wild, J., Kauristie, K., Amm, O., Bale, S. D., Balogh, A., Viljanen, A., and Opgenoorth, H. J., 2002, Coordinated ground-based and Cluster observations of large amplitude global magnetospheric oscillations during a fast solar wind speed interval, *Ann. Geophys.* **20**:405–426.

Mathie, R. A., and Mann, I. R., 2000, Observations of Pc5 field line resonance azimuthal phase speeds: A diagnostic of their excitation mechanism, *J. Geophys. Res.* **105**:10713–10728.

Mathie, R. A., Mann, I. R., Menk, F. W., and Orr, D., 1999, Pc5 ULF pulsations associated with waveguide modes observed with the IMAGE magnetometer array, *J. Geophys. Res.* **104**:7025–7036.

Menk, F. W., Waters, C. L., and Fraser, B. J., 2000, Field line resonances and waveguide modes at low latitudes 1. Observations, *J. Geophys. Res.* **105**:7747–7761.

Moldwin, M. B., Mayerberger, S., Rassoul, H. K., Collier, M. R., Lepping, R. P., Slavin, J. A., and Szabo, A., 2001, Evidence of differential magnetotail responses to small solar wind pressure pulses depending on IMF Bz polarity, *Geophys. Res. Lett.* **28**:4163–4166.

Olson, J. V., and Rostoker, G., 1978, Longitudinal phase variations of Pc 4-5 micropulsations, *J. Geophys. Res.* **83**:2481–2488.

Poulter, E. M., Allan, W., and Bailey, G. J., 1988, ULF pulsation eigenperiods within the plasmasphere, *Planet. Space Sci.* **36**:185–196.

Samson, J. C., Jacobs, J. A., and Rostoker, G., 1971, Latitude-dependent characteristics of long-period geomagnetic micropulsations, *J. Geophys. Res.* **76**:3675–3683.

Samson, J. C., and Rostoker, G., 1972, Latitude-dependent characteristics of high-latitude Pc 4 and Pc 5 micropulsations, *J. Geophys. Res.* **77**:6133–6144.

Samson, J. C., Harrold, B. G., Ruohoniemi, J. M., Greenwald, R. A., and Walker, A. D. M., 1992, Field line resonances associated with MHD waveguides in the magnetosphere, *Geophys. Res. Lett.* **19**:441–444.

Sanchez, E., Kelly, J. D., Angelopoulos, V., Hughes, T., and Singer, H., 1997, Alfvén modulation of the substorm magnetotail transport, *Geophys. Res. Lett.* **24**:979–982.

Sarafopoulos, D. V., 1995, Long-duration compressional pulsations inside the Earth magnetotail lobes, *Ann. Geophys.* **13**:926–937.

Sibeck, D. G., Baumjohann, W., Elphic, R. C., Fairfield, D. H., Fennell, J. F., Gail, W. B., Lanzerotti, L. J., Lopez, R. E., Luehr, H., Lui, A. T. Y., Maclennan, C. G., McEntire, R. W., Potemra, T. A., Rosenberg, T. J., and Takahashi, K., 1989, The magnetospheric response to 8-minute-period strong-amplitude upstream pressure variations, *J. Geophys. Res.* **94**:2505–2019.

Sibeck, D. G., Phan, T.-D., Lin, R., Lepping, R. P., and Szabo, A., 2002, Wind observations of foreshock cavities: A case study, *J. Geophys. Res.* **107**(A10):1271, doi: 10.1029/2001JA007539.

Siebert, M., 1964, Geomagnetic pulsations with latitude-dependent periods and their relation to the structure of the magnetosphere, *Planet. Space Sci.* **12**:137–147.

Southwood, D. J., 1974, Some features of field line resonances in the magnetosphere, *Planet. Space. Sci.* **22**:483–491

Stephenson, J. A. E., and Walker, A. D. M., 2002, HF radar observations of Pc5 ULF pulsations driven by the solar wind, *Geophys. Res. Lett.* **29**(9):1297, doi: 10.1029/2001GL014291.

Thomas, V. A., and Brecht, S. H., 1988, Evolution of diamagnetic cavities in the solar wind, *J. Geophys. Res.* **93**:11341–11353.

Thomas, V. A., Winske, D., and Thomsen, M. F., 1995, Simulation of upstream pressure pulse propagation through the bow shock, *J. Geophys. Res.* **100**:23481–23488.

Voelker, H., 1968, Observations of geomagnetic pulsations: Pc 3, 4, and Pi 2 at different latitudes, *Ann. Geophys.* **24**:245–252.

Yeoman, T. K., Wright, D. M., Robinson, T. R., Davies, J. A., and Rietveld, M., 1997, High spa-tial and temporal resolution observations of an impulse-driven field line resonance in radar backscatter artificially generated with the Tromsø heater, *Ann. Geophys.* **15**:634–644.

Ziesolleck, C. W. S., and McDiarmid, D. R., 1994, Auroral latitude Pc 5 field line resonances: Quantized frequencies, spatial characteristics, and diurnal variation, *J. Geophys. Res.* **99**:5817–5830.

TWO–POINT INTERBALL OBSERVATIONS OF THE LLBL

Zdeněk Němeček[1], Jana Šafránková[1], Lubomír Přech[1],
and Jiří Šimůnek[2,1]

[1]*Faculty of Mathematics and Physics, Charles University, V Holešovičkách 2, 180 00 Prague
8, Czech Republic;* [2]*Institute of Atmospheric Physics, Czech Academy of Science, Boční 1401,
141 31 Prague 4, Czech Republic*

Abstract   The low-latitude boundary layer (LLBL) is encountered as an interface between
two plasma regions – the magnetosheath and plasma sheet and thus contains
a mixture of both plasma populations.   Several mechanisms have been discussed as candidates for a formation of the LLBL. These mechanisms can be
divided into magnetic reconnection between the magnetospheric and magnetosheath magnetic fields, impulsive penetration of magnetosheath plasma, and
viscous/diffusive mixing of plasma populations at the magnetopause. The observed fluctuations of plasma parameters inside the LLBL are attributed either
to transient nature of the phenomena forming the layer or to sweeping of defor-
mations of the magnetopause or inner edge of the LLBL along the spacecraft.

The INTERBALL-1/MAGION-4 satellite pair separated by several thousands
of kilometers crossed the LLBL region in different local times and their two-
point observations allow us to distinguish between spatial and temporal changes.
The present paper surveys results achieved so far. They suggest that the most
probable source of the LLBL plasma is reconnection occurring at high latitudes.
This reconnection can supply the nightside as well as dayside LLBL during intervals of northward oriented and/ or horizontal IMF. When the IMF B_z component becomes negative, the reconnection site moves toward lower latitudes but it
can move to the subsolar point only during exceptional intervals of negative B_z
dominated IMF.

Key words: LLBL; plasma depletion layer; reconnection; plasma mantle; magnetopause.

1.   INTRODUCTION

The magnetopause is a boundary dividing two worlds – interplanetary space
where the magnetic field of solar origin is frozen into the solar wind and the

J. –A. Sauvaud and Z. Němeček (eds.),
Multiscale Processes in the Earth's Magnetosphere:From Interball to Cluster, 111-130.
© 2004 *Kluwer Academic Publishers. Printed in the Netherlands.*

magnetosphere where plasma processes are controlled by the Earth's magnetic field. These two magnetic fields are separated by a current sheet, the magnetopause. Currents flowing along the magnetopause compensate the Earth's magnetic field in the outer space. Since the conductivity of the magnetopause current layer is high, but finite, a source which is able to drive these currents is required. A long time before the first magnetopause observations, Chapman and Ferraro (1931) suggested that a small portion of the solar wind would cross the Earth's magnetic field and the resulting electric potential (now known as the cross-tail potential) would be this source. However, a penetration of the solar wind plasma into the magnetosphere implies a presence of solar wind plasma on magnetospheric field lines in a layer adjacent to the magnetopause. The penetration process (regardless of its physical nature) could modify the original solar wind distribution and mass composition but we can expect that the penetrating plasma would conserve, to a certain extent, solar wind characteristics and thus we would be able to distinguish it from the magnetospheric plasma which was already present on these field lines. The plasma entry should be permanent and thus there should be a process (processes) which returns the penetrating plasma into the solar wind immediately or transports it into the inner magnetosphere and then to the solar wind.

The above, very simplified considerations led us to a conclusion that somewhere at the magnetopause should be a layer which plays a key role in the magnetosphere formation. It would be located on magnetospheric field lines and occupied by a plasma with parameters intermediate between those of solar wind and magnetosphere. These properties can be considered as a definition of this layer. Such layer has been found everywhere along the magnetopause (e.g., Eastman et al., 1976) at low latitudes and thus it has been named the Low-Latitude Boundary Layer (LLBL).

2.   PROPERTIES OF THE LLBL

We can demonstrate basic features of the LLBL using one example of the LLBL - magnetopause crossing during relatively stable conditions in the solar wind (Němeček et al., 2002). The LLBL was crossed by INTERBALL-1 and MAGION-4 satellites at (6.3; 11.0; -0.3) R_E of GSM coordinates between 1530 and 1730 UT on April 1, 1996. Figure 1 combines the data of both spacecraft which moved essentially along the same orbit. The magnetopause current layer can be identified as a jump of the magnetic field magnitude and its rotation through a large angle ($\sim 90^\circ$) observed by INTERBALL-1 at $\sim$1639 UT (second panel). However, if we compare the magnetic field data in the first panel with the ion flux data shown as a blue line in the third panel, we should

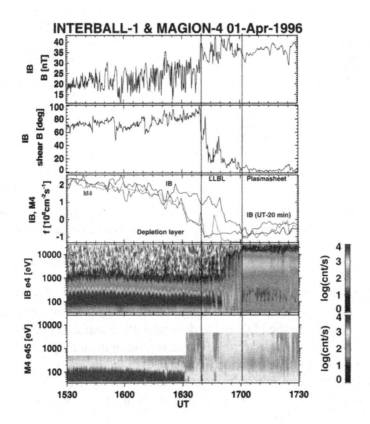

Figure 1. A comparison of INTERBALL-1 (IB) and MAGION-4 (M4) observations of the low-latitude dayside magnetopause crossing (panels from top to bottom: IB magnetic field strength; magnetic shear estimated from IB; IB and M4 ion fluxes; IB electron energy spectra; M4 electron spectra).

note that there is no significant change of the ion flux at that time and that it continues its gradual decrease.

A similar ion flux profile was observed by MAGION-4 about 20 minutes earlier and it is shown by the red line in Figure 1. The INTERBALL-1 data shifted by 20 minutes backward (black line) coincide except for temporal fluctuations which will be discussed later, with MAGION-4 measurements and thus we can consider such profile as a typical LLBL profile.

The layer in front of the magnetopause is named the Depletion Layer (DP) or Plasma Depletion Layer (PDL) and it was a subject of extended discussion in course of last years. This layer was predicted by 3-D MHD simulations (e.g., Wu, 1992) as well as by analytical treatments (Midgley and Davis, 1963; Lees, 1964; Zwan and Wolf, 1976). From these considerations, it was expected that IMF orientations perpendicular to the Earth-Sun line favor a PDL formation,

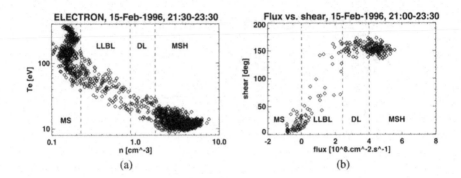

Figure 2.   Plots of electron temperature as a function of the electron density for the 2-hour interval involving all magnetopause boundary layers (a) and the ion flux as a function of magnetic shear (b) - adapted from Němeček et al. (2003a).

whereas a radial interplanetary magnetic field (IMF) does not. Since the PDL occurs in the subsolar region where a significant energy transfer is expected, it is important to define the PDL and to determine the conditions under which is formed. The PDL is characterized by a reduction in the total plasma density, a decrease in the particle pressure, an increase in the magnetic field pressure that balances the total pressure, and an increase in the p_{perp}/p_{par} pressure anisotropy (Crooker et al., 1979; Paschmann et al., 1978; Song et al., 1990; Fuselier et al., 1991; 1994).

Paschmann et al. (1993) performed a superposed epoch analysis of 22 dayside magnetopause crossings when the magnetic shear across the magnetopause was low ($< 30^o$) to obtain average profiles of the plasma parameters and magnetic field near the magnetopause. They found clear indications of a PDL and magnetic field pile-up region. In further studies, Anderson and Fuselier (1993) reported a PDL for all orientations of the magnetosheath magnetic field, although the density decrease and field increase were smallest when the magnetic field was southward. Most of these studies concern to the subsolar magnetopause but Phan et al. (1997) have shown the same properties of the flank magnetopause. The plasma parameters in the LLBL itself usually exhibit large fluctuations. Even under very steady conditions shown in Figure 1 one can identify large fluctuations of the ion flux inside the LLBL. The same fluctuations can be seen in electron energy spectrograms shown in two bottom panels. On the other hand, Anderson et al. (1997) demonstrated that the plot of electron temperature versus electron density exhibits a good organization with the temperature being nearly inversely proportional to the density. A similar organization was shown for the ion density and temperature by Vaisberg et al. (2001). Němeček et al. (2003a) suggested that even the magnetic field orientation represented as a shear angle between the current and magnetospheric field orientations is well correlated with the ion flux. These properties are demonstrated in Figure 2 by electron temperature versus density and magnetic shear vs ion flux plots. The data were taken from the INTERBALL-1 observation of a very disturbed magnetopause crossing and demonstrate a good correlation of depicted parameters on ~2-hour interval. A detail analysis of this event can be found in Němeček et al. (2003a).

Although the profiles of the LLBL depicted in previous figures are rather typical, Šafránková et al. (1997a) have shown an example of the crossing of the subsolar low-latitude region registered by the INTERBALL-1/MAGION-4 satellite pair that exhibits neither a depletion layer in front of the magnetopause nor a region of the LLBL plasma behind it.

Summarizing the above findings, we can note several conflicting properties of the LLBL. Energy spectra of LLBL ions and electrons usually contain two distinct populations appearing simultaneously or at different times. The mentioned populations can be seen in Figure 1. On the other hand, a gradual change of the particle energy (temperature) with the density suggests a smooth profile of the layer. The LLBL can be probably very broad (Sauvaud et al., 1997; Farrugia et al., 2000) under some circumstances but it can nearly (or fully) disappear (Šafránková et al., 1997a).

3. THE LLBL LOCATION

We have defined the LLBL as a layer with plasma characteristics intermediate between those of magnetosheath and plasma sheet on magnetospheric (open or closed) field lines. However, there are many such regions or layers in the magnetosphere. Figure 3 (left part) shows a schematics of the magnetospheric tail cross-section. One can find two regions which meet our definition - LLBL and plasma mantle. Looking at the cross-section of the magnetosphere at the $X - Z_{GSM}$ plane shown in Figure 3 (right part), we can find the entry layer, cleft, cusp proper, and plasma mantle, all of them containing the plasma with similar properties. As our drawings suggests, the entry layer, exterior cusp, cleft, and LLBL are different names for the same layer usually used for different parts of it. We will call it LLBL throughout the paper. The cusp proper and plasma mantle can be considered as independent regions but we are going to show that they are a part of the same process.

We have already shown an example of the dayside LLBL crossing in Figure 1. However, the LLBL is the magnetopause layer and all magnetopause field lines map onto small regions of dayside parts of auroral ovals in both hemispheres as shown in Figure 4. The top part of the figure shows the location of different regions of the dayside auroral region as has been derived from a huge number of the DMSP passes through this region and an example

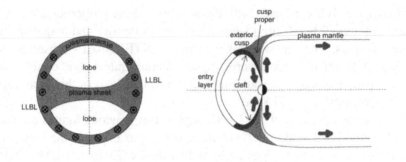

Figure 3.   A sketch of the magnetospheric tail cross-section (left panel) and the same magnetosphere cross-section in the $X - Z_{GSM}$ plane (right panel).

of DMSP data is shown at the bottom part of Figure 4. The DMSP crosses the LLBL twice, once at the morning and once at the afternoon sectors. LLBL ions can be distinguished from those of the cusp proper observed between ∼0854 and 0855 UT because they are not so dense and their energy is higher (Newell and Meng, 1988). The precipitation patterns between ∼0855 and 0857 UT were classified as a plasma mantle. However, according to the scheme in Figure 3, the plasma in low-altitude plasma mantle would generally proceeds upward in low altitudes and thus the DMSP energy spectrometers (Hardy et al., 1984) looking only toward the local zenith cannot reliably show the plasma mantle particles. We have depicted the geomagnetic coordinates of this DMSP crossing by a heavy line in the top part of the figure. This line does not cross the plasma mantle but the sketch shows average locations of regions and the whole projection is very sensitive to actual upstream conditions (e.g., see Měrka et al., 2002).

A part of this region was at the same time crossed by the MAGION-4 spacecraft equipped with a more complex spectrometer (Němeček et al., 1997). The spacecraft moved from dawn to a local noon as the projection along magnetic field lines (according to Tsyganenko and Stern (1996)) in the left part of Figure 5 shows. As indicated in the figure, the spacecraft crossed the LLBL, cusp proper, and finally the plasma mantle. The corresponding ion energy spectra are plotted in the right part of the figure. The panels correspond to channels with a different orientation with respect to the satellite spin axis. This axis was roughly oriented along the Sun-Earth line. We will concentrate on the middle panel. This analyzer was perpendicular to the spin axis and scanned nearly a full range of pitch angles. The first feature which we would note looking at this panel is the change of the energy of the particles. It decreases in several steps until 0730 UT and than rises again. We will discuss this feature later. The other distinct feature is a periodic modulation of the ion counts in the mantle region suggesting the ordered flow. A detail analysis in Němeček et

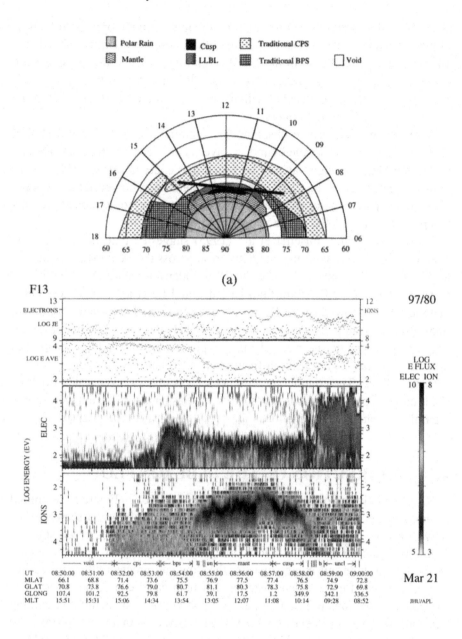

Figure 4. Mapping of magnetospheric regions onto the auroral oval (Courtesy of P. Newell). The DMSP pass of these regions is distinguished by the heavy straight line (a). Ion and electron energy spectrograms measured by DMSP F13 on March 21, 1997 (adapted according to http://sd-www.jhuapl.edu/Programs/) (b).

al. (2003a) shows that these ions are moving upward along open field lines as Figure 3 predicts for the plasma mantle at these altitudes. On the other hand, a lack of such modulation in the LLBL region reveals that this plasma is on closed field lines because there is no distinct flow in any direction. This opens a new set of questions connected with the problem of the LLBL or a part of the LLBL on closed/open field lines.

Based on *in situ* particle observations, Ogilvie et al. (1984) and Hall et al. (1991) suggested that the boundary layer is on a combination of open and closed field lines. Song et al. (1993) found that under northward IMF conditions, the inner part of the LLBL is on closed field lines, however, the topology of the outer part was not so clear, although there were some indications that this portion of the LLBL has been on closed field lines, too.

On the other hand, Fuselier et al. (1995) suggested that the LLBL is on open field lines even when the magnetic shear across the local magnetopause is low. Song and Russell (1992) assumed that under northward IMF conditions, plasma enters the LLBL from high latitudes and Le et al. (1996) concluded that the subsolar low-shear LLBL is on a combination of open and closed field lines. Similar results by Paschmann et al. (1993) indicated that under conditions of low magnetic shear, the changes in plasma thermal and flow properties may be attributed to a transition from open interplanetary to closed geomagnetic field lines. The continuous presence of a boundary layer inside the low-shear magnetopause confirms that solar wind plasma enters the magnetosphere regardless of the field orientation.

4.   A DIRECTION OF THE LLBL ION FLOW

We would like to point out that observations like those in Figure 5 are limited to the dayside magnetosphere, usually to morning or afternoon sectors. The flow direction in tail parts of the LLBL is obviously distinct as we show in Figure 6. Since the magnetic field is dominated by the B_X component inward of the low-latitude tail magnetopause, the flow can proceed either in $+X$ or $-X$ directions. Figure 6 demonstrates that both cases can be observed. The top panels show ions proceeding tailward and bottom panels those streaming sunward. In the magnetosheath which is in left part of both figures, the tailward ions (top panels) prevail. The magnetopause was crossed at ∼1347 UT in Figure 6a. After the magnetopause crossing, the magnetosheath-like tailward streaming population is mixed with plasma sheet ions. These ions are generally out of the analyzer energy range but their low-energy part is seen in the top of both panels with a similar intensity.

On the other hand, the event depicted in the right part of the figure is completely different. The magnetopause crossing at ∼0049 UT is accompanied

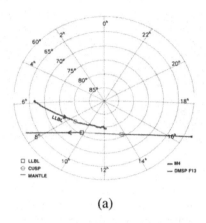

(a)

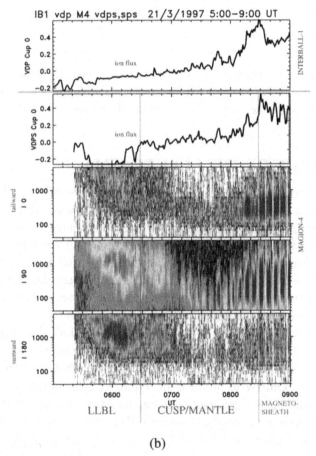

(b)

Figure 5. Projection of simultaneous MAGION-4 and DMSP passes through the auroral region onto the Earth's surface (a) and MAGION-4 ion energy spectrograms registered during this pass (b). The DMSP data are presented in Figure 4.

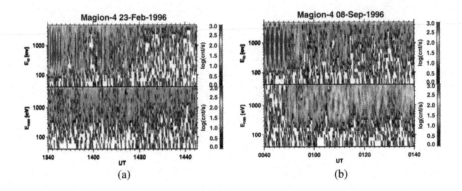

Figure 6. Two different crossings of the LLBL without (a) and with (b) distinct sunward flow (in both parts from top to bottom: energy spectra of tailward (E_{i0}) and sunward (E_{i180}) streaming ions).

with a change of the direction of the low-energy magnetosheath-like population. It is registered predominantly in a sunward direction and only hot magnetospheric ions are measured in the tailward direction.

5.   ORIGIN OF THE LLBL PLASMA

Several sources of the LLBL plasma and mechanisms of the LLBL creation were suggested over the years. We are not able to discuss all of them at the present paper and thus we limit ourself to three representatives.

The impulsive penetration of the magnetosheath (solar wind) plasma through the magnetopause was suggested e.g., by Lemaire and Roth (1978). This mechanism can explain a simultaneous presence of magnetosheath and magnetospheric populations on the same magnetic field line, as well as an intermittent occurrence of magnetosheath-like plasma (Figure 6a). Nevertheless, such mechanism should lead to low-energy plasma blobs detached from the magnetopause and to a presence of negative density gradients when the satellite deeper in the LLBL observes a higher density than the satellite at the magnetopause. Otherwise products of impulsive penetration cannot be distinguished from pressure pulses modulating the magnetopause surface. Sibeck at al. (2000) made a survey of all two-point INTERBALL-1/MAGION-4 magnetopause observations and they did not find any such example of negative gradients. They concluded that the impulsive penetration is very improbable. We would like to add that such mechanism cannot explain the LLBL flow reversal shown in Figure 6b because the penetrating plasma should conserve its original (tailward) motion.

Other mechanism mixing magnetosheath and magnetospheric populations can be diffusion (e.g., Eastman and Hones, 1979). This mechanism can simply explain a smooth change of plasma parameters across the LLBL (Figure 2) but cannot overcome the problem that the diffusion rate is too low to provide a required amount of the magnetosheath plasma in the layer. Moreover, the magnetosheath plasma entering the boundary layer via diffusion cannot gain the momentum which would turn its bulk motion sunward.

Both above mechanisms remain a magnetic field geometry nearly unchanged and thus they are rather static. By contrast, reconnection or merging of interplanetary and magnetospheric magnetic field lines which were suggested as a source of the LLBL plasma are dynamic processes (e.g., Sonnerup et al., 1981; or Song and Russell, 1992; for present review see Onsager and Scudder, 2003). Even in a case of very steady conditions, the reconnected magnetic flux should be replaced by new field lines from the inner magnetosphere and thus reconnection drives a magnetospheric convection. The reconnected field lines are open to interplanetary space. In order to keep an equilibrium, i.e., to keep the amount of the closed magnetic flux roughly constant, new reconnection is required with the same overall rate. This equilibrium condition cannot predict where this new reconnection will take place. However, one end of the reconnected field line is embedded in the magnetosheath flow and is rapidly blown tailward and thus this reconnection can be expected in a distant magnetotail.

The concept of magnetic reconnection was originally suggested by Petschek (1964) for a case of antiparallel magnetic fields. We will leave out the discussion on a possibility of other mutual orientations of magnetic fields, usually called as component reconnection (Fuselier et al., 1997; Chandler et al., 1999). It was shown by Anderson and Fuselier (1993) that the reconnection rate strongly depends on the angle between the magnetic fields on both sides of the magnetopause (usually named as the shear angle) and exhibits a sharp peak near 180°. If we consider magnetic reconnection as a principal mechanism for the solar wind plasma entry onto magnetospheric field lines, it should act at the places where both magnetic fields are roughly antiparallel. From it immediately follows that the probability (or effectiveness) of magnetic reconnection in a given place of the magnetopause would strongly depend on the IMF orientation because the magnetospheric magnetic field can be considered to be constant with respect to the IMF variability. Nevertheless, there is as a minimum one (usually two) place on the dayside magnetopause where IMF and magnetospheric field are antiparallel for any IMF orientation if the effect of the IMF draping is taken into account.

Probably the best understood situation is the case of a strongly southward oriented IMF. Schematic drawing in Figure 7 shows that the draped IMF line would cover nearly the whole dayside magnetopause and magnetic fields would be antiparallel nearly everywhere along this line. However, the reconnection

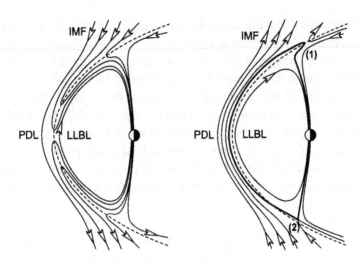

Figure 7.   A sketch of the LLBL formation for southward (left part) and northward (right part) orientations of IMF - adapted from Němeček et al. (2003a).

rate depends on plasma parameters and thus we can expect that reconnection would start at the most probable place - at the subsolar point - and thus the magnetic field geometry will evolve as Figure 7 shows. The magnetosheath flow supported by magnetic tension will push the kinks of reconnected lines toward the poles. The magnetosheath plasma accelerated by reconnection will reach low altitudes at the equatorward edge of the region creating a footprint called the LLBL or cleft. Later on, the bending of the magnetic field line is not so strong and allows the magnetosheath plasma to enter freely low altitudes, and to create the cusp proper. However, a significant part of the entering plasma is reflected back by a stronger field in low altitudes and proceeds upward and finally tailward as the corresponding field line is already lying on the tail part of the magnetopause. This part of the boundary layer is usually named the plasma mantle.

Reconnection is not limited to the subsolar point but it rather occupies a broad region of the dayside equatorial magnetopause (Maynard et al., 1997). However, there is no possibility for the magnetosheath plasma to enter onto magnetospheric lines outside the reconnection region and it can explain absence of the LLBL in the tail region. Our scheme of the plasma motion expects all boundary regions on open field lines under southward oriented IMF.

When IMF points to the north, the antiparallel reconnection site moves to high latitudes, tailward of the cusp. Since the situation in both hemispheres is similar, we can expect two conjugated reconnection sites as shown in Figure 7 (right part). The draped IMF line reconnects at point (1). Reconnection supplies not only the cusp but a whole dayside part of this line and thus we can

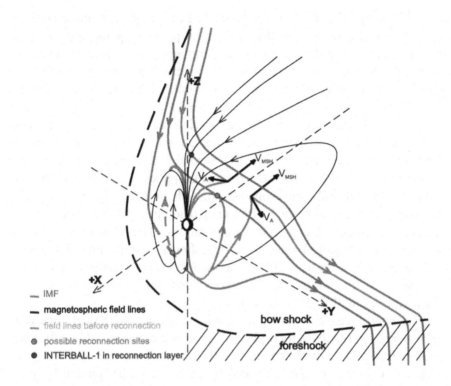

Figure 8. A 3-D sketch of the magnetic field lines formed by reconnection duskward of the northern cusp. The reconnected field line is depicted in green, directions of a motion of the kinks created by reconnection are shown by arrows (adapted from Němeček et al. (2003b)). See text for a detailed description.

observe the open LLBL covering the whole dayside magnetopause. However, such line can re-reconnect at the point (2) and one would observe the LLBL plasma on closed field lines. This scenario was suggested by Song and Russell (1992) and observed by Sandholt et al. (1999). We would like to point out that features shown above can be observed only during prolonged intervals of the unchanging IMF orientation. If IMF turns from north to south, reconnection in the subsolar region starts earlier than high-latitude reconnection is terminated and thus we can observe a mixture of plasma from two sources in the LLBL. The time needed for the equilibrium to be set up can be as long as 20 minutes (e.g., Šimůnek et al., 2003).

However, the periods of purely northward or southward IMF orientation are rather rare because IMF B_Z is usually a minor component. A scenario of the LLBL creation for IMF dominated by horizontal components was described in Nemecek et al. (2003b) and it is depicted in Figure 8. The figure presents the situation when the magnetosheath magnetic field points duskward above the cusps. The antiparallel fields can be found duskward of the cusp in the northern hemisphere and dawnward of the southern cusp. These points are shown as green dots. Providing that the tilt of the Earth dipole is small (near solstices), the situation in both hemispheres would be symmetric and we can expect equal reconnection rate at both hemispheres. We will concentrate our attention on the northern hemisphere. The newly reconnected field line is shown as green in Figure 8. Reconnection divides the line into two parts. One part points duskward of the northern cusp and creates an ionospheric projection of the LLBL in the northern hemisphere. The second part of the reconnected field line ends duskward of the southern cusp and supplies the LLBL precipitation in this region. Since the same process proceeds in the southern hemisphere, we can expect two spots of the LLBL precipitation in both hemispheres, one at dawn and the other in dusk parts of the auroral oval. This scenario is confirmed by a statistical study of Měrka et al. (2002) showing that the probability of an observation of the cusp-like (LLBL) plasma peaks at two locations separated in magnetic local time. Moreover, the authors found that this separation increases with an increasing IMF B_Y component.

Up to now, we have discussed a behavior of the part of the field line from the reconnection site to the ionosphere. The other part is directed into the magnetosheath and thus it is blown tailward with a magnetosheath speed (v_{MSH}) as shown in Figure 8. Kinks created by reconnection are pushed by magnetic tension dawn/duskward. The result is that the dawnward moving line proceeds toward higher latitudes and creates the high-latitude boundary layer. The duskward moving line gradually moves to the southern hemisphere and crosses all latitudes. The plasma on this field line is a magnetosheath plasma with an addition of a faster component originated during reconnection and thus this region would be classified as the LLBL. These lines would cover a significant part of the nightside low-latitude magnetopause during their evolution.

It should be noted here that non-zero tilt angle breaks the north/south symmetry. A magnetosheath speed above the cusp tilted toward the Sun (summer cusp) is usually subsonic and reconnection proceeds as described above. On the other hand, the speed above the winter cusp is often superalfvénic. Since reconnection can accelerate the plasma up to Alfvén speed it is not able to turn ions into the cusp and we would observe only a very weak ion precipitation. This scenario was confirmed by multisatellite observation of precipitation patterns (Němeček et al., 2003b) as well as by analysis of aurora (Sandholt and Farrugia, 2002).

Figure 8 is too schematic to reveal all peculiarities of the plasma flow along reconnected field lines. For this reason, Figure 9 shows a view from the dusk side. The magnetosheath plasma enters the field line on the dusk when IMF B_Y is positive and thus the spacecraft in a location 2 would observe the antisunward flow but the spacecraft in the location 1 would observe a sunward

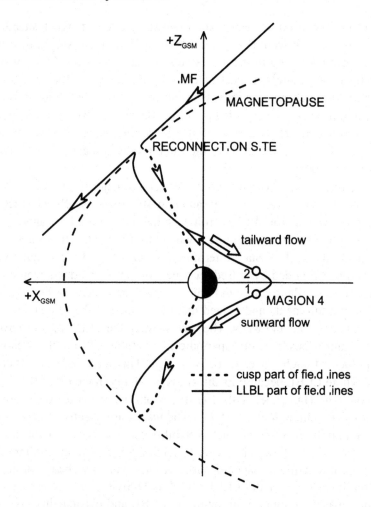

Figure 9. A schematics of observed ion flow direction during a change of the IMF B_Y sign - adapted from Němeček et al. (2003a).

streaming plasma. If the IMF B_Y component changes its sign, the situation would reverse. However, an observation of such situation is limited to a region near the terminator because a curvature radius of field lines is too small further tailward and does not allow the reversal of ion flow direction.

The process can be even more complicated. Vaisberg et al. (2004) analyzed in detail fast measurements of ion distribution function in the dayside LLBL and have shown that the LLBL flow can consist of narrow short-living and often counterstreaming ion beams. The authors attributed this effect to multiple reconnection on the same field line. As we noted above, conditions favorable for reconnection can occur at a relatively large spot on the magnetopause and thus we cannot exclude that the same magnetospheric line would reconnect in several points nearly simultaneously due to magnetosheath fluctuations. However, it would be noted that the temporal and spatial resolutions of the spectrometer used in the case study of Vaisberg et al. (2004) were exceptionally good. Standard spectrometers would smooth a fine structure and show highly turbulent plasma only. Consequently, we cannot judge if this multiple reconnection is a frequent feature of the dayside LLBL or if it is an effect limited to a range of upstream conditions. Nevertheless, an observer out of the reconnection region would observe the same products regardless of a number of reconnection points.

As we have shown, reconnection can explain a majority of LLBL observational features. However, one contradiction still remains. Steady reconnection can provide a smooth LLBL profile as that in Figure 2 but cannot explain the fluctuations like those in Figure 6 which are frequently, if not always, observed inside the LLBL. Such intermittent occurrence of the magnetosheath plasma can be explained by pulsed reconnection but its products would hardly exhibit profiles similar to those in Figure 2. Šafránková et al. (1997b) analyzed two-point measurements and argued that the surface of the low-latitude magnetopause is unstable and wavy. The waves sweep the LLBL profile across the spacecraft and it observes quasiperiodic fluctuations. The authors estimated the amplitude of such waves to be $\sim 1\ R_E$. This amplitude is sufficient to scan the whole LLBL because its thickness is usually in the range 0.1 R_E near local noon (e.g., Haerendel et al., 1978) to about 0.5 R_E at the dawn and dusk terminators (Paschmann et al., 1993). The waves are usually attributed to the Kelvin-Helmholtz instability. Such instability requires a velocity shear across the boundary but the flow velocity and direction are very often similar on both sides of the low-latitude magnetopause. A large velocity shear can be found across the inner edge of the LLBL and thus Ogilvie and Fitzenreiter (1989) concluded that the boundary dividing the LLBL and plasma sheet is Kelvin-Helmholtz unstable. On the other hand, as it has been shown both experimentally (Šafránková et al., 1997b) and theoretically (Farrugia et al., 1998), both of these boundaries move often in accord. We can thus conclude that the surface waves are probably a major source of observed fluctuations. These waves would enhance diffusion process (Book and Sibeck, 1995) and contribute to creation of the LLBL profile.

6. CONCLUSION

Both satellites of the INTERBALL-Tail project crossed the LLBL about 200 times in different local times during their active life. Not all of these crossings were analyzed until present but even this partial analysis contributes significantly to our knowledge of the LLBL.

We can conclude that:

- Dayside reconnection provides a consistent explanation of the LLBL formation.

- This reconnection proceed in the subsolar region during southward IMF but it moves toward higher latitudes for other IMF orientations.

- The LLBL is generally on open field lines.

- A presence of the LLBL on closed field lines requires multiple reconnection.

- Surface waves contribute significantly to formation of a smooth profile of plasma parameters across the LLBL.

Nevertheless, many open questions remain to be answered. Among them, we can name:

- The LLBL thickness and its dependence on upstream parameters.

- A stability of LLBL boundaries.

- A contribution of other processes to the input of the solar wind plasma into the LLBL.

- A creation of the LLBL on closed field lines for arbitrary IMF orientations.

We think that many of these problems can be elucidated by a careful analysis of CLUSTER II data. Constellation of the four spacecraft is very suitable for a study of boundary processes. Several studies based on CLUSTER II observations and presented during the NATO Advanced Research Workshop entitled Multiscale Processes in the Earth's Magnetosphere: From INTERBALL to CLUSTER (Prague, September 9-12, 2003) suggest that these expectations are justified.

ACKNOWLEDGMENTS

The present work was supported by the Czech Grant Agency under Contracts No. 205/02/0947 and 205/03/0953 and by the Research Project MSM 113200004.

REFERENCES

Anderson, B. J., and Fuselier, S. A., 1993, Magnetic pulsations from 0.1 to 4.0 Hz and associated plasma properties in the Earth's subsolar magnetosheath and plasma depletion layer, *J. Geophys. Res.* **98**(A2):1461–1479.

Anderson, B. J., Phan, T.-D., and Fuselier, S. A., 1997, Relationship between plasma depletion and subsolar reconnection, *J. Geophys. Res.* **102**(A5):9531–9542.

Book, D. L., and Sibeck, D. G., 1995, Plasma transport through the magnetopause by turbulent interchange processes, *J. Geophys. Res.* **100**:9567–9573.

Crooker, N. U., Eastman, T. E., and Stiles, G. S., 1979, Observations of plasma depletion in the magnetosheath at the dayside magnetopause, *J. Geophys. Res.* **84**:869–874.

Chandler, M. O., Fuselier, S. A., Lockwood, M., and Moore, T. E., 1999, Evidence of component merging equatorward of the cusp, *J. Geophys. Res.* **104**:22623–22633.

Chapman, S. and Farraro, V. C. A., 1931, A new theory of magnetic storms: Part I — Initial phase, *Terr. Magn. Atmos. Electr.* **36**:171–186.

Eastman, T. E., Hones Jr., E. W., Bame, S. J., and Asbridge, J. R., 1976, The magnetospheric boundary layer: Site of plasma, momentum and energy transfer from the magnetosheath into the magnetosphere, *Geophys. Res. Lett.* **3**:685–688.

Eastman, T. E., and Hones Jr., E. W., 1979, Characteristics of the magnetospheric boundary layer and magnetopause layer as observed by IMP 6, *J. Geophys. Res.* **84**:2019.

Farrugia, C. J., Gratton, F. T., Bender, L., Biernat, H. K., Erkaev, N. V., Quinn, J. M., Torbert, R. B., and Dennisenko, V. V., 1998, Charts of joint Kelvin-Helmholtz and Rayleigh-Taylor instabilities at the dayside magnetopause for northward IMF, *J. Geophys. Res.* **103**(A4):6703–6727.

Farrugia, C. J., Gratton, F. T., Contin, J., Cocheci, C. C., Arnoldy, R. L., Ogilvie, K. W., Lepping, R. P., Zastenker, G. N., Nozdrachev, M. N., Fedorov, A., Sauvaud, J.-A., Steinberg, J. T., and Rostoker, G., 2000, Coordinated Wind, Interball-tail, and ground observations of Kelvin-Helmholtz waves at the near-tail, equatorial magnetopause at dusk: January 11, 1997, *J. Geophys. Res.* **105**:7639–7667.

Fuselier, S. A., Klumpar, D. M., Shelley, E. G., Anderson, B. J., and Coates, A. J., 1991, He^{2+} and H^+ dynamics in the subsolar magnetosheath and plasma depletion layer, *J. Geophys. Res.* **96**(A12):21095–21104.

Fuselier, S. A., Anderson, B. J., Gary, S. P., and Denton, R. E., 1994, Inverse correlations between the ion temperature anisotropy and plasma beta in the Earth's quasi-parallel magnetosheath, *J. Geophys. Res.* **99**(A8):14931–14936.

Fuselier, S. A., Anderson, B. J., and Onsager, T. G., 1995, Particle signatures of magnetic topology at the magnetopause: AMPTE/CCE observations, *J. Geophys. Res.* **100**(A7):11805–11821.

Fuselier, S. A., Anderson, B. J., and Onsager, T. G., 1997, Electron and ion signatures of field line topology at the low-shear magnetopause, *J. Geophys. Res.* **102**(A3):4847–4863.

Haerendel, G., Paschmann, G., Sckopke, N., Rosenbauer, H., and Hedgecock, P. C., 1978, The frontside boundary layer of the magnetosphere and the problem of reconnection, *J. Geophys. Res.* **83**:3195.

Hall, D. S., Chaloner, C. P., Bryant, D. A., Lepine, D. R., and Triakis, V. P., 1991, Electrons in the boundary layers near the dayside magnetopause, *J. Geophys. Res.* **96**(A5):7869–7891.

Hardy, D. A., Schmitt, K. L., Gussenhoven, M. S., Marshall, F. J., Yeh, H. C., Shumacher, T. L., Hube, A., and Pantazis, J., 1984, Precipitating electron and ion detectors (SSJ/4) for the block 5D/flights 6-10 DMSP satellites: Calibration and data presentation, *Rep. AFGL-TR-84-0317*, Air Force Geophys. Lab.

Le, G., Russell, C. T., Gosling, J. T., and Thomsen, M. F., 1996, ISEE observations of low-latitude boundary layer for northward interplanetary magnetic field: Implications for cusp reconnection, *J. Geophys. Res.* **101**(A12):27239–27249.

Lees, L., 1964, Interaction between the solar wind plasma and the geomagnetic cavity, *AIAA J.* **2**:2065.

Lemaire, J., and Roth, M., 1978, Penetration of solar wind plasma elements into the magnetosphere, *J. Atmos. Terr. Phys.* **40**:331.

Maynard, N. C., Weber, E. J., Weimer, D. R., Moen, J., Onsager, T., Heelis, R. A., and Egeland, A., 1997, How wide in magnetic local time is the cusp? An event study *J. Geophys. Res.* **102**:4765–4776.

Měrka, J., Šafránková, J., and Němeček, Z., 2002, Cusp-like plasma in high altitudes: A statistical study of the width and location of the cusp from MAGION-4, *Ann. Geophys.* **20**:311–320.

Midgley, J. E., and Davis, J., 1963, Calculation by a moment technique of the perturbation of the geomagnetic field by the solar wind, *J. Geophys. Res.* **68**:5111.

Němeček, Z., Fedorov, A., Šafránková, J., and Zastenker, G., 1997, Structure of the low–latitude magnetopause: MAGION–4 observation, *Ann. Geophys.* **15**:553–561.

Němeček, Z., Šafránková, J., Přech, L., and Sauvaud, J.-A., 2003a, The structure of magnetopause layers at low latitudes: INTERBALL contribution to the topic, in *Geophysical Monograph Series, Volume 133, AGU*, Earth's Low-latitude Boundary Layer, ed. by Patrick T. Newell and Terry Onsager, 71–82.

Němeček, Z., Šafránková, J., Přech, L., Šimůnek, J., Sauvaud, J.-A., Stenuit, H., Fedorov, A., Fuselier, S. A., Savin, S., Zelenyi, L., Berchem, J., 2003b, Structure of the outer cusp and sources of the cusp precipitation during intervals of a horizontal IMF, *J. Geophys. Res.* **108**:1420, doi: 10.1029/2003JA009916.

Newell, P. T., and Meng, C.-I., 1988, The cusp and the cleft/boundary layer: Low-altitude identification and statistical local time variation, *J. Geophys. Res.* **93**:14549–14556.

Ogilvie, K. W., Fitzenreiter, R. J., and Scudder, J. D., 1984, Observations of electron beams in the low-latitude boundary layer, *J. Geophys. Res.* **89**(NA12):723–732.

Ogilvie, K. W., and Fitzenreiter, R. J., 1989, The Kelvin-Helmholtz instability at the magnetopause and inner boundary-layer surface, *J. Geophys. Res.* **94**:15113–15123.

Onsager, T. G., and Scudder, J. D., 2003, Low-latitude boundary layer formation by magnetic reconnection, in: *Earth's Low-latitude Boundary Layer*, Patrick T. Newell and Terry Onsager, eds., *Geophysical Monograph Series* 133, AGU, pp. 111–120.

Paschmann, G., Sckopke, N., Haerendel, G., Papamastorakis, I., Bame, S. J., Asbridge, J. R., Gosling, J. T., Hones Jr., E. W., and Tech, E. R., 1978, ISEE plasma observations near the subsolar magnetopause, *Space Sci. Rev.* **22**:717–737.

Paschmann, G., Baumjohann, W., Sckopke, N., Phan, T. D., and Luhr, H., 1993, Structure of the dayside magnetopause for low magnetic shear, *J. Geophys. Res.* **98**(A8):13409–13422.

Petschek, H. E., 1964, Magnetic field annihilation, in: *AAS-NASA Symposium on the Physics of Solar Flares*, NASA Spec. Publ., SP-50, pp. 425.

Phan, T. D., Larson, D., McFadden, J., Lin, R. P., Carlson, C., Moyer, M., Paularena, K. I., McCarthy, M., Parks, G. K., Reme, H., Sanderson, T. R., and Lepping, R. P., 1997, Low-latitude dusk flank magnetosheath, magnetopause, and boundary layer for low magnetic shear: Wind observations, *J. Geophys. Res.* **102**(A9):19883–19895.

Šafránková, J., Zastenker, G., Němeček, Z., Fedorov, A., Simerský, M., and Přech, L., 1997a, Small scale observation of the magnetopause motion: Preliminary results of the INTERBALL project, *Ann. Geophys.* **15**(5):562–569.

Šafránková, J., Němeček, Z., Přech, L., Zastenker, G., Fedorov, A., Romanov, S., Šimůnek, J., and Sibeck, D., 1997b, Two-point observation of magnetopause motion: INTERBALL project, *Adv. Space Res.* **20**(4/5):801–807.

Sandholt, P. E., Farrugia, C. J., Cowley, S. W. H., and Denig, W. F., 1999, Capture of magnetosheath plasma by the magnetosphere during northward IMF, *Geophys. Res. Lett.* **26**(18): 2833–2836.

Sandholt, P. E., and Farrugia, C. J., 2002, Monitoring magnetosheath-magnetosphere interconnection geometry from the aurora, *Ann. Geophys.* **20**(5):629–637.

Sauvaud, J.-A., Koperski, P., Beutier, T., Barthe, H., Aoustin, C., Throcaven, J. J., Rouzaud, J., Penou, E., Vaisberg, O., and Borodkova, N., 1997, The INTERBALL- Tail ELECTRON experiment: Initial results on the low-latitude boundary layer of the dawn magnetosphere, *Ann. Geophys.* **15**:587–595.

Sibeck, D. G., Přech, L., Šafránková, J., and Němeček, Z., 2000, Two-point measurements of the magnetopause: INTERBALL observations, *J. Geophys. Res.* **105**(A1): 237–244.

Šimůnek, J., Němeček, Z., and Šafránková, J., 2003, Configuration of the outer cusp after an IMF rotation, *Adv. Space Res.* **31**:1395–1400.

Song, P., Russell, C. T., Gosling, J. T., Thomsen, M. F., and Elphic, R. C., 1990, Observations of the density profile in the magnetosheath near the stagnation streamline, *Geophys. Res. Lett.* **17**:2035–2038.

Song, P., and Russell, C. T., 1992, Model of the formation of the low-latitude boundary layer for strongly northward interplanetary magnetic field, *J. Geophys. Res.* **97**(A2):1411–1420.

Song, P., Russell, C. T., Fitzenreiter, R. J., Gosling, J. T., Thomsen, M. F., Mitchell, D. G., Fuselier, A. S., Parks, G. K., Anderson, R. R., and Hubert, D., 1993, Structure and properties of the subsolar magnetopause for northward IMF: Multiple-instrument observations, *J. Geophys. Res.* **98**(A7):11319–11337.

Sonnerup, B. U. O., Paschmann, G., Papamastorakis, I., Sckopke, N., Haerendel, G., Bame, S. J., Asbridge, J. R., Gosling, J. T., and Russell, C. T., 1981, Evidence for magnetic field reconnection at the Earth's magnetopause, *J. Geophys. Res.* **86**(NA12):49–67.

Tsyganenko, N. A., and Stern, D. P., 1996, A new-generation global magnetosphere field model based on spacecraft magnetometer data, *ISTP Newsletter* **6**:21.

Vaisberg, O. L., Smirnov, V. N., Avanov, L. A., Waite, J. H., Burch, J. L., Gallagher, D. L., and Borodkova, N. L., 2001, Different types of low-latitude boundary layer as observed by Interball Tail probe, *J. Geophys. Res.* **106**:13067–13090.

Vaisberg, O. L., Avanov, L. A., Moore, T. E., and Smirnov, V. N., 2004, Ion velocity distributions within LLBL and their possible implication to multiple reconnection, *Ann. Geophys.* **22**:213–236.

Wu, C. C., 1992, MHD flow past and obstacle: Large-scale flow in the magnetosheath, *Geophys. Res. Lett.* **19**(2):87–90.

Zwan, B. J., and Wolf, R. A., 1976, Depletion of the solar wind plasma near a planetary boundary, *J. Geophys. Res.* **81**:1636–1648.

CLUSTER: NEW MEASUREMENTS OF PLASMA STRUCTURES IN 3D

C. P. Escoubet[1], H. Laakso[1] and M. Goldstein[2]

[1]*ESA/ESTEC, SCI-SH, Keplerlaan 1, 2200 AG Noordwijk, The Netherlands;* [2]*NASA/GSFC, Greenbelt, USA*

Abstract: After 2.5 years of operations, the Cluster mission is fulfilling successfully its scientific objectives. The mission, nominally for 2 years, has been extended 3 more years, up to December 2005. The main goal of the Cluster mission is to study in three dimensions the small-scale plasma structures in the key plasma regions in the Earth's environment: solar wind and bow shock, magnetopause, polar cusps, magnetotail, and auroral zone. During the course of the mission, the relative distance between the four spacecraft will vary from 100 km up to a maximum of 18,000 km to study the physical processes occurring in the magnetosphere and its environment at different scales. The inter-satellites distances achieved so far are 600, 2000, 100, 5000 km and recently 250 km. The latest results, which include the derivation of electric currents and magnetic curvature, the analysis of surface waves, and the observation of reconnection in the tail and in the cusp will be presented. We will also present the description of the access to data through the Cluster science data system and several public web servers, and the future plans for a Cluster archive.

Key words: solar wind; bow shock; magnetopause; polar cusp; magnetotail; auroral zone.

1. INTRODUCTION

Past multi-spacecraft missions have been investigating the role of small-scale structures in the Sun-Earth connection, in particular ISEE 1 and 2 (see e. g. Russell, 2000 for a review) and more recently Interball-1/Magion-4 (see e.g. Zelenyi et al, 2000 for a review). Cluster is following on this work with two additional spacecraft to study these structures in three dimensions.

The Cluster mission was first proposed more than 20 years ago in response to an ESA call for proposals for the next series of scientific missions (Haerendel et al., 1982). At that time the proposal selected a mother

131

J. –A. Sauvaud and Z. Němeček (eds.),
Multiscale Processes in the Earth's Magnetosphere:From Interball to Cluster, 131-147.
© 2004 *Kluwer Academic Publishers. Printed in the Netherlands.*

and three daughter spacecraft that were replaced later on by four identical spacecraft for economic reasons. The assessment study was conducted in 1983 to prove the feasibility of the mission concept. Subsequently, the Phase-A study was conducted jointly with NASA during 1984-1985. At the end of 1985 the Cluster mission was presented to the scientific community and in February 1986, the STSP programme, combining both Cluster and Soho missions, was selected by ESA Science Programme Committee. After a joint ESA/NASA Announcement of Opportunity issued in March 1987, the 11 instruments making up the scientific payload were selected in March 1988. It took about 7 years for four spacecraft to be built and tested and made ready for launch. Unfortunately the launch with the first Ariane 5 on 4 June 1996 was a failure as the rocket exploded 47 s after lift-off destroying the four spacecraft.

After the shock of seeing their work of 8 years annihilated in less that 1 min, the principal investigators and the project teams rolled up their sleeves and investigate how the mission objectives could be recovered. After ten science working team meetings and a few extraordinary ESA committees meeting, the Cluster scientists convinced the ESA Science Programme Committee (SPC) that it was essential for the European scientific community to rebuild the four spacecraft. This was agreed by the SPC in April 1997. Cluster II was born.

As the Cluster II spacecraft and instruments were essentially a rebuild of the original Cluster it took less than half the time to rebuild them (about 3 years). When the first Soyuz blasted off from Baikonur Cosmodrome on 16 July 2000, we knew that Cluster was well on the way to recovery from the previous launch setback. However, it was not until the second launch on 9 August 2000 and the proper injection of the second pair of spacecraft into orbit that we knew that the Cluster mission was back on track. In fact, the experimenters said that they knew they had a mission only after switching on their last instruments on the fourth spacecraft.

In a first section, the Cluster mission will be described with a focus on the orbit and the separation distances achieved so far and planned in the future. In a second part the instrumentation will be briefly described. Then a few examples of Cluster observations will be presented in a third part and finally the data distribution through the Cluster Science Data System and the plans for the Cluster active archive will be presented in the fourth section. The purpose of this paper is not to review what has been done in the past in the various magnetospheric missions but to present highlights of the Cluster results. The reader can find more references in the referenced papers.

2.   MISSION

The scientific objectives of the Cluster mission are to study the small-scale structures and the turbulence in the key regions of the magnetosphere. Such regions are:
– the solar wind
– the bow shock
– the magnetopause
– the polar cusp
– and the magnetotail
In addition the temporal variations of structures observed in the
– auroral zone
– mid-altitude polar cusp
– plasmasphere
can be studied for the first time as the spacecraft are following each other as a "string of pearls" near perigee.

To perform these objectives, the Cluster spacecraft have been placed in a 4x19.6 R_E polar orbit (Figure 1). The spacecraft have slightly different orbits to form a perfect tetrahedron in key regions of space such as the Northern polar cusp, Southern polar cusp and plasmasheet (Figure 1).

During the first two years of the mission, the separation distance was changed approximately every 6 months (Table 1 and Figure 2). It was decided to start with small distances (down to 100 km) and then to increase it toward the end of the mission (up to 18000 km). All measurements at small distances have to be done first since after 18000 km, the remaining fuel will no longer allow to substantially change the separation distance. The Cluster mission has been extended an additional 3 years from January 2003 to December 2005 to cover more separation distances and spend more time in all key regions. After early 2003, the constellation manoeuvres are done only once a year to decrease the operational costs and to decrease the downtime of the instrument during the manoeuvres. This was possible with the innovative manoeuvre method used by the Flight Dynamics Team at the European Space Operations Centre (ESOC). This method allows to have a tetrahedron at the same time in the tail and in the Northern polar region, therefore covering both the tail and the Northern cusp 6 months later.

The small distances in the magnetotail in Aug. 2003 were not in the initial planning of the Cluster mission, but were recommended afterwards by the International Space Science Institute substorm working group. The small scales in the tail are necessary to investigate the processes that produce geomagnetic substorms. There are two competing models: magnetic reconnection and current disruption. The existence of a small " diffusion " region where the plasma is rapidly accelerated is expected in the first model, while a disruption of cross-tail current is expected in the second model. Both phenomena have a scale size of approximately 500 km, which will require a spacecraft separation distance of a few hundred kilometers to be studied. The acquisition of data at this separation distance is occurring right now (dashed line in Figure 2).

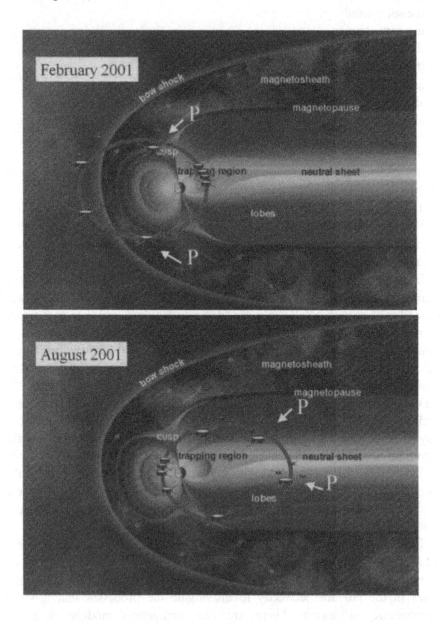

Figure 1. Regions of the magnetosphere crossed by the Cluster spacecraft. The upper panel shows the orbit in February and the lower panel in August. The perfect tetrahedron locations are marked with a "P".

Table 1. Spacecraft separation distances

Year	Phase	Separation (km)
2001	Cusp	600
2001	Tail	2000
2002	Cusp	100
2002	Tail	4000
2003	Cusp	5000
2003	Tail	100-700
2004	Cusp	100-700
2004	Tail	10000
2005	Cusp	10000-18000
2005	Tail	18000

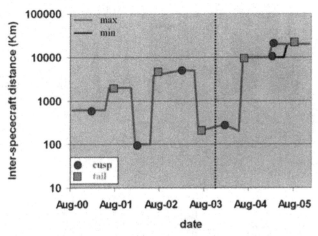

Figure 2. Separation distances during the course of the mission.

In addition, the mission data return has also been augmented by adding a second ground-station in Maspalomas (Spain). At the beginning of the mission, due to the very large amount of data produced by Cluster, the baseline data return was limited to 50% of the orbit. After a few months of operations, it was however realized that many highly bursty phenomena were missed due to their un-predictable behavior (sudden storm commencement, substorms, storms) and large scientific regions (e.g. magnetotail and North and South cusp) could not be fully observed. In February 2002, the ESA SPC agreed to extend both the data coverage and the mission. The 100% coverage started in June 2002. An example of the data return before and after data coverage extension is shown in Figure 3. In 2001, the data acquisition focused on the boundaries (Figure 3 left panel) while in 2003 the orbit is fully covered (Figure 3 right panel). With the full coverage we clearly see that the field direction in the solar wind is correlated with the rotation of the Sun such that the patterns are repeated every 28 days (this is the effect of the tilt of the magnetic pole of the Sun with respect to the Sun rotation axis). A few data gaps occur when constellations or attitude manoeuvres are performed.

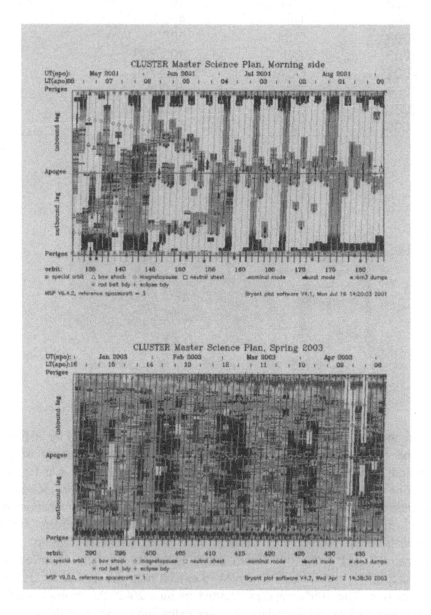

Figure 3. Master Science Plan during Spring 2001 (upper panel) and Spring 2003 (lower panel). Each vertical dashed line represents an orbit from perigee (bottom) to apogee (middle) and then to perigee (top). The Magnetic field direction (from FGM) in the GSE x-y plane is shown in color (blue=sunward, red=anti-sunward).

3.   INSTRUMENTATION

Each Cluster spacecraft contains a complete suite of instruments to measure magnetic fields, electric field, electromagnetic waves, and particles (Table 2). In addition a potential control instrument keep the spacecraft potential close to a few Volts positive (typically 7 V) in very tenuous plasma. More details on the payload can be found in Escoubet et al. (2001).

Table 2. The 11 instruments on each of the four Cluster spacecraft

Instrument	Principal Investigator
ASPOC (Spacecraft potential control)	K. Torkar (IWF, A)
CIS (Ion composition, 0<E<40 keV)	H. Rème (CESR, F)
EDI (Plasma drift velocity)	G. Paschmann (MPE, D)
FGM (Magnetometer)	A. Balogh (IC, UK)
PEACE (Electrons, 0<E<30 keV)	A. Fazakerley (MSSL, UK)
RAPID (High energy electrons and ions, 20<Ee<400 keV, 10<Ei<1500 keV)	P. Daly (MPAe, D)
DWP* (Wave processor)	H. Alleyne (Sheffield, UK)
EFW* (Electric field and waves)	M. André (IRFU, S)
STAFF* (Magnetic and electric fluctuations)	N. Cornilleau (CETP, F)
ASPOC (Spacecraft potential control)	K. Torkar (IWF, A)
WHISPER * (Electron density and waves)	P. Décréau (LPCE, F)

*Wave experiment consortium (WEC).

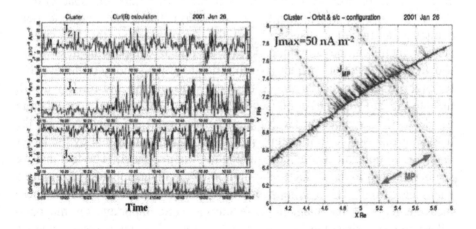

Figure 4. Electric currents measured by the four Cluster spacecraft on 26 January 2001. The left panels show the components of the current in GSE and div B. The right panel shows the electric current plotted along the orbit in the XYGSE plane. The model magnetopause is indicated for low and high solar wind dynamic pressure (Dunlop et al., 2003).

4.   CLUSTER RESULTS

Results from the first few months of operations were presented in previous papers (Escoubet et al., 2001; Escoubet and Fehringer, 2003). In this paper we will focus on specific phenomena that requires four spacecraft to be fully studied like electric currents, surface waves, magnetic curvature and reconnection in the tail, and finally the observation of the proton aurora with IMAGE and Cluster. It is not the goal of this paper to describe each result in details but just to give a brief overview of the examples presented.

4.1   The electric current

One of the first scientific objectives of the Cluster mission is to measure physical quantities that can only be measured with four spacecraft, for instance the electric current derived from the Ampère law μ_0 **J** = **curl B**. The electric current can, in principle, be measured by electron and ion detectors but the precision of the particle sensors, their limited energy range and the spacecraft potential effect on low energy particles often introduce inaccuracy in the measurement. To achieve a good measurement of **curl B**, the magnetic field needs to be measured with very high accuracy. For this reason a very stringent magnetic cleanliness programme was conducted during the development of the Cluster mission and the magnetic field could be measured with a very high accuracy (Balogh et al., 1997, 2001). An example of the current measurement at the magnetopause using the four spacecraft is shown in Figure 4 (from Dunlop et al., 2003). Using the magnetic field computed at the four spacecraft and the Ampere's law, the full current density vector can be computed. The Jx, Jy and Jz components in GSE are shown as well as div B (which give an estimate of the validity of the method: for high div B, the results should be taken with caution and for low div B, the currents are well measured (see Dunlop et al., 2002 and reference therein for details)). As expected the Magnetopause current are flowing along the magnetopause surface and mainly in the XY plane as indicated by the larger X and Y components of the current (Figure 4 upper panel). The fact that we see bursts of currents is due to the motion of the magnetopause and not to a changing current. The maximum current obtained is about 50 nA m^{-2}. These measurements of the magnetopause currents are important for magneto-spheric physics since the force produced by these currents is responsible for plasma jettings during magnetic reconnection. Furthermore, large currents on the magnetopause may be one of the driver to initiate the reconnection process. More studies have reported measurements of the electric current in various regions of the magnetosphere such as in a flux transfer event (Robert and Roux, 2003) or in the magnetotail (Slavin et al., 2003).

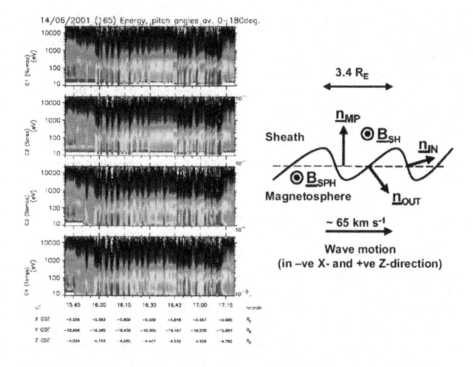

Figure 5. Left: Electron color spectrogram on the four Cluster spacecraft on 14 May 2001 during a magnetopause crossing. High fluxes below 100 eV in red indicates the magnetosheath and low fluxes above 100 eV indicate the magnetosphere. Right: sketch of the wave which propagates along the model magnetopause (dashed line) at about 65 km/s (from Owen et al., 2003).

4.2   The surface waves

In a first approximation, the magnetopause can be considered as a smooth and regular surface. However, it is known that the solar wind is very often inhomogeneous and could deform locally the magnetopause. For instance reconnection can produce a bulge of plasma propagating away from the X point or Kelvin-Helmholtz (KH) instability could form waves propagating along the magnetopause. Such example of waves has been observed with Cluster, especially when the spacecraft are on the flanks of the magnetosphere where it and can stay a long time near the magnetopause at apogee. An example is shown in Figure 5 (from Owen et al., 2003). The magnetopause, defined by the sharp drop of the high-energy electrons (around 1 keV) is crossed about 20 times in the interval of 1.5 hour considered. The spacecraft were separated by 2000 km and each spacecraft give a magnetopause crossing at a slightly different time. Using the timing analysis technique (e. g. Dunlop and Woodward, 1998; Owen et al., 2001), the time of crossing and the position of the spacecraft are giving the normal and the speed of the boundary. The inbound and outbound magnetopause crossings give a different normal, suggesting that a wave, with a wave length of about 3.4 R_E, is passing across the spacecraft at a speed of about 65 km/s. Furthermore the wave is steeper on the leading edge than on the trailing edge, consistent with KH theory (Owen et al., 2003). Surface waves have also been observed on the dusk flank of the magnetosphere by Cluster (Gustaffsson et al., 2001) or other spacecraft (e.g., Kivelson and Chen, 1995; Safrankova et al., 1997; Fairfield et al, 2000) that are consistent with these results.

4.3   The magnetic curvature

The magnetic field in the tail is elongated such that the magnetic curvature is small (radius of curvature is large) in the lobes and large (radius of curvature small) in the neutral sheet. The radius of curvature has been shown to play a key role in the motion of particles in the tail. This parameter can be computed for the first time using the measurement of the magnetic field at the four spacecraft position. An example of a neutral sheet crossing is shown in Figure 6 (from Shen et al., 2003). The spacecraft were first in the Northern lobes and crossed the neutral sheet around 09 UT. The radius of curvature (bottom panel in Figure 6) is below 2 R_E in the neutral sheet between 09:00 and 09:30 and then very large, around 10 R_E, in the lobes (before 08 or after 10). In addition Shen et al. (2003) have shown that the curvature radius is changing during the phase of the substorms: it is very small (less than 0.5 R_E) during the growth phase and expansion phase and larger during the recovery (above 0.8 R_E).

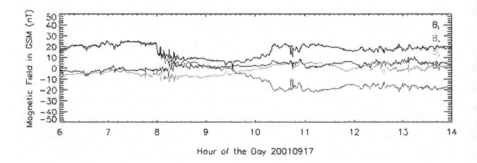

Figure 6. Magnetic field (top) and radius of curvature of the magnetic field (bottom) on 17 September 2001. Bx, BY, Bz and Bt are shown in red, green, blue and black (from Shen et al., 2003).

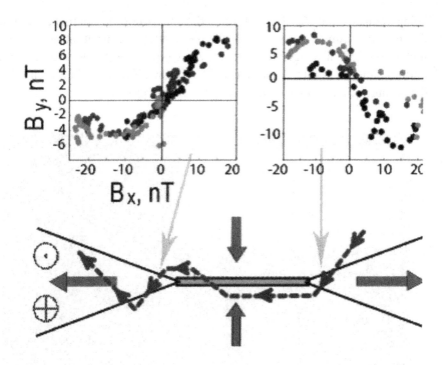

Figure 7. Two crossings of the neutral sheet on 1 October 2001. The top diagram show the magnetic field measured on Cluster and the sketch below show the reconnection event and the respective trajectory of Cluster (dashed red line) (from Runov et al., 2003).

4.4   Magnetic reconnection in the magnetotail and in the cusp

Magnetic reconnection is a universal physical process that transfers magnetic energy into plasma energy, subsequently accelerating particles to very high energy. The effects of this process can be observed deep in the tail in the neutral sheet and in the external boundary of the magnetosphere, in the polar cusp. The main signatures of this process are the plasma jets, clearly visible in the ion flow, and the magnetic reconfiguration.

The first example in Figure 7 shows the data collected by Cluster in the neutral sheet (from Runov et al., 2003). The top panels show By as a function of Bx during two crossings of the neutral sheet of the four spacecraft. During the first crossing (right panel) By changes from being positive at Bx < 0 (northern hemisphere) to negative at Bx > 0 (southern hemisphere). The other crossing (left panel) is characterized by positive By at Bx > 0 and negative By at Bx < 0. These results are consistent with the presence of Hall magnetic currents that are reversed on either side of the reconnection point (Sonnerup, 1979). These currents are typically thought of

being indicative of the decoupling of the ions from the magnetic field and
the electron fluid. A sketch of the reconnected current sheet structure
including the Hall magnetic currents (green) near the X-line is given in the
bottom panel of Figure 7. The accelerated plasma is observed on each side of
the reconnection point as indicated by the purple arrows.

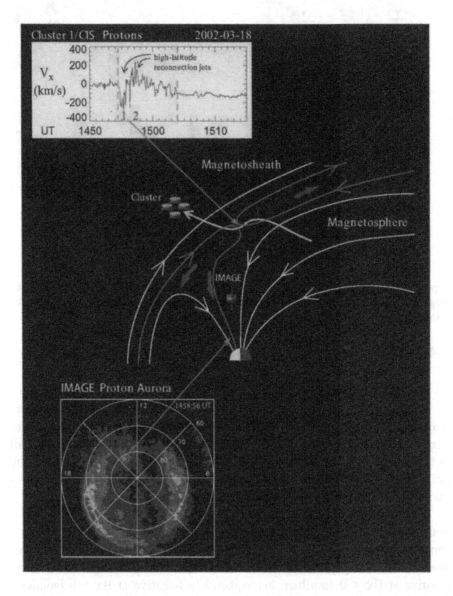

Figure 8. Cluster and IMAGE observations on 18 March 2003. Top: ion flow speed
component Vx (GSE) between 1450 and 1515 UT. Middle: sketch of the reconnection in the
lobes and the trajectory of Cluster. Bottom: Proton aurora obtained by IMAGE spacecraft at
1458 UT.

Another example of reconnection observed in the polar cusp is shown in Figure 8 (from Phan et al., 2003). Cluster was crossing the poleward boundary of the cusp and observed first a tailward flow (1), followed by an Earthward flow (2). This flows are consistent with reconnection poleward of the cusp under Northward interplanetary magnetic field. At the same time the IMAGE spacecraft looking at the precipitation of protons in the ionosphere observed a bright spot, indicative of accelerated ions, located on the same field lines. This is a first direct evidence that protons aurora in the cusp are produced by reconnection on the magnetopause.

5.   CLUSTER SCIENCE DATA SYSTEM

5.1   Cluster Science Data System

The Cluster science data system has been set-up to distribute quicklook and processed data to all Cluster Principal and Co-Investigators, as well as to the scientific community. The Cluster community consists of 11 Principal Investigators and 259 Co-Investigators from 89 laboratories located in 24 countries (see Escoubet and Fehringer, 2003 for details). To distribute the data efficiently to all users, a system of nine data centres located in Austria, China, France, Germany, Hungary, Netherlands, Sweden, United-Kingdom and United-States and interconnected with each other has been set-up. Each data centers store the full database of processed and validated data from all instruments. Data from February 2001 until end of September 2003 are available through the web at http://sci2.estec.esa.nl/cluster/csds/csds.html.

A quicklook plot is also available between a few hours to a few days after data acquisition and includes time series plots and spectrograms from most of the instruments. This is very useful for scientists to pick-up interesting events and for the Cluster project to monitor the progress of the mission. The average download rate during the year 2003 was above 5 Gbytes/month (without including the US data centre).

5.2   Public access to other data sets

The PI teams are distributing other Cluster data sets; these include particle spectrograms, high resolution data, enhanced prime parameters data or summary plots. The web links are in table 2.

Table 3. Links to Cluster additional data sets

ASPOC	http://saturn.iwf.oeaw.ac.at/acdc/acdc.html	Raw data
CIS	http://cis.cesr.fr:8000/CIS_sw_home-en.htm	6h spectrograms, 3 SC
EDI	http://edi.sr.unh.edi	Prime parameters
EFW	http://www.cluster.irfu.se/efw/data/spinfit/index.html	Ex, Ey at 4 s res.
FGM	http://www.sp.ph.ic.ac.uk/Cluster/	Prime parameters
PEACE	http://cluster2.space.swri.edu/	High resolution
RAPID	http://www.linmpi.mpg.de/english/projekte/cluster/rapid.html	High resolution
STAFF	http://www.cetp.ipsl.fr/CLUSTER/accueil/framepa.html	Spectrograms
WBD	http://www-pw.physics.uiowa.edu/plasma-wave/istp/cluster/	Spectrograms
WHISPER	http://www.whisper.cnrs-orleans.fr/	Spectrograms

5.3   Cluster active archive

The Cluster mission is successfully delivering summary and prime parameter data through the Cluster Science Data System. In February 2003, the ESA Science Programme Committee has agreed that the Cluster Project set-up a Cluster Active Archive (CAA) that will contain processed and validated high-resolution scientific data, as well as raw data, processing software, calibration data and documentation from all the Cluster instruments.   The scientific rationale underpinning this proposal is as follows:

- Maximise the scientific return from the mission by making all Cluster data available to the worldwide scientific community.
- Ensure that the unique data set returned by the Cluster mission is preserved in a stable, long-term archive for scientific analysis beyond the end of the mission.
- Provide this archive as a contribution by ESA and the Cluster science community to the International Living With a Star programme.

The CAA will be a database of high-resolution data and other allied products that will be established and maintained under the overall control of ESA.

Two important aspects of this proposal are as follows. In view of the shortage of manpower in most of the institutes processing Cluster data, ESA is supporting manpower to be deployed in institutes where the relevant expertise exists, to assist in the preparation, validation, and documentation of the high-resolution data to be deposited in the archive.

In view of the urgency in starting the programme, and recognising that much of the in-house expertise might be lost when National Agency funding to the Cluster instrument teams is expected to be greatly reduced at the end of the Cluster mission, it was imperative to start these activities as soon as possible. The design phase has started in February 2003, and will be followed by a development and implementation phase, comprising software integration and data preparation in 2004.

Processing and preparation of data to be archived will be started and will proceed in parallel, with data from all instruments entering the database at an average rate of two years of data per calendar year. The data from the year 2001 will be archived in 2004, then data from 2002-2003 in 2005, and 2004-2005 in 2006. The year 2007 is kept for reprocessing and finalization of the archive. Data from any individual experiment may be archived more or less rapidly, subject to the requirement that the archiving all data should be completed at the conclusion of CAA phase. The archive will be accessible to all scientists. Once data are included in the archive, it will be public.

6.   CONCLUSION

After 2.5 years of operations, the Cluster spacecraft have shown their full capability to make substantial advances in magnetospheric physics. For the first time plasma structures have been studied in three dimensions.

In this paper we have shown a few example of the Cluster capabilities, such as computation of electric current or magnetic field curvature and waves geometry analysis. Reconnection in the tail and at the magnetopause was observed in great details.

Next year Cluster will go for the first time to the large separation distances (10000 km) and should shed new lights on large scales phenomena such as the cusp geometry or the size of bursty bulk flow in the tail. The operations are funded up to end of 2005, however, if the spacecraft and instrument are in good health a further extension may be proposed to the ESA Science Programme Committee at the beginning of 2005.

At the end of this year, the Cluster mission will be complemented by a Chinese mission called Double Star. This mission consists of two spacecraft, one equatorial TC-1 (550 x 66,970 km, 28.5 deg. inclination) to be launched in December 2003 and the second one polar TC-2 (700 x 39,000, 90 deg. inclination) to be launched in June 2004. The orbits of Double Star have been specially designed to maximize the conjunction with Cluster in the plasma sheet and in the polar cusp. Furthermore half of the Double Star payload is made of spare or duplicate of the Cluster instruments which should allow full comparison to be made.

The latest news and the access to Cluster data through the Cluster Science Data System can be found at: http://sci.esa.int/cluster/.

ACKNOWLEDGMENTS

The authors thank all PIs and their teams who provided the Cluster data, and the JSOC and ESOC team for their very efficient Cluster operations. In particular, the authors would like to acknowledge M. Dunlop (RAL, UK) for the electric current calculation, C. Owen (MSSL,UK) for the surface waves, A. Runov, R. Nakamura and W. Baumjohann (IWF, A) for the reconnection event, C. Shen (CSSAR, China) for the magnetic field curvature computation and T. Phan (SSL, USA) for the reconnection event observed by both Cluster and IMAGE.

REFERENCES

Balogh, A., Dunlop, M. W., Cowley, S. W. H., Southwood, D. J., Thomlinson, J. G., Glassmeier, K.-H., Musmann, G., Luer, H., Buchert, S., Acuna, M. H., Fairfield, D. H., Slavin, J. A., Riedler, W., Schwingenschuh, K., and Kivelson, M. G., 1997, The CLUSTER magnetic field investigation, *Space Sci. Rev.* **79**: 65–91.

Balogh, A., Carr, C. M., Acuna, M. H., Dunlop, M. W., Beek, T. J., Brown, P., Fornacon, K.-H., Georgescu, E., Glassmeier, K.-H., Harris, J., Musmann, G., Oddy, T., and Schwingenschuh, K., 2001, The CLUSTER magnetic field investigation: Overview of in-flight performance and initial results, *Ann. Geophys.*: **19**(10/12):1207–1217.

Dunlop, M. W., and Woodward, T. I., 1998, Discontinuity analysis: Orientation and motion, in: *Analysis Methods for Multispacecraft Data, ISSI Sci. Rep. SR-001*, Götz Paschmann, Patrick W. Daly, eds., ESA Publications Division, Noordwijk, pp. 271–305.

Dunlop, M. W., Balogh, A., Glassmeier, K., Robert, P., 2002, Four-point Cluster application of magnetic field analysis tools: the curlometer, *J. Geophys. Res.*: **107**(A11):1384, doi: 10.1029/2001JA005088.

Dunlop, M. W., et al., 2003, Cluster magnetopause observation: Dynamics and current structures, *Ann. Geophys.*, submitted.

Escoubet, C. P., Fehringer, M., and Goldstein, M. , 2001, The Cluster mission, *Ann. Geophys.* **19**(10/12):1197–1200.

Escoubet, C. P., and Fehringer, M., 2003, Cluster: new view on the boundaries of the magnetosphere, in: *Proceedings of the COSPAR Colloquium on Frontiers of Magnetospheric Plasma Physics,* ISAS, Sagamihara, in press.

Fairfield, D. H., Otto, A., Mukai, T., Kokubun, S., Lepping, R. P., Steinberg, J. T., Lazarus, A. J., and Yamamoto, T., 2000, GEOTAIL observations of the Kelvin-Helmholtz instability at the equatorial magnetotail boundary for parallel northward fields, *J. Geophys. Res,* **105**(A9):21159–21174.

Gustafsson, G., André, M., Carozzi, T., Eriksson, A. I., Fälthammar, C.-G., Grard, R., Holmgren, G., Holtet, J. A., Ivchenko, N., Karlsson, T., Khotyaintsev, Y., Klimov, S., Laakso, H., Lindqvist, P.-A., Lybekk, B., Marklund, G., Mozer, F., Mursula, K., Pedersen, A., Popielawska, B., Savin, S., Stasiewicz, K., Tanskanen, P., Vaivads, A., and Wahlund, J-E., First results of electric field and density observations by Cluster EFW based on initial months of operation, *Ann. Geophys.* **19**(10/12):1219–1240.

Hearendel, G., Roux, A., Blanc, M., Paschmann, G., Bryant, D., Korth, A., and Hultqvist, B., 1982, Cluster, study in three dimensions of plasma turbulence and small-scale structure, *mission proposal submitted to ESA.*

Kivelson, G. K., and Chen, S.-H., 1995, The magnetopause: Surface waves and instabilities and their possible dynamical consequences, in: *Physics of the Magnetopause*, P. Song, B. U. Ö. Sonnerup and M. F. Thomsen, eds., Geophysical Monograph Series, Volume 90, AGU, Washington DC, pp. 257–268.

Owen, C. J., Fazakerley, A. N., Carter, P. J., Coates, A. J., Krauklis, I. C., Szita, S., Taylor, M. G. G. T., Travnicek, P., Watson, G., Wilson, R. J., Balogh, A., and Dunlop, M. W., 2001, Cluster PEACE observations of electrons during magnetospheric flux, transfer events, *Ann. Geophys.* **19**(10/12):1509–1522.

Owen, C. J., Taylor, M. G. G. T., Krauklis, I. C., Fazakerley, A. N., Dunlop, M. W., and Bosqued, J. M., 2003, Cluster Observations of Surface Waves on the Dawn Flank Magnetopause, *Ann. Geophys.* in press.

Phan, T., Frey, H. U., Frey, S., Peticolas, L., Fuselier, S., Carlson, C., Rème, H., Bosqued, J.-M., Balogh, A., Dunlop, M., Kistler, L., Mouikis, C., Dandouras, I., Sauvaud, J.-A., Mende, S., McFadden, J., Parks, G., Moebius, E., Klecker, B., Paschmann, G., Fujimoto, M., Petrinec, S., Marcucci, M. F., Korth, A., and Lundin, R., 2003, Simultaneous Cluster and IMAGE observations of cusp reconnection and auroral proton spot for northward IMF, *Geophys. Res. Lett,* **30**(10):1509, doi: 10.1029/2003GL016885.

Robert, P., Roux, A., Chanteur, G., Le Contel, O., Perraut, S., Cornilleau-Wehrlin, N., De Vulder, P., Dunlop, M. W., Balogh, A., and Glassmeier, K.-H., 2003, Estimation of the current density in a FTE, in *book of abstracts of Spatio–Temporal Analysis and Multipoint Measurements in Space.*

Russell, C. T., 2000, ISEE lessons learned for Cluster, in: *Proceedings of Cluster II workshop on Multiscale/Multipoint plasma measurements*, ESA SP 449, R. Harris, ed., ESA publications, Noordwijk, pp. 11–23.

Safrankova, J., Zastenker, G., Nemecek, Z., Fedorov, A., Simersky, M., and Prech, L., 1997, Small scale observation of magnetopause motion: preliminary results of the INTERBALL project, *Ann. Geophys.* **15**(5):562–569.

Shen, C., Li, X., Dunlop, M., Liu, Z. X., Balogh, A., Baker, D. N., Hapgood, M., and Wang, X., 2003, Analyses on the geometrical structure of magnetic field in the current sheet based on Cluster measurements, *J. Geophys. Res.* **108**(A5):1168, doi: 10.1029/2002JA009612.

Slavin, J. A., Lepping, R. P., Gjerloev, J., Goldstein, M. L., Fairfield, D. H., Acuna, M. H., Balogh, A., Dunlop, M., Kivelson, M. G., Khurana, K., Fazakerley, A., Owen, C. J., Rème, H., Bosqued, J. M., 2003, Cluster electric current density measurements within a magnetic flux rope in the plasma sheet, *Geophys. Res. Lett.* **30**(7):1362, doi: 10.1029/2002GL016411.

Sonnerup, B. U. O., 1979, Magnetic field reconnection, in: *Solar System Plasma Physics (vol. 3)*, C. F. Kennel, L. J. Lanzerotti, and E. N. Parker, eds., North Holland, pp. 45–108.

Runov, A., Nakamura, R., Baumjohann, W., Treumann, R. A., Zhang, T. L., Volwerk, M., Vörös, Z., Balogh, A., Glaßmeier, K.-H., Klecker, B., Rème, H., and Kistler, L., 2003, Current sheet structure near magnetic X-line observed by Cluster, *Geophys. Res. Lett.* **30**(11):1579, doi: 10.1029/2002GL016730.

Zelenyi, L., Zastenker, G., Dalin, P., Eiges, P., Nikolaeva, N., Safrankova, J., Nemecek, Z., Triska, P., Paularena, K., and Richardson, J., 2000, Variability and Structures in the solar wind – magnetosheath – magnetopause by multiscale multipoint observations, in: *Proceedings of Cluster II workshop on Multiscale/Multipoint plasma measurements*, ESA SP 449, R. Harris, ed., ESA publications, Noordwijk, pp 29–38.

CUSP PROPERTIES FOR B_Y DOMINANT IMF

Simon Wing, Patrick T. Newell, and Ching-I Meng
The Johns Hopkins University Applied Physics Laboratory, 11100 Johns Hopkins Road, Laurel, Maryland 20723-6099, USA

Abstract: Cusp properties during periods of B_y dominant IMF are investigated, since previous studies focus mostly on IMF B_z. The model-data comparisons for various IMF configurations show that the model captures the large-scale features of the particle precipitation very well, not only in the cusp region, but also in other open-field line regions such as the mantle, polar rain, and open-field line low-altitude boundary layer (LLBL). When the IMF is strongly duskward/dawnward and weakly southward, the model predicts the occurrence of a double cusp near noon: one cusp at lower latitude and one at higher latitude. The lower latitude cusp ions originate from the low-latitude magnetosheath whereas the higher latitude ions originate from the high-latitude magnetosheath. The lower latitude cusp is located in the region of weak azimuthal $\mathbf{E} \times \mathbf{B}$ drift, resulting in a dispersionless cusp. The higher latitude cusp is located in the region of strong azimuthal and poleward $\mathbf{E} \times \mathbf{B}$ drift. Because of a significant poleward drift, the higher latitude cusp dispersion has some resemblance to that of the typical southward IMF cusp. Occasionally, the two parts of the double cusp have such narrow latitudinal separation that they give the appearance of just one cusp with extended latitudinal width. From the 40 DMSP passes selected during periods of large (positive or negative) IMF B_y and small negative IMF B_z, 30 (75%) of the passes exhibit double cusps or cusps with extended latitudinal width. The double cusp result is consistent with the following statistical results: (1) the cusp's latitudinal width increases with |IMF B_y| and (2) the cusp's equatorward boundary moves to lower latitude with increasing |IMF B_y|.

Key words: double cusp; cusp latitudinal width; cusp equatorward boundary; cusp model; spatial feature.

J. –A. Sauvaud and Z. Němeček (eds.),
Multiscale Processes in the Earth's Magnetosphere:From Interball to Cluster, 149-174.
© 2004 *Kluwer Academic Publishers. Printed in the Netherlands.*

1.   INTRODUCTION

An important part of the dayside solar wind-magnetosphere interaction is magnetic merging or reconnection. As a result, the shocked solar wind ions and electrons, can and do enter the magnetosphere and some precipitate into the ionosphere. Although these particles originate in the solar wind, once they have entered the magnetosphere and ionosphere they exhibit distinctly different characteristics in energy, density, and temperature at different local times and latitudes. Observations at low altitude show that the resulting particle precipitation associated with open-field lines can generally be classified into four regions (ordered from low to high latitude for a typical southward IMF case): open-field low latitude boundary layer (LLBL), cusp, mantle, and polar rain (e.g., Newell et al., 1991b; Newell and Meng, 1995; Onsager and Lockwood, 1997). Out of these four regions, the cusp was discovered the first (Eather and Mende, 1971; Heikkila and Winningham, 1971; Frank, 1971), partly because of its higher flux and energy and partly because of its theoretical importance, and has attracted the most attention ever since.

There have been many studies on IMF control of particle cusp properties, e.g., locations, energy-latitude dispersions and longitudinal widths (e.g., Burch, 1972; Hill and Reiff, 1977; Carbary and Meng, 1986; Newell et al., 1989; Aparicio et al., 1991; Woch and Lundin, 1992; Zhou et al., 2000; Merka et al., 2000; Nemecek et al., 2003). With a few exceptions (e.g., Nemecek et al., 2003), most of these studies examine IMF B_z effects on the cusp. As a result, the relationships between the cusp and IMF B_y are not well known. This paper highlights and reviews (1) the observed cusp properties under various IMF conditions, particularly IMF B_y, (2) APL particle precipitation model calculations, which provide insights into the observations, e.g., locations of particle entries, convection electric field, energy-latitude dispersion, etc.

2.   APL OPEN-FIELD LINE PARTICLE PRECIPITATION MODEL

Self-consistent global models are not yet advanced enough to permit precise quantitative comparisons with the observations. For example, single-fluid MHD simulations cannot capture the parallel electric field arising from the charge-quasi neutrality constraints in the open-field lines in the magnetosphere. The suprathermal electrons, which populate much of polar rain, are absent in the MHD simulations.

Efforts to produce a model that can withstand detailed comparisons to low-altitude or mid-altitude cusp data advanced significantly with the work of Onsager et al. (1993). Instead of developing a global model self-consistently for the entire magnetosheath-magnetosphere-ionosphere system, Onsager et al. use an assimilative approach that combines good quality empirical models for different regions. In their model, for a given southward IMF orientation, solar wind temperature and density, ionospheric convection speed, and dipole tilt angle, the model computes the phase space density of the precipitating ions and electrons in three steps. In the first step, which assumes the magnetic moment is conserved, the ionospheric particles are traced back along the guiding centers to the magnetopause entry point using the Stern (1985) magnetic field model modified by uniform IMF penetration (cf. Cowley et al., 1991; Wing et al., 1995; Wing and Sibeck, 1997) and a simple dawn-dusk electric field. The second step is to compute the acceleration ($\mathbf{j} \cdot \mathbf{E} > 0$) or deceleration ($\mathbf{j} \cdot \mathbf{E} < 0$) imparted on the particles when they cross the magnetopause current layers from the magnetosheath to the magnetosphere. This computation is done with the aid of the de Hoffman-Teller reference frame in which $\mathbf{E} = 0$ (e.g., Hill and Reiff, 1977; Cowley and Owen, 1989). From this calculation, the model obtains the velocity that the particle originally had in the magnetosheath. Finally, it computes the phase space density of particles with that velocity using the gas-dynamics calculations of Spreiter and Stahara (1985) with the assumption that all the particles, ions and electrons, have Maxwellian distributions. Assuming conservation of phase space density along particle trajectories, the model can be used to compute the differential energy flux at the location where the particle was "detected" in the ionosphere. The original model result and DMSP data comparison shows that the southward IMF cusp can be modeled fairly well but the model electrons have a much more latitudinally extended entry and a much higher temperature in the mantle and polar rain regions (Onsager et al., 1993; Wing et al., 1996). Other problems include the cusp latitude being several degrees too high (mainly a problem with the magnetic field model) and ionospheric convection velocity (100 m s^{-1}) being several times too low.

The Johns Hopkins University Applied Physics Laboratory (APL) open-field line particle precipitation model basically uses the same approach as Onsager et al. (1993). However, we have introduced more realistic processes into the model and, as a result, we can model not just the cusp, but the entire open-field line particle precipitation region, namely open-field LLBL, cusp, mantle, and polar rain (Wing et al., 1996 and 2001; Newell and Wing, 1998). This is summarized below.

Electrons have thermal speeds far exceeding the magnetosheath flow speed and therefore can enter the magnetosphere along the open field lines across the polar cap. In contrast, ions have slower thermal speeds and therefore can only enter the magnetosphere from the regions in the magnetopause where the magnetosheath flow is subsonic (Reiff et al., 1977). Several researchers have noted that there has to be a mechanism that limits the entry of the electrons to balance the charge carried by the ions, maintaining charge quasi-neutrality in the precipitating particle populations (e.g., Reiff et al., 1977; Burch, 1985). Solar wind electrons have been observed to have thermal and suprathermal components (e.g., Feldman et al., 1978; Fairfield and Scudder, 1985). The original Onsager model mantle ions have much lower flux than in the DMSP data, but ions in the solar wind and the magnetosphere have been observed to have κ distributions (e.g., Feldman et al., 1974; Christon et al., 1989). A κ distribution resembles a Maxwellian at low energies, but approaches a power law distribution at high energies. For a given characteristic energy, a κ distribution produces a higher total flux in the ionosphere, owing to its high-energy tail. Magnetic field models have been steadily improved in the recent years, e.g., with the inclusion of Birkeland currents etc. (e.g., Tsyganenko and Stern, 1996). Finally, in much of the polar cap, the electric field frequently deviates from the dawn-dusk direction, especially when the IMF y-component dominates. Motivated by these results, we extended the original Onsager model as follows (Wing et al., 1996; Newell and Wing, 1998; Wing et al., 2001): (1) imposed charge-quasi neutrality with a self-adjusting parallel electric field; (2) included suprathermal electrons; (3) used a κ distribution for ions; (4) replaced the Stern (1985) magnetic field model with the T96 model (Tsyganenko and Stern, 1996); and (5) used the convective electric field obtained from the statistical APL convection patterns (Ruohoniemi and Greenwald, 1996).

　　Although the APL convection pattern provides an accurate electric field, it is not consistent with the T96 magnetic field model. The T96 model itself has its own deficiencies, e.g., it does not take into account the effects of IMF on the magnetopause shape and size, which in turn can affect the cusp footprint (e.g., Shue et al., 1997). The Spreiter and Stahara (1985) magneto-sheath model is a single-fluid gas-dynamic model that does not take into account the magnetic field. In addition, the model has not taken all the particle precipitation processes into account such as wave-particle interactions, non-adiabatic motions, particle diffusion across the magnetopause, etc.

3.　　SOUTHWARD IMF CUSP

　　In order to show how well the model works, the model results are compared with DMSP observations. First, we show the cases for strongly southward IMF and weakly southward IMF. These two cases are shown to demonstrate how well the model can produce the large-scale features in the cusp. Then, we proceed to show the cusp for B_y dominated IMF.

3.1   Strongly southward IMF cusp

The result of the model calculation for the strongly southward IMF is presented in Figure 1a (from Plate 2 in Wing et al. (2001)). Note that the y-axis of the ion panel displays the lowest energy at the top, the opposite from the way the electron is displayed. The input parameters to the model are: IMF (B$_x$, B$_y$, B$_z$) = (-3.4, -0.5, -12.3) nT, solar wind thermal n = 11 cm^{-3}, T_i = 1x10^5 °K, T_e = 3x10^4 °K, suprathermal (halo) electron n_s = 0.2 cm^{-3}, T_s = 1x10^6 °K, κ = 7, and the altitude of "detected" particle = 1.13 R$_E$, which corresponds to the DMSP spacecraft altitude. DMSP observations under similar solar wind and IMF conditions are shown in Figure 1b. DMSP are sun-synchronous satellites in a nearly circular polar orbit at an altitude of roughly 835 km and period of approximately 101 minutes per orbit. The SSJ4 instrumental package included on all recent DMSP flights uses curved plate electrostatic analyzers to measure ions and electrons from 32 eV to 30 keV in 19 logarithmically-spaced steps. One complete 19-point electron and ion spectrum is obtained each second. The magnetic coordinates used in our studies are the Altitude Adjusted Corrected Geomagnetic coordinates (AACGM) (Baker and Wing, 1989). The solar wind thermal electron temperature is taken to be somewhat lower than that of the ions to compensate for excessive heating in the model magnetosheath. This is because the Spreiter and Stahara (1985) model is a single fluid model, which overestimates the amount of electron heating in the magnetosheath. Since the ions carry most of the kinetic energy, upon encountering the magnetopause they are thermalized to a higher temperature than are electrons. Many large-scale features that are seen in the model can also be seen in a typical DMSP pass such as the one shown in Figure 1b. Figure 1 clearly shows that the model can successfully calculate the precipitating ion and electron fluxes for the cusp, mantle, and the open-field LLBL, which is located equatorward of the cusp (in order to focus more on the cusp, comparisons with the polar rain are not shown here, but they have been shown to compare well (Wing et al. (1996)).

The model cusp ions originate from the low-latitude magnetopause, within 7 R$_E$ from the subsolar point. This result is in agreement with the previous observational cusp studies during the period of southward IMF (e.g., Reiff et al., 1977). In the present paper, "low-latitude magnetopause" refers to the magnetopause locations where |z| <~ 5 R$_E$, "mid-latitude" refers to the region 5 R$_E$ < |z|< 10 R$_E$ and "high-latitude" refers to regions where |z| > ~10 R$_E$.

The success of the open-field line particle precipitation model strongly suggests that the same large-scale processes govern all four particle precipitation regions in the open-field line domain, namely, open field-line

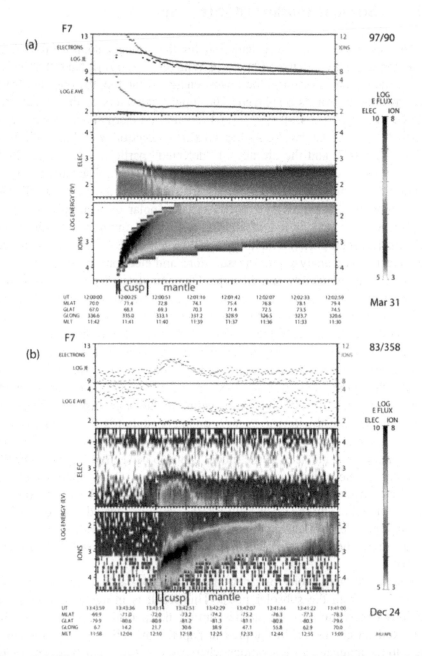

Figure 1. (a) The results of the model calculations for strongly southward IMF case and (b) a DMSP observation under similar IMF condition. The spectrogram shows log differential energy flux, in units of eV/cm^2 s sr eV, from 32 eV to 30 keV, with the ion energy scale inverted. The lower of the two line plots shows the average energy in eV for the electrons (black) and ions (orange), and the top line plot is of integral energy flux in units eV/cm^2 s sr. The red labels beneath the x-axis indicate the region types. L indicates the open-field LLBL, which is located equatorward of the cusp.

LLBL, cusp, mantle, and polar rain (Wing and Newell, 1996). Open-field LLBL is the region closest to the open/closed boundary. When the field line first becomes open, electrons having higher speeds than ions flow into the magnetosphere ahead of the ions. Charge quasi-neutrality and the resulting parallel electric field, however, limit the number of electrons that can enter. Thus, in this region, few electrons and ions are present. In the cusp, the ions have reached the ionosphere and intense fluxes of ions and electrons are usually observed. In this region, the electrons and ions can enter the magnetosphere relatively freely because the numbers of magnetosheath ions and electrons are already balanced, resulting in little or no parallel electric field. In the mantle region, fewer ions can enter as the magnetosheath flow becomes increasingly tailward and larger, whereas the magnetospheric magnetic field (and hence precipitating particle velocity) becomes more sunward, a condition which is less favorable for particle entries. In this region, $j \bullet E < 0$, which means that the magnetic stress at the magnetopause is directed to decelerate the plasma (e.g., Hill and Reiff, 1977; Cowley and Owen, 1989). Some of the solar wind thermal or core electron entries are limited by the ensuing parallel electric field that arises to maintain charge quasi-neutrality. Finally, in the polar rain region, no significant amount of ions enter the magnetosphere and the parallel electric field rises to the level where only higher energy tail end of the core electrons and the suprathermal electrons can enter the magnetosphere, by the virtue of having higher energy that can overcome the parallel electric potential.

An example of the typical parallel electric potential for these four regions for one of our model calculations (not for Figure 1) is shown in Figure 2 (from Figure 1 of Newell and Wing (1998)). This figure shows the parallel electric potential needed to maintain charge-quasi neutrality and is obtained in the model using a binary search algorithm. The parallel electric field resulting from maintaining charge quasi-neutrality of the precipitating magnetosheath ions and electrons should have implications to the ionospheric outflows. For example, the parallel electric field that prevents magnetosheath electrons from entering the magnetosphere should help increase the electron outflow and retard ion outflow.

3.2 Weakly southward IMF cusp

In the second case, the IMF is weakly southward. The input parameters to the model remain the same as before except for the IMF, which has been changed to IMF $(B_x, B_y, B_z) = (-0.5, -0.5, -3)$ nT. The model output and DMSP observations under similar IMF conditions are shown in Figures 3a and 3b, respectively (from Plate 3 in Wing et al. (2001)). Figure 3 shows that again the model seems to be able to capture the macro-scale features that are seen in the observations. The DMSP cusp in Figure 3b was obtained not under the same exact solar wind and IMF conditions as in Figure 3a. (To facilitate comparisons between the model results, the solar wind input parameters are kept the same and only the IMFs change in Figures 1a and 3a). These two factors contribute the discrepancy in the model results and the DMSP example in Figure 3.

One of the main differences between this and the previous IMF case is that the cusp location moves to higher latitude as IMF B_z increases, a well-documented phenomenon in many observational studies (e.g., Carbary and Meng, 1986; Newell et al., 1989; Zhou et al., 2000). The movement of the cusp location has been interpreted in terms of merging and the flux erosion on the dayside when IMF B_z turns more southward (e.g., Zhou et al., 2000). Our model does not have explicit merging, but most of this effect is captured by the magnetic cusp location in T96 (geometrical effect). In addition, the model magnetopause increases in size with increasing IMF B_z, resulting in longer field lines between the ionosphere and the magnetopause shape (Roelof and Sibeck, 1993). The longer field increases the duration of the

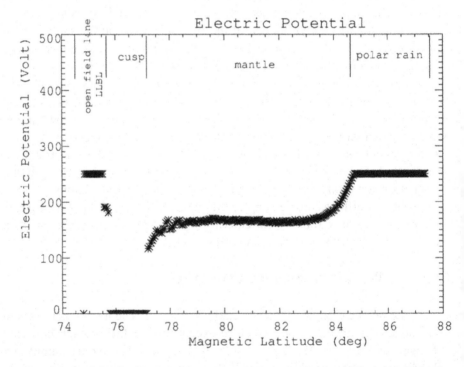

Figure 2. The model parallel electric potential between the Earth and the magnetosheath needed to retard electron entry enough to satisfy charge quasi-neutrality as a function of magnetic latitude.

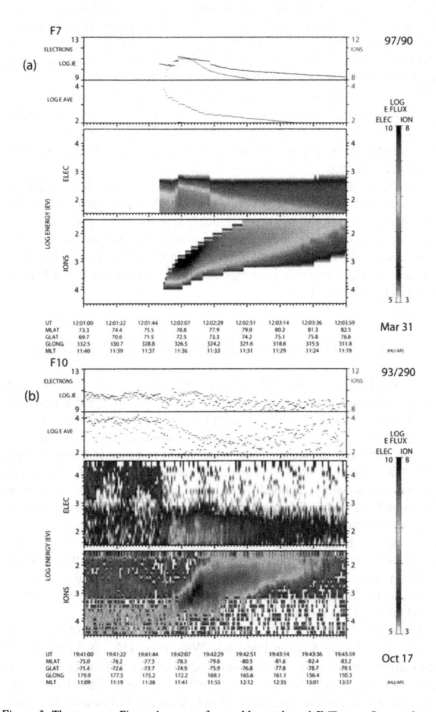

Figure 3. The same as Figure 1, except for weakly southward IMF case. See caption of Figure 1 for the descriptions of the units, scales etc.

particles undergoing **E**×**B** drift (time of flight effect). As a result, the particle cusp location is shifted more poleward of the open-closed field line separatrix compared to that in the strongly southward IMF cusp. Thus, the model predicts a wider open-field line LLBL for weakly southward IMF than that for strongly southward IMF.

The model cusp ions in the weakly southward IMF case originate in the low- to mid-latitude magnetopause/magnetosheath, $z \sim 2$–10 R_E. The higher energy cusp ions enter from mid-latitude magnetopause, $z \sim 5$–10 R_E. The entry points are at higher latitude compared to those for the strongly southward IMF case.

In both southward IMF cases, the near-noon magnetospheric magnetic field line and the **E**×**B** convection have little y-component. So, the precipitating cusp ions at noon originate approximately from the noon magnetopause at low latitude. Once they enter the magnetosphere, they undergo strong **E**×**B** poleward drift, resulting in the classical cusp dispersion in which the ion characteristic energy decreases with increasing latitude, as shown in Figures 1 and 3.

4.   CUSP FOR LARGE IMF B_Y AND SMALL IMF B_Z

For the third case, the IMF B_z is weakly negative and B_y is strongly positive. The input parameters are the same as before, except that now IMF $(B_x, B_y, B_z) = (-3.4, 12.3, -0.5)$ nT. This IMF configuration amounts to $-90°$ rotation in the y-z plane from the strongly southward IMF case while the magnitude remains unchanged. The model result is shown in Figure 4 (from Plate 4 in Wing et al. (2001)). The model predicts two cusps (double cusp) that are latitudinally separated. The lower latitude cusp has little or no dispersion (stagnant) and the higher latitude cusp exhibits dispersion that has some resemblance to the classical southward IMF dispersion. The model stops tracing whenever the particle reaches $x < -50$ R_E. This explains the sudden cutoff of the polar rain electrons in Figure 4. However, the polar rain in this region is fairly homogeneous and featureless. Had the model continued tracing tailward of $x = -50$ R_E, the resulting polar rain spectra would look just like the ones immediately preceding the cutoff.

Examples of DMSP observations when IMF B_z is small and B_y is large are shown in Figure 5 (from Plate 5 in Wing et al. (2001)). In the DMSP observations, sometimes the separation between the two cusps narrow to give the impression of just one cusp with an extended latitudinal width. However, the dispersion signatures remain the same: the lower-latitude cusp has little or no dispersion and the higher-latitude cusp has dispersion that has some resemblance to that of the southward IMF cusp.

In the model, the lower-latitude cusp ions originate from low latitude magnetopause ($-5 < z < 5$ R_E) and the higher-latitude cusp ions originate from high latitude magnetopause ($7 < z < 13$ R_E). In the APL convection pattern, the $\mathbf{E}\times\mathbf{B}$ convection in the lower latitude cusp region is weak and directed dawnward, whereas in the higher latitude cusp region, it is strong and directed dawnward and poleward (see Figure 2 of Wing et al. (2001)). Thus, the model satellite traveling in the meridional direction near noon encounters ions from two magnetosheath sources. The first population is associated with the ions that enter from the low-latitude magnetopause near noon meridian and then undergo little $\mathbf{E}\times\mathbf{B}$ dawnward convection, nearly perpendicular to the satellite path. This results in the dispersionless ion signature in the lower-latitude cusp in Figures 4 and 5. The second population is associated with ions that enter at the high-latitude magnetopause eastward of the satellite location. Upon entering the magnetopause, the ions $\mathbf{E}\times\mathbf{B}$ convect strongly westward and poleward. Because of a significant poleward convection, the model satellite "observes"

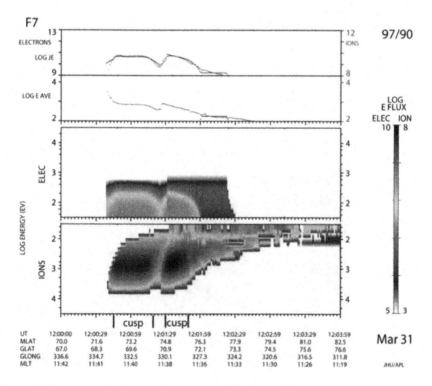

Figure 4. The same as Figure 1, except for strongly duskward and weakly southward IMF case. The calculation result shows two cusp regions that are latitudinally separated (double cusp). The model stops tracing at $x < -50$ R_E, which explains the sudden cut off of the polar rain electron spectra. See caption of Figure 1 for the descriptions of the units, scales etc.

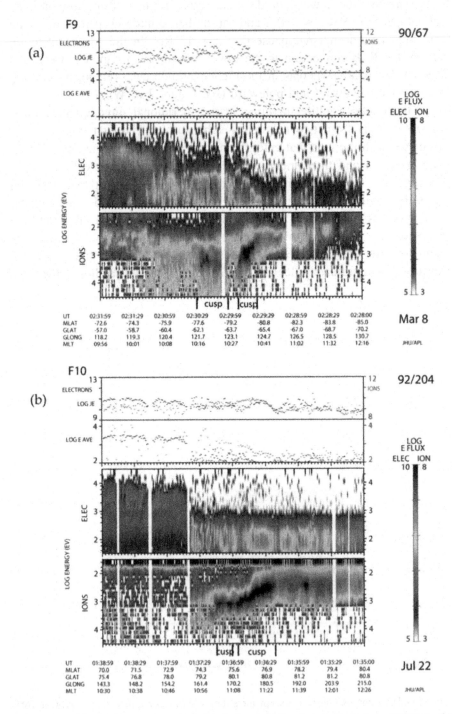

Figure 5. DMSP double cusp events (a, b) during periods of strongly duskward IMF. In (b) the lower latitude and the higher latitude cusp appear to form one cusp with extended latitudinal width. See caption of Figure 1 for the descriptions of the units, scales etc.

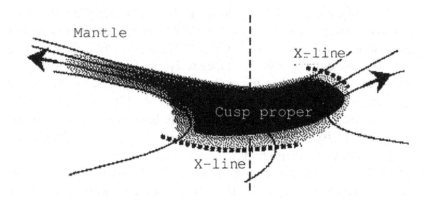

Figure 6. With two simultaneous merging sites, an ionospheric satellite traveling in a meridional trajectory near noon (dashed line) could encounter discontinuous cusp ion dispersions and two sources of ion population. The lower latitude cusp ions are associated with the field lines that have recently merged at low-latitude near noon magnetopause. The higher latitude cusp ions are associated with the field lines that have recently merged at high-latitude post-noon and then convect westward to pre-noon magnetopause. The schematic diagram is for a steady, large and positive IMF B$_y$ (adapted from Figure 5 in Weiss et al., 1995).

dispersion that is similar to the classical southward IMF dispersion. Our model does not have explicitly merging processes. If all magnetosheath ion entries are the result of merging, then the result here suggests that merging simultaneously occurs at the high- and low-latitude magnetopause. This scenario is depicted in Figure 6, which is adapted from Figure 5 in Weiss et al. (1995).

Recently several observational studies at mid- and low-altitudes with Polar, Fast, and DMSP satellites report the discontinuous cusp as a spatial feature rather than temporal feature (e.g., Trattner et al., 1999; 2002; Su et al., 2001; Pitout et al., 2002). Evidence for latitudinally separated cusp during B$_y$ dominant IMF has also been presented in a recent high-altitude cusp study (Merka et al., 2002). Our model can provide the framework to interpret these results. Trattner et al. (1999; 2002) report observations of multiple cusps (more than 2) under steady solar wind and IMF. However, in their study, they do not distinguish among the cusp, the mantle, and the open-field line LLBL regions, but rather all these three regions are lumped together as cusp. In our classification scheme (Wing et al., 1998 and 2001), some of their cusps would be labeled as mantle or open-field line LLBL. As an example, Figure 7 shows a Polar TIMAS observation for an event discussed in Trattner et al. (2002). During this period, IMF was southward and duskward, with the z-component comparable to the y-component, average $(B_x, B_y, B_z) = \sim(-3, 3.5, -3.5)$ nT. They identified 4 cusp regions as indicated by the solid horizontal lines. Not only is the double cusp featured prominently in this event, but it has the same dispersion signature predicted by our model, namely the lower-latitude cusp has little or no dispersion and the higher-latitude cusp has some dispersion. The lower and higher latitude cusps are labeled cusp 1 and 2, respectively, in Figure 7. The region poleward of cusp 2, which has lower fluxes and energies, would be called the mantle in our classification scheme (Wing et al., 1998 and 2001). The region equatorward of cusp 1, which has slightly higher energy, may be the open-field line LLBL, although this is hard to ascertain without the accompanying electron spectrogram.

We would also like to distinguish our double cusp events from the event presented in Coleman et al. (2001). Their event shows a discontinuity in the mantle (not cusp) region, which they attribute to merging locations in the northern and southern hemisphere that is consistent with the anti-parallel merging. The discussion in this paper pertains only to the cusp.

Not all the features in the DMSP observations match the model results because the model still needs to incorporate a number of processes, as mentioned in section 2. Nonetheless, the model seems to be able to capture the large-scale features in the observations.

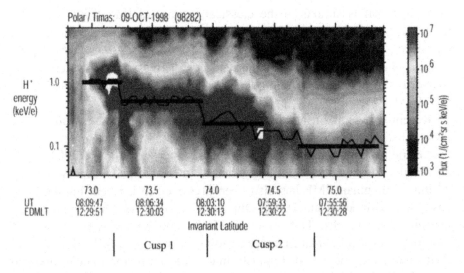

Figure 7. An example of double cusp in Polar TIMAS flux measurements (1/(cm²·s·sr·keV/e)). Consistent with the model prediction, the lower latitude cusp (cusp 1) has little or no dispersion whereas the higher latitude cusp (cusp 2) shows the classical southward IMF dispersion. The region poleward of cusp 2 has lower fluxes and energies, which would be classified as mantle. The region equatorward of cusp 1, which has higher energies, may be the open-field line LLBL, although this is hard to ascertain without the accompanying electron spectrogram. From Plate 2 of Trattner et al. (2002).

5.   THE FREQUENCY OF DOUBLE CUSP OCCURRENCE IN THE DMSP DATA SET

The DMSP database for the period of 1985–1995 was searched for cusp events when the IMF has a large *y*-component and a small negative *z*-component. The automated algorithm that identifies auroral oval boundaries and structures based on DMSP particle precipitation data developed by Newell et al. (1991a; 1991b) was used to search for these events. IMF was obtained from the IMP-8 15-s database provided by the NASA NSSDC website. The solar wind propagation delay from IMP8 to the ionosphere is estimated rather crudely as Δt = ballistic propagation of solar wind to the magnetopause standoff distance (*x* = 10 R$_E$) +5 min propagation in the magnetosheath (e.g., Lockwood et al., 1989; Ridley et al., 1998) +3.5 min for 1 keV ions to travel along the field line from the magnetopause to the ionosphere (15 R$_E$) (e.g., Carlson and Torbert, 1980). The database was divided into two classes: IMF B$_y$<<0 (toward sector) and IMF B$_y$>>0 (away sector).

The criteria for selecting IMF B$_y$>>0 events are: (a) IMF −4 nT ≤ B$_z$ ≤ 0 nT and B$_y$ ≥ 8 nT; and (b) the IMF has been relatively stable so that (a) is satisfied for at least 15 minutes. The latter requirement attempts to restrict events to those in a quasi-steady state. The search returns a total of 22 cusp events. From these 22 events, 16 events or 73% of the total events show double cusps or latitudinally extended cusps while six events do not.

The criteria for selecting IMF B$_y$<<0 events are the same as above except that the IMF B$_y$ condition is reversed: (a) IMF −4 nT ≤ B$_z$ ≤ 0 nT and B$_y$ ≤ −8 nT; and (b) the IMF has been relatively stable so that (a) is satisfied for at least 15 minutes. There are 18 cusp events that satisfy the IMF criteria. Of these 18 events, 14 events show double cusps or latitudinally extended cusps and four do not. This amounts to 77% of the total events with double cusps or latitudinally extended cusps.

In all, double cusps or latitudinally extended cusps appear fairly frequently, approximately 75% of the time, when IMF has a large (positive or negative) *y*-component and a small negative *z*-component.

6.   IMF CONTROL OF THE CUSP LOCATION AND LATITUDINAL WIDTH

There have been many statistical studies of the IMF and solar wind control of cusp properties e.g., locations, boundaries etc. (e.g., Carbary and Meng, 1986; Newell et al., 1989; Aparicio et al., 1991; Zhou et al., 2000). However, none of these studies has examined the IMF B$_y$ control of the latitudinal cusp width or the cusp's equatorward boundary. As discussed above, IMF B_y should have some influence on these two cusp properties. For example, the cusp latitudinal width increases for the type of double cusp events shown in Figure 5b, e.g., when the latitudinal separation between the two cusps is very narrow. With years of DMSP data available, these two cusp properties can now be determined statistically.

For selecting the cusp events, we used the same DMSP automated cusp identification algorithm in the case study above to search cusp events in the DMSP data for the period of one solar cycle, 1985–1995 (Newell et al., 1991a; 1991b). Upon inspection of several double cusp events, it is found that this automated algorithm works reasonably well most of the time. However, it sometimes identifies cusps with low energy flux as LLBL. Although this inevitably introduces noise into the data set, there has been no perfect automated cusp identification algorithm. NASA NSSDC provides the IMP-8 simultaneous hourly averaged solar wind and IMF data. With this method and database, 2259 cusp events were identified. The cusp's equatorward boundary and latitudinal width are correlated with the IMF B_y and B_z. In each case, the data are divided according to the sign of the IMF components. Thus, there are four cases to be considered.

6.1   Cusp's equatorward boundary

It is well established that the cusp latitudinal location correlates well with IMF B_z. The same result holds with our data and methodology, which uses computer algorithms to search the DMSP data for cusp events for the period of one solar cycle. The result can be seen in Figure 8a, which includes 2177 data points. The results of the linear least square fits (for all these points) are: cusp equatorward boundary (ceb) = (0.78 ± 0.03) IMF B_z + (77.3 ± 0.1) °Λ and ceb = $(6\times10^{-4}\pm0.04)$ IMF B_z + (77.9 ± 0.1) °Λ for southward and northward IMF respectively. The correlation coefficients are 0.55 and 5×10^{-4} for southward and northward IMF respectively. The near-zero correlation coefficient of the latter simply reflects the nearly constant locations of the cusp's equatorward boundary latitude during periods of northward IMF, as can also be seen in the scatter plot in Figure 8a. These results are in very good agreement with the previous result of ceb = 0.76 IMF B_z + 77.0 °Λ and ceb = 0.11 IMF B_z + 77.2 °Λ, respectively (Newell et al., 1989) and are comparable to ceb = 0.86 IMF B_z + 79.5 ° Λ and ceb = 0.07 IMF B_z + 79.2 °Λ, respectively (Zhou et al., 2000). The latter results were obtained with mid-altitude POLAR satellite observations, which may explain the slight location shift. The decrease of the cusp latitude with decreasing IMF B_z during periods of southward IMF B_z has been interpreted as the effect of merging and flux erosion on the dayside (e.g., Aubry et al., 1970; Zhou et al., 2000). In contrast, IMF B$_z$ does not control much of the cusp's equatorward boundary during periods of northward IMF.

The above relationship between IMF B$_z$ and the cusp's equatorward boundary is obtained when all cusp events are included. If the cusp events with large |IMF B$_y$| are removed from the data, then the cusp equatorward boundary moves to higher latitude. There are 798 such cusp events which were chosen with the IMF B$_y$ criterion: -3 nT $\leq$ IMF B$_y$ $\leq$ 3nT. The result of the linear least square fits are ceb = (0.81±0.05) IMF B$_z$ + (77.7±0.2) °Λ and ceb = (0.04±0.06) IMF B$_z$ + (78.1±0.2) °Λ for southward and northward IMF respectively. Their correlation coefficients are 0.54 and 0.04 respectively. This difference is statistically significant, e.g., for IMF B$_z$ = 0, the difference of the cebs (77.7–77.3) is larger than the uncertainty ($\sqrt{0.1^2 + 0.2^2}$). The poleward shift, resulting from the removal of large |IMF B$_y$| events, ranges from 0.1° to 0.4° as IMF B$_z$ increases from -10 to 0 nT. Thus, the poleward shift is greater for weakly southward IMF than for strongly southward IMF. This shift is consistent with the removal of the double cusp events. However, there could be other factors at work simultaneously, such as the effect of merging and flux removal (discussed next).

Merging also occurs during periods of large IMF B$_y$ and the ensuing flux erosion is expected to move the cusp's latitudinal location equatorward as in the case for southward IMF. We selected 1337 cusp events with small IMF B$_z$, -3nT < IMF B$_z$ < 3 nT. The results of the linear least square fits of IMF B$_y$ versus the equatorward boundary of the cusp are shown in Figure 8b. The linear least square fit results in ceb = (0.12±0.05) IMF B$_y$ + (77.3±0.2) and ceb = (-0.14±0.04) IMF B$_y$ + (77.7±0.2) for negative and positive IMF B$_y$ respectively. The correlation coefficient is 0.10 and -0.13 for IMF B$_y$ < 0 and IMF B$_y$ > 0 respectively. The slopes are much smaller than that for southward IMF case. The cusp equatorward boundary moves slightly equatorward when IMF B$_y$ increases in magnitude but this effect is much weaker than the southward IMF effect. The small correlation coefficients indicate the presence of rather large scatter in the data distribution but they are statistically significant considering the size of the data set, namely 635 and 696 points for IMF B$_y$ < 0 and IMF B$_y$ > 0 respectively. A t-test indicates that the probability that IMF B$_y$ and ceb are uncorrelated is <1% (e.g., Pugh and Winslow, 1966). Furthermore, this result is consistent with the poleward shift of ceb in Figure 8a when large |IMF B$_y$| events are removed, as discussed in the previous paragraph.

The relationships between IMF and ceb can be illustrated more easily by the plots of their medians. The median values of ceb in 2 nT IMF B$_z$ and B$_y$ bins are indicated by horizontal bars in Figure 8a and 8b respectively. The correlation coefficients of these medians are 0.99, -0.57, 0.93, -0.98 for IMF B$_z$ <0, IMF B$_z$ >0, IMF B$_y$ <0, IMF B$_y$ >0, respectively.

Cusp Equatorward Boundary

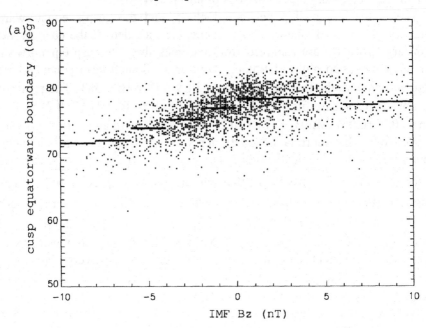

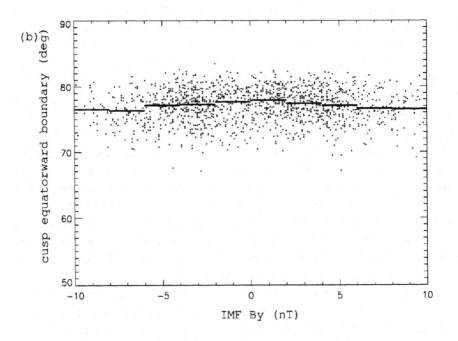

Figure 8. Cusp equatorward boundary as a function of (a) IMF B_z and (b) IMF B_y. The medians in (a) 2 nT IMF B_z and (b) IMF B_y bins are indicated by horizontal bars.

The cusp's equatorward boundary is clearly more affected by IMF B$_z$ than IMF B$_y$. The larger effect of IMF B$_z$ over IMF B$_y$ is typical for many cusp properties.

6.2 Cusp's latitudinal width

In contrast to the cusp equatorward latitude, the effect of IMF B$_y$ is at least as strong as that of IMF B$_z$ on the cusp's latitudinal width near noon meridian, $11 \leq$ MLT ≤ 13. Figure 9 shows that the cusp latitudinal width increases with |IMF B$_y$| and |IMF B$_z$|. The misclassification of weak cusps (cusps with lower fluxes) as open-field LLBL partly contributes to the large scatter in the figure. In this study, the cusp's latitudinal width is obtained within ~1–2 min from an individual cusp observation made by a DMSP pass that reaches 81°Λ or higher. This requirement helps eliminate passes that just graze the cusp; e.g., the statistical location of the cusp is well below 81° Λ (e.g., Newell et al., 1989).

In Figure 9b, all the events have been selected so that they have weakly southward IMF component, -3 nT $\leq$ IMF B$_z \leq 0$ nT which restricts the number of events to 396. The medians of the cusp's latitudinal width (clw) in 2 nT IMF B$_y$ bins are plotted as horizontal lines in Figure 9b. The medians are computed only for bins that contain five data points or more. The least square fits of the medians are clw = (-0.06 ± 0.004) IMF B$_y$ + (0.5 ± 0.02) degree and clw = (0.04 ± 0.02) IMF B$_y$ + (0.5 ± 0.1) degree for IMF B$_y$<0 and IMF B$_y$>0 respectively. The correlation coefficients are -0.99 and 0.76, respectively. This result is consistent with our case study above which shows that at times the double cusp, associated with large |IMF B$_y$|, forms a single cusp with extended latitudinal width, e.g., Figure 5b.

Figure 9a shows that the effect of IMF B$_z$ on the cusp latitudinal width. The results of the least square fit of the medians are clw = (-0.03 ± 0.01) IMF B$_z$ + (0.6 ± 0.05) and clw = (0.04 ± 0.02) IMF B$_z$ + (0.4 ± 0.1) for southward and northward IMF, respectively. Their correlation coefficients are -0.91 and 0.84 for southward and northward IMF, respectively. Again, only bins containing five or more data points are included in the median calculations. Figure 9a shows a similar trend as Figure 5 of Zhou et al. (2000), especially if their extremely small (IMF B$_z$<-8) and large IMF B$_z$ (IMF B$_z$> 5) bins, which contain much fewer points, are excluded from their figure. In any case, the scatter is very large in both studies. The previous study may have included mantle precipitation in their operational definition of "cusp" and thus determine a larger cusp size for southward IMF. In this study, the large scatter may be partly due to the misclassification of cusps having low energy flux as open-field line LLBL as well as other factors such as dipole tilt etc. Also, the usage of the hourly-averaged IMF may contribute to the noise.

Cusp Latitudinal Width

(a)

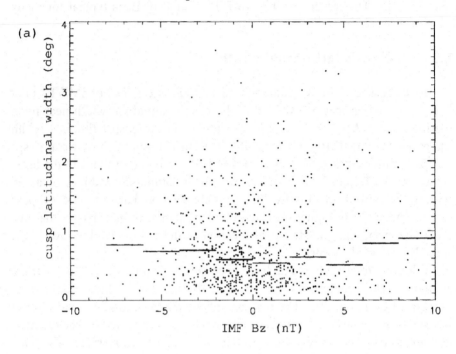

(b)

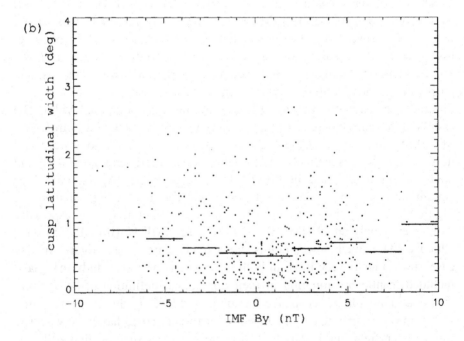

Figure 9. Cusp latitudinal width as a function of (a) IMF B_z and (b) IMF B_y. The medians in (a) 2 nT IMF B_z and (b) IMF B_y bins are indicated by horizontal bars.

7.   SUMMARY

The discontinuous cusp ion signature has long been associated with the discontinuity in the IMF, solar wind, and/or merging rate (e.g., FTE, pulsed or bursty or intermittent injections etc.) (e.g., Lockwood and Smith, 1989; 1992; Smith et al., 1992; Escoubet et al., 1992; Lockwood et al., 1995; Boudouris et al., 2001).

However, Trattner et al. (1999) present discontinuous cusps events in which two satellites crossed the same open flux tubes at different times and yet observed similar discontinuous cusp structures. These observations led them to conclude that (1) the discontinuous cusps can be a stable spatial feature that can persist up to 1.5 hours and (2) the presence of the discontinuous cusp does not necessarily indicate the temporal nature of the merging parameters.

It turns out that some of these spatial discontinuous cusps favor certain IMF orientations (Wing et al., 2001). During periods of stable and B_y dominant IMF, two cusps (double cusp) are frequently observed by DMSP satellites with the following properties:

1. The two cusps are separated latitudinally.
2. The lower latitude cusp ion exhibits little or no energy-latitude dispersion (stagnant) whereas the higher latitude cusp ion exhibits the classical southward IMF dispersion (see Figure 5). Sometimes the latitudinal separation of the two cusps in the double cusp narrows to the point of giving the impression of just one cusp with extended latitudinal width. This may result from several factors, e.g., the seasonal variation which changes the latitudinal locations of the northern and southern merging sites, satellite trajectories etc. (e.g., Rodger et al., 2000; Weiss et al., 1995).
3. In the DMSP data, the double cusp is observed in 30 out of 40 events (75% of the events) during periods of B_y dominant IMF.

The formation of the double cusp was actually predicted by APL open-field line particle precipitation model (Wing et al., 2001). In the model:

1. The lower-latitude cusp ions originate from the pre-noon low-latitude magnetopause, $z \sim -5$ to 5 R_E. In this region, the ions undergo a moderate $\mathbf{E} \times \mathbf{B}$ azimuthal drift during their flight from the magnetopause to the ionosphere. This results in a dispersionless or stagnant cusp.
2. The higher-latitude cusp ions originate from higher-latitude magnetosheath regions ($z \sim 7$–13 R_E) eastward/westward of the satellite, depending on the orientation of the IMF B_y. Once the ions enter the magnetosphere, they undergo strong $\mathbf{E} \times \mathbf{B}$ azimuthal and poleward drift, resulting in the classical dispersion.

The lower-latitude dispersionless cusp perhaps corresponds to the "stagnant" or "weak IMF B_z" cusp in the cusp classification scheme developed by Yamauchi and Lundin (1994). In their study, stagnant cusps occur most frequently during periods of weak IMF B_z. With a weak z-component, the IMF orientation may be dominated by the y-component. If this is the case, then their results can be explained in terms of merging locations and $E \times B$, as discussed here. The results here strongly suggest that in addition to IMF B_z, IMF B_y plays an equally important role in determining cusp morphology.

Both the statistical widening of the cusp's latitudinal width and shifting of equatorward boundary to a lower latitude during periods of large IMF B_y are consistent with the formation of double cusps (separation between the two cusp can sometimes narrow to give the impression of just one cusp with an extended latitudinal width).

The model does not explicitly include merging processes. However, if the magnetosheath ion entries are assumed to result from merging, then the result here suggests that merging simultaneously occurs at low- and high-latitude magnetopause during periods of large |IMF B_y| and small IMF B_z. Merging at low-latitudes during periods of non-southward IMF orientation has been previously reported. For example, observations in the vicinity of the magnetopause indicated that in the vicinity of the subsolar region merging occurs at modest magnetic shear, ranging from 60° to 180°, but at high-latitudes merging occurs when the magnetic shear is larger, > 135° (e.g., Gosling et al., 1990; 1991). A recent cusp study reports simultaneous merging at both low- and high-latitudes during periods of northward IMF (e.g., Fuselier et al., 2000).

ACKNOWLEDGMENTS

The DMSP SSJ4 instrument was designed and built by Dave Hardy and colleagues at the AFRL. Fred Rich has been helpful in our acquiring this data, as has the World Data Center in Boulder, Colorado. Simon Wing was supported by NASA Grant NAG5-10971 and NSF Grants ATM-9819705 and ATM-0344690. Patrick Newell was supported by AFOSR Grant F49620-00-1-0172.

REFERENCES

Aparicio, B., Thelin, B., and Lundin, R., 1991, The polar cusp from a particle point of view: A statistical study based on Viking data, *J. Geophys. Res.* **98**:14,023–14,031.

Aubry, M. P., Russell, C. T., and Kivelson, M. G., 1970, Inward motion of the magnetopause before a substorm, *J. Geophys. Res.* **75**:7018–1031.

Baker, K. B., and Wing, S., 1989, A new magnetic coordinate system for conjugate studies at high latitudes, *J. Geophys. Res.* **94**:9139–9143.

Boudouridis, A., Spence, H. E., and Onsager, T. G., 2001, Investigation of magnetopause reconnection models using two colocated, low-altitude satellites: A unifying reconnection geometry, *J. Geophys. Res.* **106**:29,451–29,466.

Burch, J. L., 1972, Precipitation of low energy electrons at high latitudes: Effects of interplanetary magnetic field and dipole tilt angle, *J. Geophys. Res.* **77**:6696.

Burch, J. L., 1985, Quasi-Neutrality in the polar cusp, *Geophys. Res. Lett.* **12**:469–472.

Carbary, J. F., and Meng, C.-I., 1986, Relations between the interplanetary magnetic field B_z, AE index, and cusp latitude, *J. Geophys. Res.* **91**:1549–1556.

Carlson, C. W., and Torbert, R. B., 1980, Solar wind ion injection in the morning auroral oval, *J. Geophys. Res.* **85**:2903–2908.

Christon, S. P., Williams, D. J., Mitchell, D. G., Frank, L. A., and Huang, C. Y., 1989, Spectral characteristics of plasma sheet ion and electron populations during undisturbed geomagnetic conditions, *J. Geophys. Res.* **94**:13,409–13,424.

Coleman, I. J., Chisham, G., Pinnock, M., and Freeman, M., 2001, An ionospheric convection signature of antiparallel reconnection, *J. Geophys. Res.* **106**:28,995–29,007.

Cowley, S. W. H., and Owen, C. J., 1989, A simple illustrative model of open flux tube motion over the dayside magnetopause, *Planet. Space Sci.* **37**:1461–1475.

Cowley, S. W. H., Morelli, J. P., and Lockwood, M., 1991, Dependence of convective flows and particle precipitation in the high-latitude dayside ionosphere on the x and y components of the interplanetary magnetic field, *J. Geophys. Res.* **96**:5557–5564.

Eather, R. H., and Mende, S. B., 1971, Airborne observations of auroral precipitation patterns, *J. Geophys. Res.* **76**:1746.

Escoubet, C. P., Smith, M. F., Fung, S. F., Anderson, P. C., Hoffman, R. A., Baasinska, E. M., and Bosqued, J. M., 1992, Staircase ion signature on the polar cusp: A case study, *Geophys. Res. Lett.* **19**(17):1735–1738.

Fairfield, D. H., and Scudder, J. D., 1985, Polar rain: Solar coronal electrons in the Earth's magnetosphere, *J. Geophys. Res.* **90**:4055–4068.

Feldman, W. C., Asbridge, J. R., Bame, S. J., and Montgomery, M. D., 1974, Interplanetary solar wind streams, *Rev. Geophys. Space Phys.* **12**:715.

Feldman, W. C., Asbridge, J. R., Bame, S. J., Gosling, J. T., and Lemons, D. S., 1978, Characteristic electron variations across simple high-speed solar wind streams, *J. Geophys. Res.* **83**:5285–5295.

Frank, L. A., 1971, Plasma in the Earth's polar magnetosphere, *J. Geophys. Res.* **76**:5202–5219.

Fuselier, S. A., Trattner, K. J., and Petrinec, S. M., 2000, Cusp observations of high- and low-latitude reconnection for northward interplanetary magnetic field, *J. Geophys. Res.* **105**:253–266.

Gosling, J. T., Thomsen, M. F., Bame, S. J., and Elphic, R. C., 1990, Plasma flow reversals at the dayside magnetopause and the origin of asymmetric polar cap convection, *J. Geophys. Res.* **95**:8073–8084.

Gosling, J. T., Thomsen, M. F., Bame, S. J., and Elphic, R. C., 1991, Observations of reconnection of interplanetary and lobe magnetic field lines at high-latitude magnetopause, *J. Geophys. Res.* **96**:14,097–14,106.

Harel, M., Wolf, R. A., Spiro, R. W., Reiff, P. H., Chen, C.-K., Burke, W. J., Rich, F. J., and Smiddy, M., 1981, Quantitative simulation of a magnetospheric substorm, 1, Model logic and overview, *J. Geophys. Res.* **86**:2217–2241.

Heikkila, W. J., and Winningham, J. D., 1971, Penetration of magnetosheath plasma to low altitudes through the dayside magnetospheric cusps, *J. Geophys. Res.* **76**:833.

Heppner, J. P., and Maynard, N. C., 1987, Empirical high-latitude electric field models, *J. Geophys. Res.* **92**:4467–4489.

Hill, T. W., and Reiff, P. H., 1977, Evidence of magnetospheric cusp proton acceleration by magnetic merging at the dayside magnetopause, *J. Geophys. Res.* **82**:3623–3628.

Lockwood, M., Sandholt, P. E., Cowley, S. W. H., and Oguti, T., 1989, Interplanetary magnetic field control of dayside auroral activity and the transfer of momentum across the dayside auroral magnetopause, *Planet. Space Sci.* **37**(11):1347–1365.

Lockwood, M., and Smith, M. F., 1989, Low-altitude signatures of the cusp and flux transfer events, *Geophys. Res. Lett.* **16**(8):879–882.

Lockwood, M., and Smith, M. F., 1992, The variation of reconnection rate at the dayside magnetopause and cusp ion precipitation, *J. Geophys. Res.* **97**(A10):14,841–14847.

Lockwood, M., Davis, C. J., Smith, M. F., Onsager, T. G., and Denig, W. F., 1995, Location and characteristics of the reconnection X-line deduced from low-altitude satellite and ground observations, Defense Meteorological Satellite Program and European Incoherent Scatter data, *J. Geophys. Res.* **100**(A11):21803–21813.

Menietti, J. D., and Burch, J. L., 1988, Spatial extent of the plasma injection region in the cusp-magnetosheath interface, *J. Geophys. Res.* **93**:105–113.

Merka, J., Safrankova, J., Nemecek, Z., Savin, S., Skalsky, A., 2000, High-altitude cusp: Interball observation, *Adv. Space Res.* **25**:1425–1434.

Merka, J., Safrankova, J., and Nemecek, Z., 2002, Cusp-like plasma in high altitudes: a statistical study of the width and location of the cusp from Magion-4, *Ann. Geophys.* **20**:311–320.

Maynard, N. C., Denig, W. F., and Burke, W. J., 1995, Mapping ionospheric convection patterns to the magnetosphere, *J. Geophys. Res.* **100**:1713–1721.

Nemecek, Z., Safrankova, J., Prech, L., Simunek, J., Sauvaud, J. A., Fedorov, A., Stenuit, H., Fuselier, S. A., Savin, S., Zelenyi, L., and Berchem, J., 2003, Structure of the outer cusp and sources of the cusp precipitation during intervals of a horizontal IMF, *J. Geophys. Res.* **108**(A12):1420, doi: 10.1029/2003JA009916.

Newell, P. T., and Meng, C.-I., 1988, Hemispherical asymmetry in cusp precipitation near solstices, *J. Geophys. Res.* **93**:2643–2648.

Newell, P. T., Meng, C.-I., Sibeck, D. G., and Lepping, R., 1989, Some low-altitude dependencies on the interplanetary magnetic field, *J. Geophys. Res.* **94**:8921–8927.

Newell, P. T., and Meng, C.-I., 1989, Dipole tilt angle effects on the latitude of the cusp and the cleft/LLBL, *J. Geophys. Res.* **94**:6949–6953.

Newell, P. T., Burke, W. J., Meng, C.-I., Sanchez, E. R., and Greenspan, M. E., 1991a, Identification and observations of the plasma mantle at low altitude, *J. Geophys. Res.* **96**:35–45.

Newell, P. T., Wing, S., Meng, C.-I., and Sigilito, V., 1991b, The auroral oval position, structure, and intensity of precipitation from 1984 onward: An automated on-line data base, *J. Geophys. Res.* **96**:5877–5882.

Newell, P. T., Burke, W. J., Sanchez, E. R., Meng, C.-I., Greenspan, M. E., and Clauer, C. R., 1991c, The low-latitude boundary layer and the boundary plasma sheet at low-altitude: Prenoon precipitation regions and convection reversal boundaries, *J. Geophys. Res.* **96**:21,013–21,023.

Newell, P. T., and Meng, C.-I., 1995, Magnetopause dynamics as inferred from plasma observations on low-altitude satellites, in: *Physics of Magnetopause*, Geophys. Monogr. Ser. **90**, P. Song, B. U. Ö. Sonnerup, and M. F. Thomsen, eds., AGU, Washington D. C., pp. 407–416.

Newell, P. T., and Wing, S., 1998, Entry of solar wind plasma into the magnetosphere: Observations encounter simulation, in: *Geospace Mass and Energy Flow: Results from the International Solar-Terrestrial Physics Program*, Geophys. Monogr. Ser. **104**, J. L. Horwitz, D. L. Gallagher, and W. K. Peterson, eds., AGU, Washington D. C., pp. 73–84.

Onsager, T. G., Kletzing, C. A., Austin, J. B., and MacKiernan, H., 1993, Model of magnetosheath plasma in the magnetosphere: Cusp and mangle particles at low-altitudes, *Geophys. Res., Lett.* **20**:479–482.

Onsager, T. G., and Lockwood, M., 1997, High-latitude particle precipitation and its relationship to magnetospheric source regions, *Space Sci. Rev.* **80**:77–107.

Pitout, F., Newell, P., and Buchert, S., 2002, Simultaneous high- and low-latitude reconnection: ESR and DMSP observations, *Ann. Geophys.* **20**:1311–1320.

Pugh, E. M., and Winslow, G. H., 1966, *The Analysis of Physical Measurements*, Addison-Wesley, Reading, Mass., pp. 188–199.

Reiff, P. H., Hill, T. W., and Burch, J. L., 1977, Solar wind plasma injection at the dayside magnetospheric cusp, *J. Geophys. Res.* **82**:479–491.

Reiff, P. H., Burch, J. L., and Spiro, R. W., 1980, Cusp proton signatures and the interplanetary magnetic field, *J. Geophys. Res.* **85**:5997–6005.

Ridley, A. J., Lu, G., Clauer, C. R., and Papitashvili, V. O., 1998, A statistical study of the ionospheric convection response to changing interplanetary magnetic field conditions using the assimilative mapping of ionospheric electrodynamics technique, *J. Geophys. Res.* **103**:4023–4039.

Roelof, E. C., and Sibeck, D. G., 1993, Magnetopause shape as a bivariate function of interplanetary magnetic field B~z~ and solar wind dynamic pressure, *J. Geophys. Res.* **98**:21,421–21,450.

Rodger, A. S., Coleman, I. J., and Pinnock, M., 2000, Some comments on transient and steady-state reconnection at the dayside magnetopause, *Geophys. Res. Lett.* **27**:1359–1362.

Ruohoniemi, J. M., and Greenwald, R. A., 1996, Statistical patterns of high-latitude convection obtained from Goose Bay HF radar observations, *J. Geophys. Res.* **101**:21,743–21,763.

Shue, J.-H., Chao, J. K., Fu, H. C., Russell, C. T., Song, P., Khurana, K. K., and Singer, H. J., 1997, A new functional form to study the solar wind control of the magnetopause size and shape, *J. Geophys. Res.* **102**:9497–9511.

Smith, M. F., and Lockwood, M., 1996, Earth's magnetopsheric cusps, *Rev. Geophys.* **34**:233–260, 1996.

Spreiter, J. R., and Stahara, S. S., 1985, Magnetohydrodynamic and gasdynamic theories for planetary bow waves, in: *Collisionless Shocks in the Heliosphere: Reviews of Current Research*, Geophys. Monogr. Ser. **35**, B. T. Tsurutani and R. G. Stone, eds., AGU, Washington, D. C., pp. 85–107.

Su, Y.-J., Ergun, R. E., Peterson, W. K., Onsager, T. G., Pfaff, R., Carlson, C. W., and Strangeway, R. J., 2001, FAST auroral snapshot observations of cusp electron and ion structures, *J. Geophys. Res.* **106**:25595–25600.

Trattner, K. J., Fuselier, S. A., Peterson, W. K., Sauvaud, J.-A., Stenuit, H., and Dubouloz, N., 1999, On spatial and temporal structures in the cusp, *J. Geophys. Res.* **104**(A12):28411–28421.

Trattner, K. J., Fuselier, S. A., Peterson, W. K., and Carlson, C. W., 2002, Spatial features observed in the cusp under steady solar wind conditions, *J. Geophys. Res.* **107**(A10):1288 doi: 10.1029/2001JA000262.

Stern, D. P., 1985, Parabolic harmonics in magnetospheric modeling: The main dipole and the ring current, *J. Geophys. Res.* **90**(NA11):10,851–10,863.

Tsyganenko, N. A., and Stern, D. P., 1996, Modeling the global magnetic field of the large-scale Birkeland current systems, *J. Geophys. Res.* **101**(A12):27187–27198.

Voigt, G.-H., 1974, Calculation of the shape and position of the last closed field line boundary and the coordinates of the magnetopause neutral points in a theoritical magnetospheric field model, *J. Geophys.* **40**:213–228.

Weiss, L. A., Reiff, P. H., Weber, E. J., Carlson, H. C., Lockwood, M., and Peterson, W. K., 1995, Flow-aligned jets in the magnetospheric cusp: Results from the Geospace Environment Modeling Pilot program, *J. Geophys. Res.* **100**:7649–7659.

Wing, S., Newell, P. T., Sibeck, D. G., and Baker, K. B., 1995, A large statistical study of the entry of interplanetary magnetic field Y-component into the magnetosphere, *Geophys. Res. Lett.* **22**:2083–2086.

Wing, S., Newell, P. T., and Onsager, T. G., 1996, Modeling the entry of magnetosheath electrons into the dayside ionosphere, *J. Geophys. Res.* **101**:13,155–13,167.

Wing, S., and Sibeck, D. G., 1997, Effects of interplanetary magnetic field z component and the solar wind dynamic pressure on the geosynchronous magnetic field, *J. Geophys. Lett.* **102**:7207–7216.

Wing, S., Newell, P. T., and Ruohoniemi, J. M., 2001, Double cusp: Model prediction and Observational Verification, *J. Geophys. Res.* **106**:25,571–25,593.

Woch, J., and Lundin, R., 1992, Magnetosheath plasma precipitation in the polar cusp and its control by the interplanetary magnetic field, *J. Geophys. Res.* **97**:1421–1430.

Xue, S., Reiff, P. H., and Onsager, T. G., 1997, Mid-altitude modeling of cusp ion injections under steady and varying conditions, *Geophys. Res. Lett.* **24**:2275–2278.

Yamauchi, M, and Lundin, R., 1994, Classification of large-scale and meso-scale ion dispersion patterns observed by Viking over the cusp-mantle region, in: *Physical signatures of magnetospheric boundary layer processes*, J. A. Holtet and A. Efeland, eds., Kluwer Academic Pubs., The Netherlands, pp. 99–109.

CEP AS A SOURCE OF UPSTREAM ENERGETIC IONS

Jiasheng Chen and Theodore A. Fritz

Center for Space Physics, Boston University,
725 Commonwealth Avenue, Boston, MA 02215, USA

Abstract   The cusp energetic particles (CEP) have been observed in the dayside high-altitude cusp region, showing orders of the magnitude increase of ion intensities with energies from 20 keV up to 10 MeV. Associated with these charged particles are large diamagnetic cavities with significant fluctuations of the local magnetic field strength. The CEPs may provide a answer to a long-standing unsolved fundamental issue about the origin of upstream energetic ions. On June 28, 1999, the WIND spacecraft (near the forward libration point) observed a sudden increase (by more than one order of magnitude) of the solar wind pressure at about 4:45 UT and an upstream ion event at 5:23-5:45 UT, the INTERBALL-1 spacecraft located just upstream of the bow shock in the morningside measured an upstream ion event from 5:16 UT to 6:00 UT, the GEOTAIL spacecraft in the afternoonside near the bow shock detected an upstream ion event from 5:50 UT to 6:16 UT, while the POLAR satellite at 7 hours of magnetic local time detected an energetic particle event in the high-altitude region associated with turbulent diamagnetic cavities from 5:12 UT to 6:30 UT. Energetic oxygen ions of both ionospheric and solar wind origin were observed by the POLAR spacecraft during this event period. The energetic ions and the associated turbulent magnetic field are very similar to what was found in the high-altitude dayside cusp region. It is argued that the bow shock is not the main source of energetic ions in these upsteam events since their energy spectra are independent of the solar wind velocity and their intensities are independent of the bow shock geometry and solar wind pressure. The event onset was first detected in the cusp by POLAR at 5:12 UT, then near the bow shock in the morningside by INTERBALL-1, and then in far the upstream by WIND. The measured energetic ion intensity decreased with increasing distance from the cusp before 5:42 UT. At 5:50 UT, GEOTAIL detected the event onset that showed an energy dispersion, suggesting a drift effect. These observational facts together with the IMF directions suggest that these upstream energetic ions most likely came from the cusp.

Key words: cusp; energetic particles; bow shock; upstream events.

J. –A. Sauvaud and Z. Němeček (eds.),
Multiscale Processes in the Earth's Magnetosphere:From Interball to Cluster, 175-194.

1. INTRODUCTION

The origins of the > 40 keV/e ions in upstream particle events have been an outstanding issue and have remained unresolved and controversial. There are three possible source regions for the upstream energetic ions: (1) the bow shock, (2) leakage from the outer radiation belt, and (3) the high-altitude dayside cusp. In the case of the first source energetic ions could be energized by the shock drift acceleration at the quasi-perpendicular bow shock or by the Fermi mechanism at the quasi-parallel bow shock (e.g., Lin et al., 1974; West and Buck, 1976; Gosling et al., 1978; Anderson, 1981; Terasawa, 1981; Lee et al., 1981; Lee, 1982; Kudela et al., 2000; Meziane et al, 1999, 2002; Freeman and Parks, 2000). The energetic ions could also be energized by the dipolization process in the geomagnetic tail during magnetic storms and substorms (Lezniak and Winckler, 1970; Quinn and Southwood, 1982; Aggson et al., 1983; Delcourt et al., 1990; Lopez et al., 1990; Hesse and Birn, 1991). These ions then drift westward to form the outer radiation belt. Those energetic ions in the outer radiation belt may encounter the magnetopause and escape through a tangential discontinuity or a rotational discontinuity (Speiser et al., 1981; Scholer et al., 1981; Anagnostopoulos et al., 1986, 1998; Sibeck et al., 1987; Paschalidis et al., 1994; Karanikola et al., 1999; Kudela et al., 2002). The energetic ions could also be energized in the cusp diamagnetic cavities by the resonate interactions of these ions with the turbulent ultra-low frequency (ULF) electromagnetic power (Chen and Fritz, 1998). Some of the cusp energetic ions may escape into the upstream through open field lines (Chen and Fritz, 2002, 2003).

To determine which is the dominant source region for the upstream energetic ions, particle and field data from multiple spacecraft are used. The energetic ions measured simultaneously by the INTERBALL-1, the WIND, the GEOTAIL and the POLAR spacecraft showed different onset and intensities at different locations during the June 28, 1999 high solar wind pressure period, providing important information on the origins of the upstream energetic ions, when the spacecraft were magnetically connected with the possible energetic ion source regions.

The energetic ion data were obtained from ion detectors onboard the various spacecraft. The energetic particle experiment DOK-2 onboard INTERBALL-1 was designed to measure ions over the energy range of 25-850 keV (Kudela et al., 1995). The Energetic Particle and Ion Composition (EPIC) instrument onboard GEOTAIL was designed to measure ions over the energy range of 8 keV/e (keV per charge) to 6 MeV (Williams et al., 1994). The 3-Dimensional Plasma and Energetic Particle (3DP) instrument onboard WIND was designed to make measurements of full 3-dimensional distribution of ions from 20 keV to 11 MeV (Lin et al., 1995). The Imaging Proton Sensor (IPS) onboard PO-

LAR was designed to measure 3-dimensional proton angular distributions over the energy range of 20 keV to 10 MeV (Blake et al., 1995). The Charge and Mass Magnetospheric Ion Composition Experiment (CAMMICE) onboard POLAR was designed to measure the charge and mass composition over the energy range of 1 keV/e to 60 MeV, to determine the fluxes of various ion species and their relative abundances and to determine the incident charge state of these ions (Chen et al., 1997). Earlier versions of the CAMMICE instruments have been described in detail by Fritz et al. (1985) and Wilken et al. (1992). The inter-calibration among different ion instruments on board INTERBALL-1, POLAR, and GEOTAIL has been performed carefully (see Appendix).

2. CUSP ENERGETIC PARTICLES

The cusp energetic particle (CEP) events were discovered by the POLAR spacecraft in 1996 (Chen et al., 1997, 1998; Fritz et al., 1999a). They are defined as follows: (1) a decrease in magnetic field magnitude in the dayside cusp, (2) a more than one order of magnitude increase in intensity for the 1-10 keV ions, and (3) a more than three sigma increase above background for > 40 keV ion (dominated by protons) intensity. One example is shown in Figure 1. On May 13, 1999, the POLAR spacecraft observed an extremely large diamagnetic cavity in the high-altitude cusp region in the morningside (Fig. 1). From 14 to 23:20 UT on 5/13/99, the local magnetic field strength showed a large decrease with strong field turbulence (bottom panel of Fig. 1). The middle panel is the plot of the time profiles of the lower energy (1-10 keV/e) O^{+6} (dotted line) and (1-18 keV/e) He^{++} (solid line), while the top panel is of the higher energy (55-200 keV/e) He^{++}. These two panels exhibit orders of magnitude enhancement of the ion intensities. It is noted that the POLAR spacecraft has an orbital period of about 18 hours. Figure 1 shows that on May 13, 1999, POLAR was in a cusp diamagnetic cavity (CDC) during half of its orbit period, indicating that this CDC was extremely large. This CDC has a size of about 6 R_E in the latitudinal and/or radial directions, much larger than expected.

The CEP events were not only observed by POLAR as mentioned above, but are also observed by CLUSTER. Figure 2 displays the March 5, 2001 CEP event observed by the CLUSTER spacecraft (C3), when this spacecraft was crossing through the dayside high-altitude northern cusp region. The position of the CLUSTER satellites in GSE coordinates is labled at the bottom of Figure 2. The shaded area represents a data gap. A large CDC (bottom panel) was observed by CLUSTER at 7:35-10:45 UT, corresponding to a significant increase of the energetic ion fluxes (middle panel) and an enhancement of the energetic

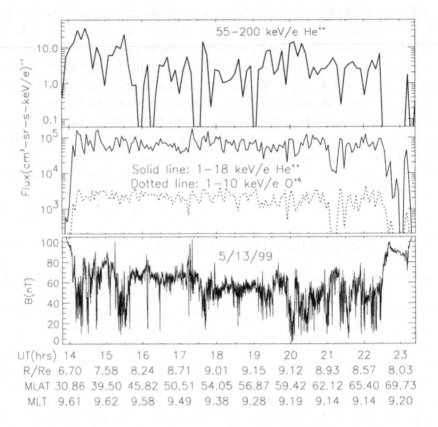

Figure 1. The cusp energetic particle events observed by POLAR on 13 May 1999. The panels show the variation of the 55-200 keV/e He^{++} flux (top panel), the 1-18 keV/e He^{++} (solid line) and 1-10 keV/e O^{+6} (dotted line) fluxes (middle), and the magnetic field (bottom) versus time, respectively. The distance of POLAR from the Earth (in Re), the magnetic latitude (MLAT), and the magnetic local time (MLT) are shown at the bottom.

electron flux (top panel). Note that at 8:50-10:00 UT the 27-95 keV proton flux was about three to four orders of magnitude higher than that after 10:45 UT when CLUSTER went into the magnetosheath. Figure 2 further shows that the enhancement of the cusp energetic electron fluxes are much less than that of the cusp energetic ion fluxes. This result is the same as that observed by POLAR (Chen et al., 1998; Chen and Fritz, 2000, 2002).

In fact, the CEP events are very common in the high-altitude dayside cusp regions and are always there day after day. Around solar minimum at POLAR launch through the end of 1997, there were about 300 cusp crossings, in which 279 or 93% of the crossings were identified as CEP events (Fritz et al., 1999b). In April 1999 when closer to solar maximum, there were 40 cusp crossings and all of them were identified as CEP events (Fritz et al., 2003a); in May 1999, there were 35 cusp crossings and again all of them were CEP events.

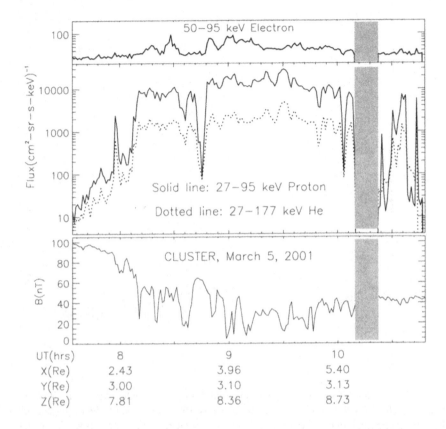

Figure 2. The CEP event observed by CLUSTER on 5 March 2001. The panels show the variation of the 50-95 keV electron flux (top panel), the 27-95 keV proton (solid line) and 27-177 keV helium (dotted line) fluxes (middle), and the magnetic field (bottom) versus time, respectively. The distance of CLUSTER from the Earth (in Re) are shown at the bottom in GSE coordinates.

3. SOLAR WIND PRESSURE AND POLAR ION OBSERVATIONS

Figure 3 displays the measurements by the WIND and POLAR spacecraft during the June 28, 1999 high solar wind pressure period, when the WIND spacecraft was near the forward libration point. The top panel of Figure 3 shows a peak solar wind pressure value of about 47 nPa measured by the Solar Wind Experiment (SWE) (Ogilvie et al., 1995) on WIND; this value is more than one order of magnitude higher than the normal one. POLAR detected about one to two orders of magnitude enhancements of 0.8-1.1 MeV ion (panel b) and 1-18 keV/e He^{++} (panel c) fluxes throughout the period when the local geomagnetic field (GMF) strength measured by POLAR showed diamagnetic cavities with large variations (panel d). At about 5:12 UT the GMF strength

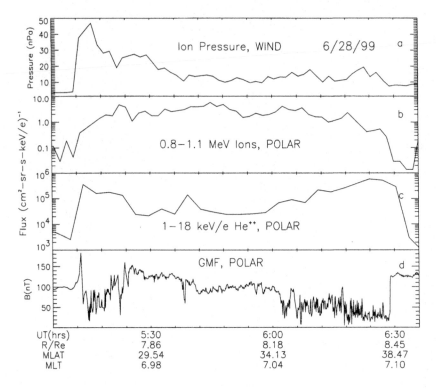

Figure 3. WIND observation: (a) The solar wind pressure, together with POLAR observations: (b) The 0.8-1.1 MeV ion flux, (c) the 1-18 keV He^{++} flux, and (d) the magnitude of the local magnetic field, during the June 28, 1999 high solar wind pressure period. The distance of POLAR from the Earth (in Re), the magnetic latitude (MLAT), and the magnetic local time (MLT) are shown at the bottom of the figure. Corrections have been made for the propagation time from WIND to POLAR.

(bottom panel) increased from about 100 nT to 180 nT, and then decreased to about 18 nT, corresponding to the significant incease of the solar wind pressure (top panel). In Figure 3, the WIND data are displaced by about 27 minutes from WIND local observations to match in location to the POLAR observations. An inspection on Figure 3 indicates that the event period observed by POLAR (bottom three panels) is corresponding to the high solar wind pressure period observed by WIND, but the intensity of the charged particles is not one to one corresponding to the changing solar wind pressure.

The composition of the ions measured by POLAR show that both the energetic (about 70-200 keV/e) $O^{\leq +2}$ of ionospheric origin and energetic (55-200 keV/e) He^{++} of solar wind origin (top panel of Fig. 4) increased significantly during this event period. It is noted that the time-intensity profiles of the energetic ions (panel b of Fig. 3 and top panel of Fig. 4) are different from the thermalized lower energy solar wind ions (1-18 keV/e He^{++} in panel c of Fig. 3 and 1-10 keV/e $O^{\geq +3}$ in middle panel of Fig. 4). The bottom panel

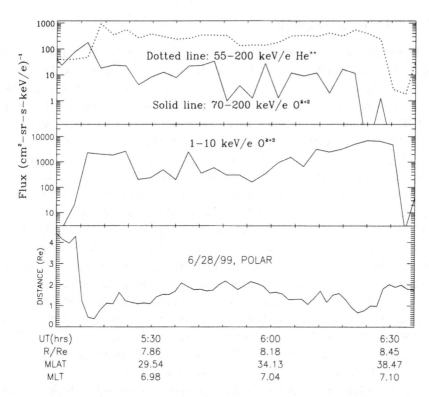

Figure 4. POLAR observations: The 70-200 keV/e $O^{\leq+2}$ (solid line) and 55-200 keV/e He^{++} (dotted line) fluxes (top panel), the 1-10 keV/e $O^{\geq+3}$ flux (middle panel), and POLAR distance from the magnetopause (bottom panel) during the June 28, 1999 high solar wind pressure period. The distance of POLAR from the Earth (in Re), the magnetic latitude (MLAT), and the magnetic local time (MLT) are shown at the bottom of the figure.

of Figure 4 plots the distance of POLAR from the magnetopause, where a positive value indicates POLAR being inside the magnetopause in the magnetosphere and a negative value indicates POLAR outside the magnetopause in the magnetosheath. The magnetopause is obtained from the model of Shue et al. (1998). This panel suggests that POLAR was about one to two Re from the magnetopause inside the magnetosphere at all times during the event. Both the ion and field data in Figures 3 and 4 showed three basic features similar to the cusp energetic particle (CEP) events reported previously in the normal cusp (Chen et al. 1997, 1998; Fritz et al., 1999a); the three basic features are: (1) the diamagnetic cavities with large field fluctuations, (2) the more than one order of magnitude increase in intensity for lower energy solar wind plasma, and (3) the significant increase of higher energy charged particles. Figures 3 and 4 thus suggest that POLAR observed a CEP event during the high solar wind pressure period.

4.   IMF AND GMF

Figure 5 compares the three components of the interplanetary magnetic field (IMF) (top two panels) measured by WIND and the geomagnetic field (GMF) (bottom two panels) observed by POLAR during June 28, 1999 high solar wind pressure period, where the corrections for the solar wind time delay from WIND (27 minutes) to POLAR have been made. No obvious correlations of the GMF with the IMF are found. One interesting point of Figure 5 is that at 5:15-5:42 UT and 6:12-6:30 UT the IMF B_y component was positive, while the cusp diamagnetic cavities (CDC) were observed at 7 hours magnetic local time (morningside) in the northern hemisphere (see bottom of Fig. 4). According to the prediction of the current MHD models a positive IMF B_y would move the dayside northern cusp duskward into afternoonside (e.g., Crooker et al., 1998), so that under the positive IMF B_y conditions the observation of the CDC in the morningside is unexpected and not predicted by the existing MHD models. This is a newly recognized property of the high-altitude dayside cusp region, something for which there is as yet no quantitative model. Now, assuming this prediction of the current MHD models is correct, one would expect that the northern CDC should also exist in the afternoonside under the positive IMF B_y conditions on June 28, 1999. This suggests from another point of view a very large CDC along the longitudinal direction.

Figure 6 further compares the IMF components measured in the far upstream region by WIND with that measured near the bow shock in the morningside by INTERBALL-1 during the event period. In spite of about 200 Re difference in distance between WIND and INTERBALL-1, the IMF conditions measured by these spacecraft were similar.

5.   UPSTREAM ION EVENT, Θ_{BN}, AND TIMING

During the high solar wind pressure period (shown in Fig. 3), INTERBALL-1 located just upstream of the bow shock in the morningside (X=22.4Re, Y=-15.7Re, Z=6.25Re in GSE) observed an upstream ion event from 5:16 UT to 6:00 UT on that day. The ion fluxes measured by INTERBALL-1 in the upstream are plotted as dotted lines in panel 3 from top of Figure 7 for 65-89 keV and in bottom panel for 477-821 keV. For comparison, the ion fluxes measured simultaneously by POLAR in the cusp are plotted as solid lines in these two panels with similar energy ranges. Three important features can be seen in these two panels: (1) The onset of the event (increase of energetic ion intensity) was detected earlier in the cusp than in the region upstream from the bow shock; (2) the time interval of the event was longer in the cusp than in the upstream; and (3) the energetic ion flux was higher in the cusp than in the upstream.

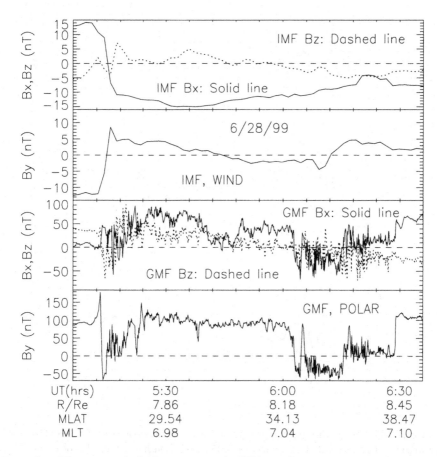

Figure 5. Three magnetic field components (in GSM coordinates) measured by WIND (top two panels) and by POLAR (bottom two panels) at 5:06-6:36 UT on June 28, 1999.

The top panel of Figure 7 is the solar wind speed versus time measured by WIND. It is noticed that during this upstream event period (5:16-6:00 UT) the solar wind speed was rather stable with a value of about 900 km/s while there was a two orders of magnitude change of the energetic ion flux measured by INTERBALL-1. The panel 2 from top of Figure 7 is a plot of Θ_{Bn} versus time for INTERBALL-1. The Θ_{Bn} is the angle between the IMF direction and the bow shock normal, where the bow shock normal is determined from the model of Formisano (1979). The bow shock is called quasi-parallel if the Θ_{Bn} is less than 45°, and quasi-perpendicular if this angle is larger than 45°. The panel shows that on June 28, 1999, INTERBALL-1 was magnetically connected with the quasi-parallel bow shock before the onset (5:16 UT) of this upstream event until 5:37 UT and was magnetically connected with the quasi-perpendicular bow shock after 5:37 UT. Such magnetic connections with the bow shock can also be seen from the direct IMF measurement by INTERBALL-1 shown in

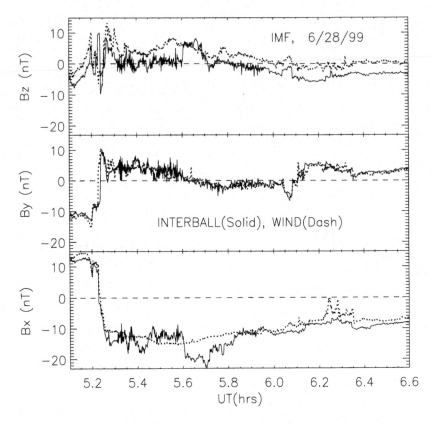

Figure 6.   Three magnetic field components (in GSE coordinates) measured by WIND (dashed lines) and by INTERBALL-1 (solid lines) at 5:06-6:36 UT on June 28, 1999. Corrections have been made for the propagation time from WIND to INTERBALL-1.

Figure 6. Before 5:37 UT the ULF (ultra-low frequency) waves were present with large fluctuations (Fig. 6), indicating that INTERBALL-1 was magnetically connected with the quasi-parallel bow shock at the time; in contrast, after 5:37 UT the fluctuations were much smaller and INTERBALL-1 was either magnetically connected with the quasi-perpendicular bow shock or not magnetically connected with the bow shock. At 5:37-6:00 UT INTERBALL-1 still observed significant energetic ion fluxes even though it was in the quasi-perpendicular bow shock geometry.

The simultaneous energetic ion observations by the four spacecraft (WIND, dashed line; GEOTAIL, dotted line; INTERBALL-1, thick dot-line; POLAR, solid line), from 5:09 to 6:33 UT and over an energy range of about 110-400 keV, are compared in Figure 8 during the June 28, 1999 high solar wind pressure period, where no time delay corrections were made for the 100-400 keV ions transported from the WIND, INTERBALL-1, and GEOTAIL to the PO-LAR. During this period, the INTERBALL-1 spacecraft was upstream from

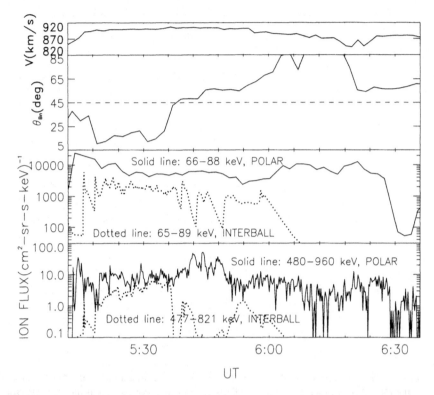

Figure 7. The time profiles of the solar wind speed (top panel), the Θ_{Bn} determined for INTERBALL-1 (panel 2 from top) and the ion fluxes measured by POLAR (solid line) and INTERBALL-1 (dotted line) over two energy intervals ($\sim$ 65-89 keV and 480-900 keV) (bottom two panels) at 5:12-6:36 UT on June 28, 1999.

the bow shock in the morningsidelocated at (X=22.4Re, Y=-15.7Re, Z=6.25Re) in GSE coordinates, the GEOTAIL was near the bow shock in the afternoonside at $\sim$ (10.8Re, 26.7Re, -2.2Re), and the WIND was near the forward libration point at $\sim$ (209Re, -22.5Re, -2.6Re). As shown in Figure 8, The event onset was observed first by POLAR in the cusp at 5:12 UT, then by INTERBALL-1 at 5:16 UT, then by WIND at 5:23 UT, and then by GEOTAIL at 5:50 UT. The onset time delay (7 minutes) from INTERBALL-1 (5:16 UT) to WIND (5:23 UT) was not due to the IMF connection to the bow shock because the IMF measured by WIND were very similar from 5:19 to 5:36 UT and was dominated by the IMF B_x component (Fig. 6). In fact, an 115 keV proton takes about 4.24 minutes from INTERBALL-1 (22.4Re) to reach WIND (209Re) for a simple time of flight, and 7 minutes is expected if taking into account the proton spiral movement along the IMF field line.

The case for GEOTAIL is more complicated. Figure 9 is a plot of GEOTAIL observations, which show the ion-intensity profiles over four energy ranges (top panel) and three IMF components (bottom three panels). The bottom three

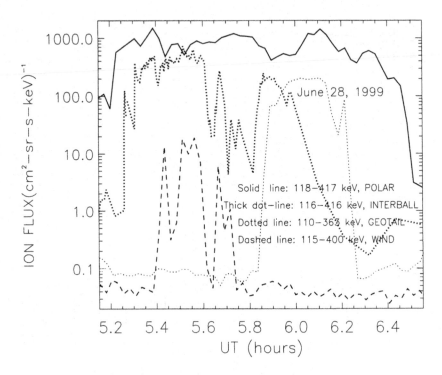

Figure 8.   The time profiles of the ∼ 110-400 keV ion fluxes measured by the WIND (dashed line), the GEOTAIL (dotted-line), the INTERBALL-1 (thick dot-line), and the POLAR (solid line) spacecraft during 5:09-6:33 UT on June 28, 1999.

panels of Figure 9 show the ULF waves with large fluctuations at about 5:56-6:14 UT, suggesting that GEOTAIL was connected to the quasi-parallel bow shock at the time; however, the top panel of Figure 9 reveals that GEOTAIL observed the ion event onset at 5:50 UT before connecting to the quasi-parallel bow shock at 5:56 UT. Therefore, this upstream ion event observed by GEO-TAIL was independent of the bow shock geometry. Before 5:48 UT, GEOTAIL did not observe an enhancement of energetic ion flux since at the GEOTAIL location (X=10.8Re, Y=26.7Re) the IMF (B_x <0, B_y >0 shown in the bottom two panels of Fig. 9) did not connect to the possible ion source region (bow shock or magnetosphere). Another feature of Figure 9 is the energy dispersion shown in the arrival time (top panel) for the ion event measured by GEOTAIL.

6.   THE CUSP SOURCE

In order to be accelerated at a quasi-parallel bow shock, the ions need to interact with the bow shock many times and to stay there for an extended period. The higher the energy obtained, the longer the time required. By analyzing 33

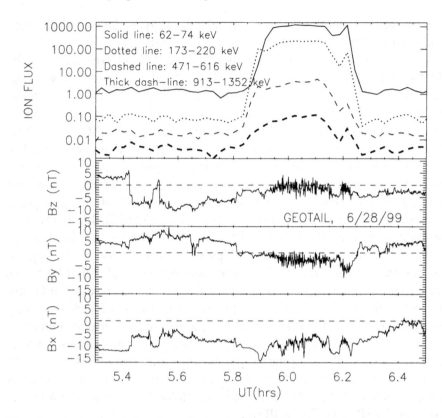

Figure 9. The energetic ion fluxes (in ions/cm^2-sr-s-keV) over four energy intervals (top panel) and three magnetic field components (bottom three panels) measured by GEOTAIL at 5:18-6:30 UT on June 28, 1999.

diffuse ion events upstream from the bow shock, Ipavich et al. (1981) found an inverse velocity dispersion signature in every event. The onset time in the upstream ion event measured by INTERBALL-1 for different energy ranges were almost the same (no any obvious inverse velocity dispersion) (Figs. 7 and 8). The onset time measured by GEOTAIL showed a normal velocity dispersion not an INVERSE velocity dispersion, indicating a ion drift effect not bow shock acceleration effect (Fig. 9). Furthermore, if the quasi-parallel bow shock was the source, the resulting spectral index of the ion energy spectrum should depend only on the solar wind velocity, and the spectral amplitude should be related to the bow shock geometry and solar wind density (or pressure if the solar wind velocity is constant) (e.g., Trattner et al., 2003; Sheldon et al., 2003). The energetic ion intensities in the event are independent of the bow shock geometry and the solar wind pressure (Fig. 3, 7, and 9). Even though the solar wind velocity was rather stable from 5:18 UT to 5:54 UT, the energetic ion fluxes measured by INTERBALL-1 showed a harder energy spectrum at 5:36 UT and a softer spectrum at 5:18 UT (Fig. 7). These observational facts can rule out the bow shock as the main source of the energetic ions in the upstream event observed by INTERBALL-1 and by GEOTAIL.

From 5:00-6:30 UT on 6/28/99, the 50-400 keV ion fluxes, measured by the LANL1994_084 (near local noon), the LANL1991_080 (near dawn), and the LANL1989_046 (near dusk) geostationary satellites, were rather stable. At the time, INTERBALL-1 was in the morningside upstream from the bow shock and was 6.25Re north above equatorial plane (X=22.4Re, Y=-15.7Re, Z=6.25Re), and the IMF B_z measured by INTERBALL-1 was near zero. Under these conditions INTERBALL-1 would not observe outer radiation belt energetic ions leaked from the duskside magnetopause. In other words, the leakage from the equatorial outer radiation belt was also not the main source of energetic ions in this upsteam event observed by INTERBALL-1. The story of the upstream ion event observed by GEOTAIL is different. The energey dispersion shown in the top panel of Figure 9 indicates a drift effect, suggesting that the leakage from the equatorial outer radiation belt could have been the source of the energetic ions in the upstream event observed by GEOTAIL at 5:50-6:16 UT. A further comparison indicates that the AE index was very small before 5:12 UT. At about 5:12 UT, the AE index increased from about 100 nT to 800 nT, suggesting a substorm started at this time. If this substorm was the source of the energetic ions in the upstream event observed by GEO-TAIL, then the 38 minutes (= 5:50 - 5:12) onset time delay was too long for a MeV ion to drift from the substorm in the midnight region to the GEOTAIL position (10.8Re, 26.7Re, -2.2Re).

One possible source is the production of energetic ions in the cusp. The cusp energetic particles (CEP) with energies from 40 keV to 8 MeV have been reported previously (Chen et al., 1997, 1998; Fritz et al., 1999a). Recently, extremely large diamagnetic cavities with a size of as large as 6 Re have been observed in the dayside high-altitude cusp regions and were always there day after day (Fritz et al., 2003a). These diamagnetic cavities were associated with energetic ions and strong magnetic field turbulence. The intensities of the cusp energetic ions were observed to increase by as large as four orders of the magnitude, and their seed populations were a mixture of ionospheric and solar wind particles (Chen and Fritz, 2001; Fritz et al., 2003a, 2003b). The cusp is connected to the nightside equatorial plane through the Shabansky orbits followed by drifting energetic particles (Shabansky and Antonova, 1968; Shabansky, 1971; Antonova and Shabansky, 1975) along closed field lines. If the charged particles start in the cusp they have almost complete access to the equatorial plasma sheet on the nightside in outer magnetosphere (Fritz and Chen, 1999; Fritz et al., 2000). A 38 minute onset in the time delay is expected if a MeV ions starts in the high-altitude dayside cusp, drifts through nightside to the magnetopause in the duskside, and then escapes along open field line to the GEOTAIL position. The two observational facts (event onsets and energetic ion intensities) suggest that the upstream energetic ions measured by INTERBALL-1 may be interpreted as leakage of the cusp energetic ions along open field lines. The 6/28/99 upstream event was observed by WIND when the IMF was radial (Fig. 6), which is similar to what has been reported by Desai et al. (2000) who investigated 1225 upsteam events. Most of the events cannot be satisfactorily explained by either magnetospheric leakage or Fermi acceleration models (Desai et al., 2000). These upstream events may be explained by the CEP events since the characteristics of the energetic ions in the upstream ion events (Desai et al., 2000) are similar with that in the CEP events (Chen and Fritz, 2002).

7.   CONCLUSIONS

The origins of upstream energetic ions have been investigated during the June 28, 1999, high solar wind pressure period when both the bow shock and the magnetosphere were compressed. At 5:06-6:36 UT, the WIND spacecraft was close to the forward libration point, the INTERBALL-1 spacecraft was upstream and near the bow shock on the morningside, the GEOTAIL spacecraft was upstream and near the bow shock on the afternoonside, and the POLAR satellite was in the high-altitude dayside region. Our principal conclusions from the simultaneous observations are the following:

(1) The solar wind pressure indirectly produced a energetic ion event one to two Re inside the magnetopause.

(2) Both particle and field features suggest that the ion event detected by POLAR was a CEP event even though POLAR was at 7 hours magnetic local time.

(3) Energetic ions of both ionospheric origin and solar wind origin were observed by the POLAR during this event period.

(4) The energetic ion event was observed by WIND at 5:23-5:45 UT, by INTERBALL-1 at 5:16-6:00 UT, by GEOTAIL at 5:50-6:16 UT, and by PO-LAR at 5:12-6:30 UT.

(5) The bow shock was not the main source of the energetic ions in this upsteam event since their intensities were independent of the bow shock geometry and solar wind pressure and their energy spectra were independent of the solar wind velocity.

(6) The event onset was first detected in the cusp, then near the bow shock on the morningside, and then in the far upstream region; and the measured energetic ion intensity decreased with increasing distance from the cusp before 5:42 UT. At 5:50 UT, GEOTAIL detected the event onset that showed an energy dispersion, suggesting a drift effect.

(7) The energetic ion intensity was higher in the cusp than in the upstream.

(8) These observational facts suggest that these upstream energetic ions most likely came from a source producing them in the cusp.

ACKNOWLEDGMENTS

We want to acknowledge valuable discussions at Boston University with G. L. Siscoe, Q.-G. Zong, R. B. Sheldon, E. Foreman and to acknowledge the contribution of B. Laubscher, R. Hedges, R. Vigil, and G. Lujan on the CAMMICE sensor system at the Los Alamos National Laboratory; S. Livi, H. Sommer, and H. Steinmetz at the Max Planck Institute for Aeronomy in Germany; M. Grande and colleagues at the Rutherford Appleton Laboratory in Great Britian; J. Fennell, R. Koga, P. Lew, N. Katz, J. Roeder, and B. Crain on the data processing units at the Aerospace Corporation; and the administrative support and interest provided by D. D. Cobb at the Los Alamos National Laboratory. We are grateful to C. T. Russell for providing us the POLAR GMF data, D. J. Williams for the GEOTAIL ion data, S. Kokubun for the GEOTAIL IMF data, K. Kudela for the INTERBALL-1 ion data, M.Nozdrachev for the INTERBALL-1 IMF data, R. P. Lin for the WIND ion data, R. Lepping for the WIND IMF data and K. Ogilvie for the WIND solar wind plasma data. This research was supported by NASA grants NAG5-2578, NAG5-7677, NAG5-7841, NAG5-9562, and NAG5-11397.

APPENDIX

The inter-calibration between the instruments on POLAR has been done carefully. Examples have been published by Chen and Fritz (2001) and Fritz et al. (2003b), where ion data measured by three different instruments on POLAR were compared. Since the three instruments were calibrated independently, the agreements in the resulting spectra give good confidence in the instrumental calibrations. It is very difficult to calibrate different instruments between different spacecraft due to their different orbits. However, we have identified a time period (5/1/98 at 12:45-13:30 UT) when both POLAR and INTERBALL-1 were very close and in northern polar cap in the region of open field lines and GEOTAIL was upstream from the bow shock under normal solar wind and IMF condition. Table A.1 lists the spacecarft positions and the local magnetic field data in GSM at 13 UT on May 1, 1998.

The values in Table A.1 are also about the mean values at 12:45-13:30 UT. During this period (12:45-13:30 UT on 5/1/98) the solar wind velocity was almost constant at about 430 km/s and solar wind ion pressure was also almost constant at about 3 nPa. The POLAR pitch angle distribution data indicate that energetic ions (> 40 keV) were isotropic. The ion fluxes measured by three spacecraft are plotted in Figure A.1. It shows that within a factor of 2 the ion fluxes observed by the three spacecraft are consistent with each other over 30-700 keV, which give good confidence in the instrumental calibrations. It is noted that while the ion fluxes measured by both INTERBALL-1 and POLAR were almost the same over 60-700 keV, the ion flux measured by INTERBALL-1 was lower (within a factor of 2) than what was measured by POLAR at energies below 60 keV and was higher at energies greater than 700 keV.

Table A.1. Spacecraft Positions and Local Magnetic Fields in GSM

	POLAR	INTERBALL-1	GEOTAIL
X(Re)	4.6	2.9	16.3
Y(Re)	.96	-1.5	14.6
Z(Re)	7.7	9.0	-0.4
B_x(nT)	12	20	-5
B_y(nT)	-16	-15	4
B_z(nT)	-83	-80	0

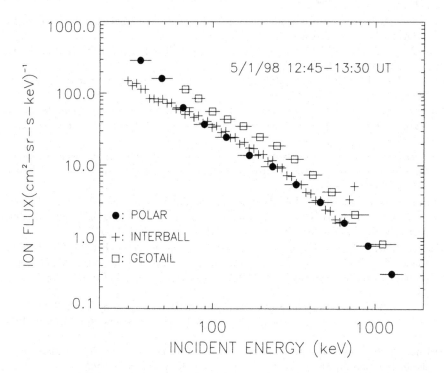

Figure A.1.   The ion fluxes measured by three spacecraft (POLAR, solid circles; INTERBALL-1, pluses; GEOTAIL, open squares) in the locations shown in Table A.1 at 12:45-13:30 UT on May 1, 1998.

REFERENCES

Aggson, T. L., Heppner, J. P., and Maynard, N. C., 1983, Observations of large magnetospheric electric fields during the onset phase of a substorm, *J. Geophys. Res.* **88**:3981–3990.

Anagnostopoulos, G. C., Rigas, A. C., Sarris, E. T., and Krimigis, S. M., 1998, Characteristics of upstream energetic (E > 50 keV) ion events during intense geomagnetic activity, *J. Geophys. Res.* **103**(A5):9251–9533.

Anagnostopoulos, G. C., Sarris, E. T., Krimigis, S. M., 1986, Magnetospheric origin of energetic (E > 50 keV) ions upstream of the bow shock: The October 31, 1977, event, *J. Geophys. Res.* **91**:3020–3028.

Anderson, K.A., 1981, Measurements of bow shock particles far upstream from the Earth, *J. Geophys. Res.* **86**(NA6):4445–4454.

Antonova, A. E., and Shabansky, V. P., 1975, Particle and magnetic field in the outer dayside geomagnetosphere, *Geomangn. Aeron.* **15**(2):297–302.

Blake, J. B., Fennell, J. F., Friesen, L. M., Johnson, B. M., Kolasinski, W. A., Mabry, D. J., Osborn, J. V., Penzin, S. H., Schnauss, E. R., Spence, H. E., Baker, D. N., Belian, R., Fritz, T. A., Ford, W., Laubscher, B., Stiglich, R., Baraze, R. A., Hilsenrath, M. F., Imhof, W. L., Kilner, J. R., Mobilia, J., Voss, D. H., Korth, A., Gull, M., Fisher, K., Grande, M., Hall, D., CEPPAD — Comprehensive energetic particle and pitch-angle distribution experiment on polar, *Space Sci. Rev.* **71**(1-4):531–562.

Chen, J., Fritz, T. A., 1998, Correlation of cusp MeV helium with turbulent ULF power spectra and its implications, *Geophys. Res. Lett.* **25**(22):4113–4116.

Chen, J., Fritz, T. A., 2000, Origins of energetic ions in CEP events and their implications, *Int. J. Geomagnetism and Aeronomy* **2**:31–44.

Chen, J., Fritz, T. A., 2001, Energetic oxygen ions of ionospheric origin observed in the cusp, *Geophys. Res. Lett.* **28**(8):1459–1462.

Chen, J., Fritz, T. A., 2002, The global significance of the CEP events, in: *Solar-Terrestrial Magnetic Activity and Space Environment*, H.N. Wang and R.L. Xu, eds., *COSPAR Colloquia Series,* 14, pp. 239–249.

Chen, J., Fritz, T. A., 2003, Cusp as a source for high-latitude boundary layer energetic ions, in: *Earth's Low-Latitude Boundary Layer*, P. Newell and T. Onsager, eds., *Geophys. Monogr. Series,* 133, pp. 283–292.

Chen, J., Fritz, T. A., Sheldon, R. B., Spence, H. E., Spjeldvik, W. N., Fennell, J. F., and Livi, S., 1997, A new, temporarily confined population in the polar cap during the August 27, 1996 geomagnetic field distortion period, *Geophys. Res. Lett.* **24**(12):1447–1450.

Chen, J., Fritz, T. A., Sheldon, R. B., Spence, H. E., Spjeldvik, W. N., Fennell, J. F., Livi, S., Russell, C. T., Pickett, J. S., and Gurnett, D. A., 1998, Cusp energetic particle events: Implications for a major acceleration region of the magnetosphere, *J. Geophys. Res.* **103**(A1):69–78.

Crooker, N. U., Lyon, J. G., and Fedder, J. A., 1998, MHD model merging with IMF By: Lobe cells, sunward polar cap convection, and overdraped lobes, *J. Geophys. Res.* **103**(A5):9143–9151.

Delcourt, D. C., Sauvaud, J.-A., and Pedersen, A., 1990, Dynamics of single-particle orbits during substorm expansion phase, *J. Geophys. Res.* **95**(A12):20853-20865.

Desai, M. I., Mason, G. M., Dwyer, J. R., Mazur, J. E., von Rosenvinge, T. T., Lepping, R. P., 2000, Characteristics of energetic ($\geq$ 30 keV/nucleon) ions observed by the Wind/STEP instrument upstream of the Earth's bow shock, *J. Geophys. Res.* **105**(A1):61–78.

Formisano, V., 1979, Orientation and shape of the Earth's bow shock in three dimensions, *Planet. Space Sci.* **27**:1151–1161.

Freeman, T. J., and Parks, G. K., 2000, Fermi acceleration of suprathermal solar wind oxygen ions, *J. Geophys. Res.* **105**(A7):15715–15727.

Fritz, T. A., and Chen, J., 1999, The cusp as a source of magnetospheric particles, *Radiation Measurements* **30**(5):599–608.

Fritz, T. A., Chen, J., and Sheldon, R. B., 2000, The role of the cusp as a source for magnetospheric particles: A new paradigm? *Adv. Space Res.* **25**(7-8):1445–1457.

Fritz, T. A., Chen, J., Sheldon, R. B., Spence, H. E., Fennell, J. F., Livi, S., Russell, C. T., and Pickett, J. S., 1999a, Cusp energetic particle events measured by POLAR spacecraft, *Phys. Chem. Erath (C)* **24**(1-3):135–140.

Fritz, T. A., Karra, M., Finkemeyer, B., and Chen, J., 1999b, Statistical studies with Polar of energetic particles near the dayside cusp, *EOS Trans. AGU* 80(46):F862.

Fritz, T. A., Chen, J., and Siscoe, G. L., 2003a, Energetic ions, large diamagnetic cavities, and Chapman-Ferraro cusp, *J. Geophys. Res.* 108(A1):1028–1036.

Fritz, T. A., Zurbuchen, T. H., Gloeckler, G., Hefti, S., and Chen, J., 2003b, The use of iron charge state changes as a tracer for solar wind entry and energization within the magnetosphere, *Annales Geophysicae* 21:2155–2164.

Fritz, T. A., Young, D. T., Feldman, W. C., Bame, S. J., Cessna, J. R., Baker, D. N., Wilken, B., Stuedemann, W., Winterhoff, P., Bryant, D. A., Hall, D. S., Fennell, J. F., Chenette, D., Katz, N., Imamoto, S. I., Koga, R., and Soraas, F., 1985, The mass composition instruments (AFGL-701-11), in: *CRRES/SPACERAD Experiment Descriptions*, M. S. Gussenhoven, E. G. Mullen, and R. C. Sagalyn, eds., Air Force Geophysics Laboratory Report AFGL-TR-85-0017 (Hanscom AFB, MA), pp. 127.

Gosling, J. T., Asbridge, J. R., Bame, S. J., Paschmannn, G., and Sckopke, N., 1978, Observations of two distinct populations of bow shock ions in the upstream solar wind, *Geophys. Res. Lett.* 5:957.

Hesse, M., and Birn, J., 1991, On dipolarization and its relation to the substorm current wedge, *J. Geophys. Res.* 96:19417–19426.

Ipavich, F. M., Galvin, A. B., Gloeckler, G., Scholer, M., Hovestadt, D., 1981, A statistical survey of ions observed upstream of the Earth's bow shock: Energy spectra, composition, and spatial variation, *J. Geophys. Res.* 86:4337–4342.

Karanikola, I., Anagnostopoulos, G. C., and Rigas, A., 1999, Characteristics of $\geq$ 290 keV magnetosheath ions, *Ann. Geophysicae* 17:650–658.

Kudela, K., Slivka, M., Rojko, J., and Lutsenko, V. N., 1995, The Apparatus DOK-2 (Project Interball), Output Data Structure and Modes of Operation, IEP SAS, Kosice, UEF 01-95, pp. 18.

Kudela, K., Slivka, M., Sibeck, D. G., Lutsenko, V. N., Sarris, E. T., Safrankova, J., Nemecek, Z., Kiraly, P., and Kecskemety, K., 2000, Medium Energy Proton Fluxes Outside the Magnetopause: Interball-1 data, *Adv. Space Res.* 25(7/8):1517–1522.

Kudela. K., Lutsenko, V. N., Sibeck, D. G., andx Slivka, D. G., 2002, Energetic ions and electrons within the magnetosheath and upstream of the bow shock: Interball-1 overview, *Adv. Space Res.* 30(7):1685–1692.

Lee, M. A., 1982, Coupled hydromagnetic wave excitation and ion acceleration upstream of the Earth's bow shock, *J. Geophys. Res.* 87(NA7):5063–5080.

Lee, M. A., Skadron, G., and Fisk, L. A., 1981, Acceleration of energetic ions at the Earth's bow shock, *Geophys. Res. Lett.* 8:401–404.

Lezniak, T. W., and Winckler, J. R., 1970, Experimental study of magnetospheric motion and the acceleration of energetic electrons during substorms, *J. Geophys. Res.* 75:7075.

Lin, R. P., Meng, C.-I., and Anderson, K. A., 1974, 30- to 100-keV proton upstream from the Earth's bow shock, *J. Geophys. Res.* 79:489.

Lin, R. P. Anderson, K. A., Ashford, S., Carlson, C., Curtis, D., Ergun, R., Larson, D., McFadden, J., McCarthy, M., Parks, G.K., Reme, H., Bosqued, J. M., Coutelier, J., Cotin, F., D'uston, C., Wenzel, K. P., Sanderson, T. R., Henrion, J., Ronnet, J. C., and Paschmann, G., 1995, A three-dimensional plasma and energetic particle investigation for the WIND spacecraft, *Space Sci. Rev.* 71:125–153.

Lopez, R. E., Sibeck, D. G., McEntire, R. W., and Krimigis, S. M., 1990, The energetic ion substorm injection boundary, *J. Geophys. Res.* 95:109–117.

Meziane, K., Lin, R. P., Parks, G. K., Larson, D. E., Bale, S. D., Mason, G. M., Dwyer, J. R., and Lepping, R. P., 1999, Evidence for acceleration of ions to ~ 1 MeV by adiabatic-like reflection at the quasi-perpendicular Earth's bow shock, *Geophys. Res. Lett.* **26**(19):2925–2928. doi: 10.1029/1999GL900603.

Meziane, K., Hull, A. J., Hamza, A. M., and Lin, R. P., 2002, On the bow shock Θ_{Bn} dependence of upstream 70 keV to 2 MeV ion fluxes, *J. Geophys. Res.* **107**(A9):1243, doi: 10.1029/2001JA005012.

Ogilvie, K. W., Chornay, D. J., Fritzenreiter, R. J., Hunsaker, F., Keller, J., Lobell, J., Miller, G., Scudder, J. D., Sittler Jr., E. C., Torbert, R. B., Bodet, D., Needell, G., Lazarus, A. J., Steinbert, J. T., Tappan, J. H., Mavretic, A., and Gergin, E., 1995, SWE, a comprehensive plasma instrument for the WIND spacecraft, *Space Sci. Rev.* **71**:55–77.

Paschalidis, N. P., Sarris, E. T., Krimigis, S. M., McEntire, R. W., Levine, M. D., Daglis, I. A., and Anagnostopoulos, G. C., 1994, Energetic ion distributions on both sides of the Earth's magnetopause, *J. Geophys. Res.* **99**:8687–8703.

Quinn, J. M., and Southwood, D. J., 1982, Observation of parallel ion energization in the equatorial region, *J. Geophys. Res.* **87**(NA12):536–540.

Scholer, M., Ipavich, F. M., Gloeckler, G., Hovestadt, D., and Klecker, B., 1981, Leakage of magnetospheric ions into the magnetosheath along reconnected field lines at the dayside magnetopause, *J. Geophys. Res.* **86**:1299–1304.

Shabansky, V. P., 1971, Some processes in the magnetosphere, *Space Sci. Rev.* **12**:299–418.

Shabansky, V. P., and Antonova, A. E., 1968, Topology of particle drift shells in the Earth's magnetosphere, *Geomangn. Aeron., Engl. Transl.* **8**:844.

Sheldon, R., Chen, J., and Fritz, T. A., 2003, Comment on "Origins of energetic ions in the cusp" by K. J. Trattner et al., *J. Geophys. Res.* **108**(A7):1302, doi: 10.1029/2002JA009575.

Shue, J.-H., Song, P., Russell, C. T., Steinberg, J. T., Chao, J. K., Zastenker, G., Vaisberg, O. L., Kokubun, S., Singer, H. J., Detman, T. R., and Kawano, H., 1998, Magnetopause location under extreme solar wind conditions, *J. Geophys. Res.* **103**(A8):17691–17700.

Sibeck, D. G., McEntire, R. W., Lui, A. T. Y., Krimigis, S. M., Zanetti, L. J., and Potemra, T. A., 1987, The magnetosphere as a source of energetic magnetosheath ions, *Geophys. Res. Lett.* **14**(10):1011–1014.

Speiser, T. W., Williams, D. J., and Garcia, H. A., 1981, Magnetospherically trapped ions as a source of magnetosheath energetic ions, *J. Geophys. Res.* **86**:723–732.

Terasawa, T., 1981, Energy spectrum of ions accelerated through Fermi process at the terrestial bow shock, *J. Geophys. Res.* **86**(NA9):7595–7606.

Trattner, K. J., Fuselier, S. A., Peterson, W. K., Chang, S.-W., Friedel, R., Aellig, M. R., 2003, Reply to comment on "Origins of energetic ions in the cusp" by R. Sheldon, J. Chen, and T. A. Fritz, *J. Geophys. Res.* **108**(A7):1303, doi: 10.1029/2002JA009781.

West Jr., H. I., and Buck, R. M., 1976, Observations of > 100-keV protons in the Earth's magnetosheath, *J. Geophys. Res.* **81**:569–584.

Wilken, B., Weiss, W., Hall, D., Grande, M., Soraas, F., and Fennell, J. F., 1992, Magnetospheric ion composition spectrometer onboard the CRRES spacecraft, *J. Spacecraft Rockets* **29**(4):585–591.

Williams, D. J., McEntire, R. W., Schlemm, C., Lui, A. T. Y., Gloeckler,G., Christon, S. P., and Gliem, F., 1994, GEOTAIL energetic particles and ion composition instrument, *J. Geomag. Geoelectr.* **46**(1):39–57.

MAGNETIC CLOUD AND MAGNETOSPHERE— IONOSPHERE RESPONSE TO THE 6 NOVEMBER 1997 CME

Alexander Z. Bochev[1] and Iren Ivanova A. Dimitrova[2]

[1] *Solar—Terrestrial Influences Laboratory, Bulgarian Academy of Sciences (BAS), 1113 Sofia, Bulgaria;* [2] *Space Research Institute, BAS, 1113 Sofia, Bulgaria*

Abstract: In the present paper, we analyze the magnetic cloud (MC) at 1 AU on November 9, 1997. The appearance of a hotter and dense part (dense filament), with a radial extent 10^6 km, immediately behind the frontal part of the MC, is a distinctive feature of the event. The INTERBALL-Auroral probe had a chance to observe field-aligned currents in the mid-altitude magnetosphere during the substorm expansion phase intensification related to the dense filament. We emphasize the appearance of unusual "N"-shape magnetic structure, duration 3 minutes, amplitude 50 nT between the field-aligned current region 1 and the magnetosphere lobe in the late evening hours. The "N"-shape structure is related to a significant amount of wave energy transfer red towards the ionosphere.

Key words: coronal mass ejection; solar wind; magnetosphere/ionosphere; wave-energy transfer.

1. INTRODUCTION

The coronal mass ejection (CME) appears to be the main real solar event responsible for geomagnetic storms (Crooker, 1994). CME events were detected from space-born coronographs in the 70-s (Gosling et al., 1974). According to their visual features, CMEs have an observable change in the coronal structure that (1) occurs on a time scale between a few minutes and several hours and (2) involves the appearance of a new, bright white-light feature in the coronograph field of view (Hundhausen, 1993). In some work it is assumed that the CME propagates at 1 AU in the form of a magnetic cloud (MC), (Burlaga et al., 1982). Considering the evolution of CME in the

J. –A. Sauvaud and Z. Němeček (eds.),
Multiscale Processes in the Earth's Magnetosphere:From Interball to Cluster, 195-204.
© 2004 *Kluwer Academic Publishers. Printed in the Netherlands.*

interplanetary space, it is also correctly to be related as "ICME" (Interplanetary coronal mass ejection). The January 6, 1997 CME, corresponding to a typical MC, appeared to be a remarkable event, which largely affected both the magnetosphere and ionosphere (Fox et al., 1998; Thomsen et al., 1998; Yermolaev et al., 1998; Sanchez et al., 1998).

Another unusual event was the 6 November 1997 CME, being related to the active region AR 8100 (Maia et al., 1999; Dermendjiev et al., 1999 ; Pick et al., 1999; Mason et al., 1999; Mazur et al., 1999; Delannée et al., 2000; Delannée and Aulanier, 1999). It is remarkable for its very large-scale extent in latitude seen by both SOHO/LASCO/C2 coronograph and Nancy radioheliograph (Maia et al., 1999). The present work aims to investigate the 9 November MC and its magnetosphere—ionosphere response. For this purpose, we analyze the magnetic field and plasma parameters in the interplanetary space at 1 AU (WIND data) during 9-10 November 1997. In order to show field-aligned current (FAC) systems, we analyze data from the three-component flux-gate magnetometer (Arshinkov et al., 1995) aboard the INTERBALL-Auroral Probe satellite (or shortly INTERBALL-AU). The latter is spin stabilised along the X-axis directed to the Sun (with a period 120 s) and with orbital parameters: apogee $4R_E$ (R_E is the Earth's radius), period 5.5 h, and inclination 65°. The magnetometer X-axis is aligned with the spin axis, and the Y and Z axes are in the spin plane of the spacecraft.

2.   MAGNETIC CLOUD AT 1 AU

We analyze the magnetic field and plasma parameters in the interplanetary space at 1 AU on 9 November 1997 using data from the WIND satellite (coordinates: X=144, Y=-44 and Z=20 in Earth's radius). The arrival of a shock is clearly seen at about 10:35 UT (Figure 1; Figure 2, panels 1-2). Note that the time in these plots has been shifted by 35 minutes to allow for the MC propagation time from WIND. The delay time corresponds to the beginning of the sharp increase of the magnetic field magnitude at GEOS-9 at 10:35 UT on 9.11.1997. The frontal boundary of a large-scale magnetic structure corresponds to Bz fast reversal from +4 nT to -5 nT at 18:00 UT (Figure 1 (2), panel 4 (1)). Then a general positive trend of Bz dominates until 20:00 UT on 10 November (the rear boundary, followed by irregular oscillations). Within these boundaries, the radial extent of the event is about 0.3 AU. An examination of the magnetic field topology was done previously (Dermendjiev et al., 1999; Bochev, 2000). As a whole, the magnetic field configuration differs from a typical cylindrical cloud (Burlaga et al., 1998). We can identify some significant differences as compared with previous studies of MC: (1) A pronounced step-like change of B from 5 to 9 nT (from

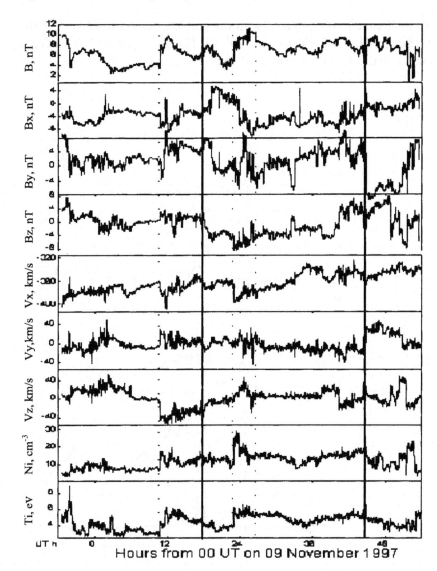

Figure 1. A plot of the magnetic field and solar wind parameters: the magnetic field module B, the Bx, By, and Bz components, the bulk velocity components (Vx, Vy, and Vz), the proton density (Ni), the ion temperature (Ti). Vertical lines indicate: the shock; magnetic cloud (MC) boundaries and the dense and hotter part of MC.

23:00 to 26:30 UT) delineates a region which is clearly indicated by considering the remaining plots: velocities, density and temperature (Figure 1). Note the appearance of a warmer and dense (Ni = 25 cm^{-3}) region immediately after the frontal portion of MC. Further in the text, we denote this part of the MC, with boundaries defined by B and Ni, as dense filament; (2) We observe a relatively higher temperature Ti = 6 eV (or 6.7×10^4 K) in the dense filament as compared with the ambient solar wind ahead of the shock where Ti = 3.5 eV (or 4.1×10^4 K), (Figure1, panel 9). The dense filament radial extent is more than 10^6 km.

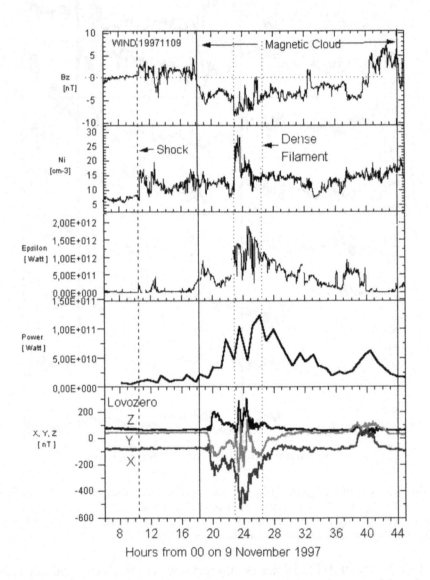

Figure 2. Panel 1 — magnetic cloud (MC) Bz magnetic field component; panel 2 — ion density Ni; panel 3 — energy coupling function epsilon (e); panel 4 — hemisphere power, and panel 5 — magnetic field components at Lovozero Observatory. Vertical lines indicate: shock (dash), magnetic cloud (solid) and dense filament (fine dash).

3. MAGNETOSPHERE—IONOSPHERE RESPONSE

We have calculated the energy coupling function ε (Appendix A) in order to make a quantitative estimation of the expected magnetosphere response (Figure 2, panel 3). The first separate enhancement ε = 5x10[11] W was on 09 November 1997 at 19 UT. Then, we observe two larger enhancements: ε = 1.2×10[12] and ε = 1.5×10[12] W at about 23:00 and 24:50 UT, respectively. The planetary Kp index increases from 1+ to 4. Ground based magnetograms from Lovozero show a substorm onset at about 19:30 UT (Figure 2, panel 5). Then an enhancement begins at about 23:00 UT. After that, we examine the Hemisphere Power (briefly Power), issued by the Space Environment Centre, NOAA, Boulder, CO (available with the time resolution of 30 or 60 minutes). This parameter is proportional to the auroral electrojet activity. As a whole, we observe a dynamic picture in form of three peaks of 0.7×10[11], 1×10[11], and 1.2×10[11] W at about 21:50, 23:30, and 26:30 UT, respectively (Figure 2, panel 4). Three aspects stand out from these considerations: (1) Before the arrival of MC, the magnetosphere was quiet. As a whole, the MC interaction with the magnetosphere provoked substorms and a magnetic storm; (2) There is an indication of the substorm expansion phase intensification with the arrival of the MC dense filament at about 23 UT on 9 November 1997; (3) A large portion of the power transferred to the magnetosphere-ionosphere corresponds to the MC dense filament.

4. RESPONSE OF FIELD-ALIGNED CURRENT (INTERBALL-AU DATA)

INTERBALL-AU had a chance to monitor FACs exactly during the interaction of the MC dense filament with the magnetosphere. Here we are inclined to emphasize on this period, taking into a consideration a magnetogram recorded on 9 November 1997, from 22:48 to 24:00 UT. To characterize the magnetic field disturbances, we have plotted for a simplicity only the Bh component residuals (measured – IGRF/1995 model of the Earth's magnetic field), Bh = $(By^2 + Bz^2)^{1/2}$, where By and Bz are spin plane components (Figure 3, panel 2). Corrected geomagnetic coordinates are shown at the bottom side. The orbit footpoints cross the polar cap, and the auroral oval at ≈ 22 MLT. Along this orbital interval, the spacecraft altitude decreases from 3.87 R_E to 2.88 R_E. In the magnetogram, a large-scale disturbance due to the FACs (Region 1 and Region 2) dominates. The disturbance amplitude is as large as 100 nT, which exceeds by a factor of three the normal quantities for these heights.

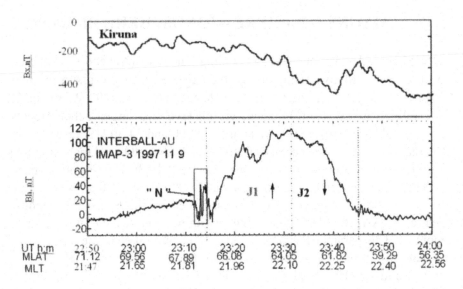

Figure 3. Panel 1 — The Bx magnetic field component during an expansion phase of a substorm in Kiruna; panel 2 — INTERBALL-AU magnetogram suggestive for upward and downward large-scale field-aligned currents, J1 and J2 (Region 1 and Region 2) and an "N"-shape structure.

Next, we try to show what is the portion of dissipated energy in the ionosphere. We assume that the FACs closure in the ionosphere appears to be the Pedersen currents under the condition of div $\mathbf{j}$ = 0. The general expression of Joule heating rate is $Q = \mathbf{j}\cdot(\mathbf{E} + \mathbf{V}\times\mathbf{B})$, where $\mathbf{j}$, $\mathbf{E}$ and $\mathbf{V}$ are ionosphere current density, electric field and velocity, respectively. Under $\mathbf{V} = 0$, equipotentiality of the magnetic field line ($\mathbf{E} = \mathbf{E}_\perp$), and predomination of the meridional orientation of the electric field $\mathbf{E}_\perp = \mathbf{E}_x$, Q will depend only on the transverse components of $\mathbf{j}$ and $\mathbf{E}$, so that $Q = i_\perp \cdot E_x$, where Pedersen current $i_\perp$ is expressed by the FAC density $i_\perp = \int j_\parallel \, dx$, where x is along the orbit. Assuming infinite current sheets for the FAC density $j_\parallel \cong \Delta B_h/(\mu_0 \, \Delta l \,)$, where ΔB_h is B_h amplitude, Δl is the current sheet width, $\mu_0 = 4\pi/10^7$. Then $i_\perp = 0.12$ A/m and assuming E = 0.080 V/m in the ionosphere, for an unit area Q = 0.0096 W/m². Accepting for the half width of the auroral oval (AO), l = 500 km in one hour MLT sector, we have for the heating rate $q = 0.26\cdot R\cdot Q\cdot l\cdot \sin\theta$ (Bochev, 2000) or $q = 1.3\times10^{10}$ W for $R = 2\times10^4$ km and $\theta = 30°$. For a quarter of AO (half of the night sector), $q = 0.8\times10^{11}$ W. In this way, we can draw that the dissipated energy of the FACs closure in the ionosphere was large; it corresponds to the Hemisphere Power, which indicates that the substorm overlaid not only on the Kiruna-Lovozero sector but probably about 2 MLT hours to the west.

Apart from the enhanced large-scale currents, a zone of intense small-scale variations in the poleward edge of Region 1 could be revealed from 23:12 UT to 23:14 UT (Figure 3, panel 2). We denote this peculiar type of perturbation "N"-shape structure. Originally, this name was used in the former investigations by the TRIAD mission (Iijima and Potemra, 1978). In this zone, peak to peak variations reach 50 nT (Figure 3, panel 2). Assuming a spatial interpretation, they could be related to a three-sheet current structure. Accordingly, high current densities $j_{||} = 2.10^{-6}$ A/m^2 would be inherent for this zone. However, an interpretation in terms of time variations seems more probable: the dense filament as a pressure pulse would compress the magnetosphere and cause the time variation in the magnetic field (Cahill et al., 1986; Potemra et al., 1988; 1989). Recently, an intriguing interpretation appeared (Keiling et al., 2000).

5. DISCUSSION AND CONCLUSIONS

We have studied the magnetic field and plasma parameters at heliocentric distance 1 AU. As one can trace in Figure 1, on 9 November, WIND registered a shock (10:30 UT) and sheath and then the magnetic field and plasma which we identify as similar to a magnetic cloud; we stress on a well distinguished warm and dense filament in the MC frontal part. The filament appears as a large pressure pulse characterized by a southward magnetic field (Bz = −8 nT) and a high ion density (Ni=25 cm^{-3}), (Figure 2, panel 1-2). It appears that the strengthening of reconnection at the nose of the magnetosphere due to these important parameters began a little before 22:56 UT. The latter time could be identified by a sudden (20 nT) increase of the GEOS-9 Bz component by the respective compressional wave in the dayside magnetosphere (data from GEOS are widely accessible). The jumps of solar wind parameters at about 23:00 UT in Figure 1 were identified as a shock in the ISTP preliminary event list. For timing of the substorm development, apart from Lovozero, we have examined other observatories too. For example, Kiruna which is nearer to the INTERBALL footpoint shows a similar record (Figure 3, panel 1). There may exists an uncertainty in obtaining the exact timing of the substorm onset, however, it is obvious that beginning from 23:08 UT there was a sharp and continuous decrease of the Bx component indicative for the expansion phase intensification. It is possible that the substorm started even at earlier UT to the west of Kiruna, where INTERBALL conjunction was. The more important for us is that both observatories sensed almost simultaneously features characterising the expansion phase of a substorm during the INTERBALL transit. We have emphasized the appearance of an unusual "N"-shape magnetic field structure between FAC Region 1 and the open field lines (magnetosphere lobe) registered by the INTERBALL magnetometer at 23:12 UT, 21.19 MLT (Figure 3, panel 2). The disturbance is mainly in the spin plane of the spacecraft (Dermendjiev et al., 1999). Note a sharper pulse in the middle of "N"-shape, time scale of 10 s (30 km), dB/dt = 5 nT/sec. We interpret these fast and large fluctuations of **B** in terms of MHD waves. This magnetic field disturbance would create a strong variation of the electric field, too: $\nabla \times \mathbf{E} = -\partial/\partial t(\mathbf{B})$, (Faraday's law). Under the assumption of N-S electric field **E**, it follows $E_x = \partial/\partial t(\mathbf{B})dz$, where dz is a vertical line and element $dz = V_A\,dt$, V_A is the Alfven speed. Assuming $V_A = 3 \times 10^6$ m/s we obtain $E_x = 0.150$ V/m. Note that V_A may be even two times larger in this region (Keiling et al., 2000). The associated energy flux between the magnetosphere and ionosphere along magnetic field lines is derived from the Poynting vector flux $\mathbf{S} = (\mathbf{E} \times \mathbf{B})/\mu o$. Under the above assumptions for the vertical component $S_z = E_x B_y/\mu o$, we find approximately $S_z = 0.006$ W/m^2 at this height. This quantity of wave energy flux grows along the magnetic field to the Earth. Its magnitude could presumably exceed the Joule heating rate in the ionosphere. Our finding is comparable with the POLAR satellite observations occurred in close temporal proximity to a substorm onset (Keiling et al., 2000). The "N"-shape may be a result of a significant amount of wave energy transfer towards the ionosphere.

ACKNOWLEDGEMENTS

Dr. Ivan S. Arshinkov from the Scientific Industrial laboratory of Special Sensors and Systems ("SDS" Lab's) — Bulgarian Academy of Sciences (BAS), Sofia, Bulgaria, was the original principal investigator of the INTERBALL-Auroral Probe/ magnetometer. Dr. Alexander Z. Bochev from the Solar-Terrestrial Influences laboratory (STIL)—BAS, Sofia, Bulgaria, has played that role since 1996. We acknowledge the WIND, GEOS-9, Lovozero and Kiruna data providers. We thank the referee for the helpful comments.

APPENDIX A

The energy coupling function of Perrault-Akasofu is $\varepsilon = V \cdot B^2 \sin^4 (\theta/2) \cdot l_o^2 \, [10^{-7}$ W/s], where V is the solar wind speed, B is the IMF magnitude, $\theta = \tan^{-1} (|By/Bz|)$ for Bz > 0, $\theta = 180 - \tan^{-1}(|By/Bz|)$ for Bz < 0, $l_o = 7$ Re (Re—Earth's radius). As originally defined, ε is the energy flux integrated across an effective magnetopause cross sectional area. Ranges of ε given by Akasofu are: $\varepsilon \geq 10^{11}$ W for a substorm onset, $10^{11} < \varepsilon < 10^{12}$ W for a typical substorm and $\varepsilon > 10^{12}$ W for triggering of a magnetic storm.

REFERENCES

Arshinkov, I. S., Zhuzgov, L. N., Bochev, A. Z., et al., 1995, Magnetic field experiment in the INTERBALL project /experiment IMAP/, *Interball Mission and Payload,* CNES, pp. 222–228.

Aschwanden, M. J., Benz, A. O., 1997, Electron densities in solar flare loops, chromospheric evaporation upflows, and acceleration sites, *Astrophys. J.* **480**(2):825–839.

Bochev, A. Z., 2000, Field -aligned currents during forcing of magnetosphere on 9-10 November 1997, *Compt. Rend. Acad. Bulg. Sci.* **53**(12):53–56.

Burlaga, L. F., Klein, L. W., Sheeley, N. R., Michels, D. J., Howard, R. A., Koomen, M. J., Schwenn, R., and Rosenbauer, H., 1982, A magnetic cloud and a coronal mass ejection, *Geophys. Res. Lett.* **9**(12):1317–1320.

Burlaga, L., Fitzenreiter, R., Lepping, R., Ogilvie, K., Szabo, A., Lazarus, A., Steinberg, J., Gloeckler, G., Howard, R., Michels, D., Farrugia, C., Lin, R. P., and Larson, D. E., 1998, A magnetic cloud containing prominence material: January 1997, *J. Geophys. Res.* **103**(A1):277–295.

Cahill, L. J., Lin, N. G., Engebretson, M. J., Weimer, D. R., and Sugiura, M., 1986, Electric and magnetic observations of the structure of standing waves in the magnetosphere, *J. Geophys. Res.* **91**(A8):8895–8907.

Gosling, G. T., Hildner, E., MacQueen, R. M., et al., 1974, Mass ejections from the Sun: A view from Skylab, *J. Geophys. Res.* **79**(31):4581–4587.

Crooker, N., 1994, Replacing the solar myth, *Nature,* **367**:595–596.

Delannée, C., Delaboudinière, J.-P., Lamy, P., 2000, Observation of the origin of CME in the low corona. *Astron. Astrophys.,* **355**:725–742.

Delannée, C., and Aulanier, G., 1999, CME associated with transequatorial loops and a bald patch flare, *Solar Phys.* **190**(1-2):107–129.

Dermendjiev, V. N., Bochev, A. Z., Koleva, K. Zh., and Kokotanekova, J. St., 1999, The white light flare and CME on November 6, 1997 and the Earth's magnetosphere response, in: *Magnetic fields and Solar Processes,* ESA SP-448, **2**:971–977.

Fox, N. J., Peredo, M., and Thompson, B. J., 1998, Cradle to grave tracking of the January 6-11, 1997 Sun-Earth connection event, *Geophys. Res. Lett.* **25**:2461–2464.

Iijima, T., Potemra, T. A., 1978, Large-scale characteristics of field-aligned currents associated with substorms, *J. Geoph. Res.* **83**:599–615.

Harra, L. K., Matthews, S. A., and van Driel-Gesztetyi, L., 2003, Evidence of flaring in a transequatorial loop on the Sun, *Astrophys. J.* **598**(1):L59–L62.

Hundhausen, A. J., 1993, Sizes and locations of coronal mass ejections, *J. Geophys. Res.* **98**:13177–13200.

Keiling, A., Wygant, J. R., Catell, C., Temerin, M., Mozer, F. S., Kletzing, C. A., Scudder, J., Russell, C. T., Lotko, W., and Streltsov, A. V., 2000, Large Alfven wave power in the plasma sheet boundary layer during the expansion phase of substorms, *Geophys. Res. Lett.* **27**(19):3169–3172.

Maia, D., Vourlidas, A., Pick, M., Howard, R., Schwenn, R., and Magalhaes, A., 1999, Radio signatures of a fast coronal mass ejection development on November 6, 1997, *J. Geoph. Res.* **104**(A6):12507–12513.

Mason, G. M., Cohen, C. M. S., Cummings, A. C., Dwyer, J. R., Gold, R. E., Krimigis, S. M., Leske, R. A., Mazur, J. E., Mewaldt, R. A., Mobius, E., Popecki, M., Stone, E. C., von Rosenvinge, T. T., and Wiedenbeck, M. E., 1999, Particle acceleration and sources in the November 1997 solar energetic particle events, *Geophys. Res. Lett.* **26**(2):141–144.

Mazur, J. E., Mason, G. M., Looper, M. D., Leske, R. A., and Mewaldt, R. A., 1999, Charge state of solar energetic particles using the geomagnetic cutoff technique: SAMPEX measurements in the November 6, 1997 polar particle event, *Geophys. Res. Lett.* **26**(2):173–177.

Pick, M., Demoulin, P., Maia, D., and Plunkett, S., 1999, Coronal mass ejections, in: *Proc. 9-th European Meeting on Solar Physics, Magnetic fields and Solar Processes,* ESA SP-448, **2**:915–926.

Potemra, T. A., Zanetti, L. J., Bythrow, P. F., Erlandson, R. E., Lundin, R., Marklund, G. T., Block, L. P., and Lindqvist, P. A., 1988, Resonant geomagnetic field oscillations and Birkelands currents in the morning sector, *J. Geophys. Res.* **93**(A4):2661–2674.

Potemra, T. A., Luhr, H., Zanetti, L. J., Takahashi, K., Erlandson, R. E., Marklund, G. T., Block, L. P., Blomberg, L. G., and Lepping, R. P., 1989, Multisatellite and ground-based observations of transient ULF waves, *J. Geophys. Res.* **94**(A3):2543–2554.

Sanchez, E. R., Thayer, J. P., Kelly, J. D., and Doe, R. A., 1998, Energy transfer between the ionosphere and magnetosphere during the January 1997 CME event, *Geophys. Res. Lett.* **25**:2597–2601.

Thomsen, M. F., Borovsky, J. E., McComas, D. J., Elphic, R. C., and Maurice, S., 1998, The magnetospheric response to the CME passage of January 10-11, 1997, as seen at geosynchronous orbit, *Geophys. Res. Lett.* **25**(14):2545–2548.

Yermolaev, Y. I., Zastenker, G. N., Nozdrachev, M. N., Skalsky, A. A., and Zelenyi, L. M., 1998, Plasma populations in the magnetosphere during the passage of magnetic cloud on January 10-11, 1997: INTERBALL/Tail observations, *Geophys. Res. Lett.* **25**(14):2565–2569.

MULTIPOINT OBSERVATIONS OF TRANSIENT EVENT MOTION THROUGH THE IONOSPHERE AND MAGNETOSPHERE

G. I. Korotova[1], D. G. Sibeck[2], H. J. Singer[3], and T. J. Rosenberg[4]

[1]*IZMIRAN, Moscow Region, 142090 Russia;* [2]*GSFC/NASA, Greenbelt, MD 20771, USA;*
[3]*SEC/NOAA, Boulder, CO 80305, USA;* [4]*IPST, UMD, College Park, MD 20742, USA*

Abstract   We present the results of a case study of transient event observed in high-latitude ground magnetograms on May 8, 1997. We use the GOES-8, GOES-9, and GOES-10 spacecraft to identify corresponding signatures in high-time resolution geosynchronous magnetometer observations. We determine the event's spatial extent and velocity and show that the direction of event motion through the noon magnetosphere and ionosphere was similar. Wind, Geotail and Interball solar wind observations indicate that the interplanetary magnetic field (IMF) orientation controls the direction of transient event motion near local noon. The transient event corresponded to the motion of the foreshock away from the subsolar bow shock.

Key words: magnetosphere; transient event motion; high-latitude ground magnetograms; IMF orientation; solar wind discontinuity.

1.   INTRODUCTION

Isolated transient events are common in high-latitude dayside ground magnetograms. They provide evidence for one or more unsteady solar wind-magnetosphere interaction mechanisms, including abrupt variations in the solar wind dynamic pressure (Sibeck et al., 1989), bursty merging at the magnetopause (Lanzerotti et al., 1986) and the Kelvin-Helmholtz instability (McHenry et al., 1988). Events produced by each of these mechanisms exhibit differing patterns for event occurrence and recurrence as a function of solar wind conditions. Recent efforts to determine the origin of the events have reached no

J. –A. Sauvaud and Z. Němeček (eds.),
Multiscale Processes in the Earth's Magnetosphere:From Interball to Cluster, 205-215.
© 2004 *Kluwer Academic Publishers. Printed in the Netherlands.*

consensus (McHenry et al., 1990; Konik et al., 1994; Korotova and Sibeck, 1995).

The locations where the events originate and their direction of motion can help in determining its relative significance. All models predict that the majority of events move antisunward. However, the bursty merging and pressure pulse models predict sunward moving events in the vicinity of local noon. According to the reconnection model, transient events should move dawnward during periods of duskward IMF ($B_y > 0$), and move duskward during periods of dawnward IMF ($B_y < 0$) (Cowley and Owen, 1989). According to the pressure pulse model, they should move dawnward during spiral IMF, duskward during orthospiral IMF (Sibeck, 1990).

The purpose of our paper is to see if we can predict the motion of events near local noon on the basis of solar wind observations and illustrate the problems and methods involved in identifying the cause of transient events. We present a case study of the transient event observed at the South Pole station on May 8, 1997. We use high-time resolution geosynchronous and MACCS data to determine the direction of transient event motion in the pre- and post-noon magnetosphere/ionosphere. We use Wind, Geotail, and Interball as solar wind monitors to identify a trigger (if any) for the generation of the transient event.

2.   CASE STUDY

Figure 1 presents GOES-8, GOES-9, GOES-10, Interball, and Geotail satellite locations in the GSE $x - y$ plane from 1700 UT to 1900 UT on May 8, 1997. Interball moved antisunward from GSE $(x, y, z) = (18.14, 5.49, -7.46)$ to $(16.51, 5.14, -8.06)$ R_E outside the early post-noon bow shock. Geotail moved antisunward from GSE $(x, y, z) = (20.60, 15.95. - 2.83)$ to $(20.20, 17.44, -2.92)$ R_E nearly parallel to post-noon bow shock. Wind moved antisunward from GSE $(x, y, z) = (204.76, - 0.28, 21.24)$ to $(204.61, -0.24, 21.20)$ R_E far upstream from the noon bow shock. GOES-8 (at 76.4° W, $LT = UT - 5$) moved from 1200 to 1400 LT, GOES-10 (at 104.5° W, $LT = UT - 7$) from 1000 to 1200 UT, and GOES-9 (at 135.3° W, $LT = UT - 9$) from 0800 to 1000 LT.

The upper panel of Figure 2 presents the H (northward) component of the South Pole (SP, $F = 75°$, $LT = UT - 3.5$) ground magnetogram at 1-s time resolution. The isolated event from 1800 to 1810 UT exhibited a monopolar signature with a peak-to-peak amplitude of ~ 80 nT at about 1804 UT.

Densely-spaced ground observations within the northern hemisphere allow us to determine the direction in which the event moved. We can use the times at which peak amplitudes were observed to determine the azimuthal velocity of the transient event through the MACCS chain.

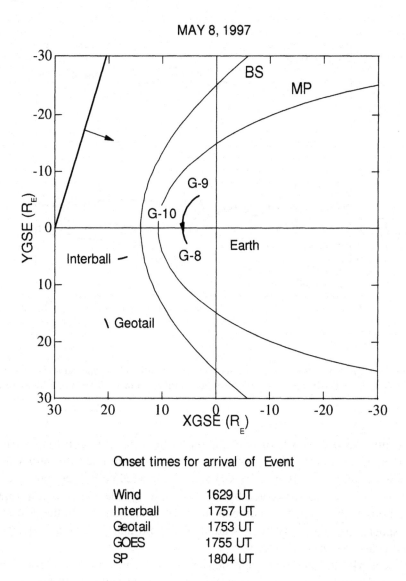

MAY 8, 1997

Onset times for arrival of Event

Wind	1629 UT
Interball	1757 UT
Geotail	1753 UT
GOES	1755 UT
SP	1804 UT

Figure 1. GOES-8, GOES-9, GOES-10, Interball, and Geotail satellite locations in the GSE xy plane shown from 1700 to 1900 UT on May 08, 1997. The solid line shows the orientation of the corresponding solar wind discontinuity for the event.

Figure 3 presents observations from 6 MACCS magnetometers at 5-s time resolution. Arrows indicate the times of prominent peaks in the X component of the magnetic field at GH, PB, CH, and CD. At the time of this event at 1805 UT, all the stations were located near noon geomagnetic local times. GH and PB bound geomagnetic latitude $\sim 78.2°$, but are separated by $\sim 12°$ in longitude. The peak in the GH X component reached PB some 65 s later, consistent

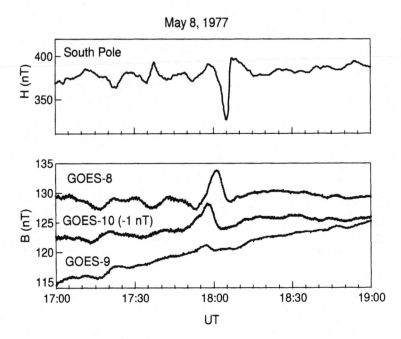

Figure 2. The panels present the H component of the South Pole magnetogram and GOES-8, GOES-9, and GOES-10 total magnetic field observations from 1700 to 1900 UT on May 8, 1997.

with sunward and duskward propagation through the pre-noon ionosphere at an azimuthal velocity of ~ 4 kms^{-1}. We also determined the propagation velocity using observations from CH and CD, located near a geomagnetic latitude of $\sim 74.3°$, but separated by $\sim 11.7°$ in longitude. As shown in Figure 3, CH observed the peak X component 45 s earlier then CD. In the post-noon ionosphere, the observations indicate antisunward and duskward propagation at a greater velocity ~ 7.8 kms^{-1}.

Past work indicates that such transient events correspond to abrupt variations in the geosynchronous magnetic field strength (Sibeck, 1993). To confirm this, the bottom panel of Figure 2 presents 0.5 s time resolution GOES-8, GOES-10, and GOES-9 observations of the total magnetic field strength. At 1755 UT, they recorded a similar positive magnetic field strength variation with peak amplitudes of about 7, 4 and 1.5 nT, respectively, corresponding to the transient event identified at South Pole. During the event, GOES-9 was located near mid-morning local time, GOES-8, and GOES-10 were at post-noon and pre-noon local times, respectively. Close examination of the traces reveals that GOES-9 observed the event before GOES-8 and GOES-10.

The peak correlation coefficient for the 15-minute data interval encompassing the event occurred for a lag of 300 s from GOES-9 to GOES-8, 185 s

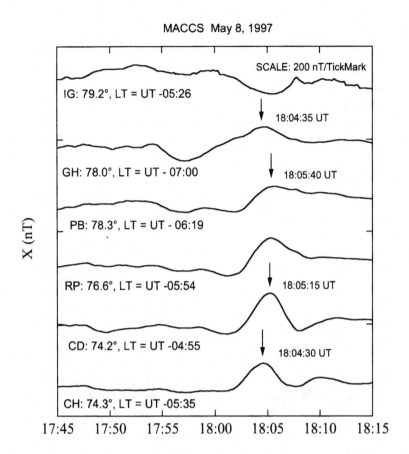

Figure 3. The X component of MACCS magnetograms from 1745 UT to 1815 UT on May 8, 1997.

from GOES-9 to GOES-10 and 119 s from GOES-10 to GOES-8 indicating duskward propagation through local noon with an azimuthal velocities of $\sim 120 - 190 \text{ kms}^{-1}$. The propagation direction of the transient event observed at geosynchronous orbit is shown by an arrow in Figure 1.

To identify solar wind triggers for the transient event, we inspected Wind, Geotail, and Interball observations. Figure 4 presents Wind Magnetic Field Investigation (MFI) and Three-Dimensional Plasma (3DP) observations with 3s and 80s time resolution, respectively. From about 1600 to 1630 UT Wind observed a cloud characterized by enhanced magnetic field strengths, depressed densities, sunward ($+B_x$), duskward ($+B_y$), and southward ($-B_z$) IMF components. At $\sim$1629 UT, the plasma observations showed an impulse-like increase of the density from 9.5 to 12.8 cm^{-3} accompanied by a sharp decrease of total magnetic field strength, a shift of the large positive B_y component to values near 0, and then a rotation of the IMF to dawnward orientations.

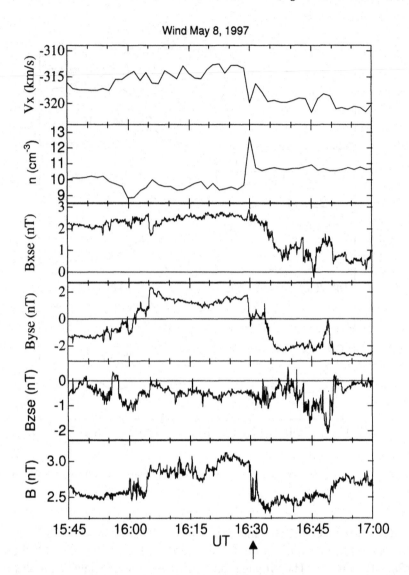

Figure 4.   Wind MFI and 3DP observations from 1545 to 1700 UT on May 8, 1997. From top to bottom, the panels present the antisunward component of velocity, number density, the GSE components of magnetic field, and the total magnetic field strength.

To test the hypothesis that the transient ionospheric/magnetospheric event could be directly associated with the arrival of the 1629 UT discontinuity at Earth, and assuming it to be tangential, we calculated its normal as the cross-product of the upstream and downstream magnetic fields for the interval from 1629:07 to 1629:49 UT. The normal pointed in the GSE $(x, y, z) = (-0.22, 0.07, -0.97)$ direction, i.e., antisunward, duskward and southward, as illustrated in Figure 1. Solar wind discontinuities with this orientation should first encounter Wind, strikes regions of the pre-noon bow shock and magne-topause, then sweep duskward past the subsolar magnetopause and encounter Interball and Geotail.

We now inspect Interball and Geotail observations to see if they provide a reasonable lag time for the solar wind features observed by Wind. Figure 5 presents Interball-1 Electron, VDP plasma, and MIF-M magnetic field ob-servations in GSE coordinates with 2-minute and 1-s time resolution, respec-tively. Comparison with Figure 4 shows that Interball observed features similar to those at Wind. The cloud seen by Wind arrived at Interball at ~1720 UT. From 1720 to 1757 UT, Interball observed pronounced wave activity indicat-ing that the strong positive IMF B_x placed it and the region upstream from the subsolar bow shock within the foreshock. At 1757 UT, the large increase in density and rotation in the IMF orientation arrived, causing an abrupt termi-nation in wave activity and removing the foreshock from the subsolar region. We should note that the GOES magnetic field was also more variable before 1800 UT (behind foreshock) than after (no foreshock). We conclude that the discontinuity producing the transient event in the magnetosphere/ionosphere arrived at Interball at 1757 UT, i. e., 1h 28 min later than at Wind. This delay is consistent with the orthospiral orientaton of the IMF and the low solar wind velocity.

Figure 6 presents Geotail magnetic field and plasma observations. We use observations from the Magnetic Field Experiment (MGF) with 3-s time resolu-tion and from the Comprehensive Plasma Instrumentation (CPI) with 52-s time resolution on the spacecraft. There are many similarities between the Geotail and Interball observations. The transition from the foreshock to non-foreshock occurred at Geotail at 1753 UT, i.e., 4 min earlier than at Interball. We con-sider the lag times for the discontinuity at Interball, Geotail, and GOES space-craft. The calculated normal to the interplanetary discontinuity points strongly southward, significantly antisunward, and slightly duskward. Although Geo-tail is located at post noon local time and Interball is located at early post-noon local time, Geotail observes the discontinuity prior to Interball. The reason for this is in the fact that the Interball lies further below the equatorial plane. The predicted lag time from Geotail to Interball can be calculated from the equation $t = \frac{d \cdot n}{V \cdot n}$, where d is the vector distance from Geotail to Interball, n is the normal to the solar wind discontinuity, and V is the solar wind velocity. Using the locations of two spacecraft at 1750 UT and a solar wind velocity of -314 km/s in the x direction, we predict a lag of 6.2 min. The observed lag is ~ 4 min, reasonably consistent given the uncertainties in normal calculation and the possibility that the discontinuity front may not be purely planar.

Our examination of the Interball and Geotail observations confirms that we have correctly identified the solar wind discontinuity responsible for the tran-sient ionospheric and magnetospheric event and correctly determined its ori-

Interball May 8, 1997

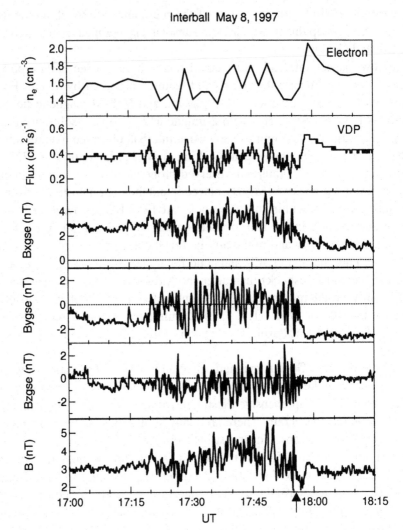

Figure 5. Interball-1 Electron, VDP, and MFI observations from 1700 to 1815 UT on May 8, 1997. From top to bottom the panels present thermal electron number density over the full energy range, integrated flux (10^9 cm^{-2} s^{-1}), the GSE components of the magnetic field, and the total magnetic field strength.

entation. As predicted by the pressure pulse model for an IMF discontinuity with an orthospiral orientation, the transient event moved duskward through local noon in both the magnetosphere and ionosphere (Korotova et al., 2000). Because the IMF maintained a southward orientation throughout the interval under study, it might be suggested that the transient event in the magnetosphere and ionosphere corresponded to a flux transfer event generated by a

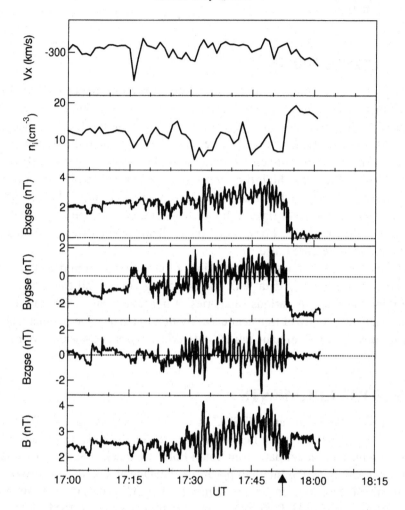

Figure 6. Geotail magnetometer and plasma moment observations from 1700 UT to 1815 UT on May 8, 1997. From top to bottom the panels show the antisunward component of the velocity, the number density, the GSE components of magnetic field and the total magnetic field strength.

burst of reconnection (in turn perhaps triggered by the solar wind pressure pulse). However, as shown in Figure 4, the IMF exhibited a duskward orientation immediately before and after the 1629 UT IMF discontinuity. Instead of moving duskward (as observed), events produced by bursty merging should move dawnward through the high-latitude northern ionosphere. The ground observations rule out an explanation in terms of bursty merging. Instead, it seems more reasonable to associate the transient magnetospheric and ionospheric events directly with the pressure pulse, not via the intermediary of a bursty of merging.

3. CONCLUSION

We present a case study demonstrating a clear relationship between transient events in simultaneous solar wind, geosynchronous, and ground observations. We used high-time resolution magnetic field data from the GOES spacecraft and MACCS chain to show that the direction of event motion through the noon magnetosphere and ionosphere was similar and controlled by the IMF orientation. This means that high-time resolution geosynchronous magnetometer observations can be used to track events even when ground data are sparse. The transient event was triggered by a solar wind pressure pulse precisely at the edge of the foreshock. Fortunately, Interball and Geotail were well situated to observe this pressure pulse.

The case study demonstrates the utility of combining ground and geosynchronous magnetometer data with corresponding solar wind observations to determine the cause of various categories of transient events. We plan to perform a statistical study of transient event motion in the dayside magnetosphere to determine the predominant cause(s) of various types of transient events.

ACKNOWLEDGMENTS

We thank the SPDF and NSSDC at GSFC for supplying the Wind, and Interball-1 plasma, magnetic field data. We are grateful to L. J. Lanzerotti for providing magnetometer data from South Pole. We thank J. Safrankova and V. Petrov for supplying the Interball plasma and magnetic field data, respectively. We thank M. Engebretson for providing the MACCS data. Part of this work was performed when G. I. Korotova was a visiting scientist at ISSI, Bern.

REFERENCES

Cowley, S. W., and Owen, C. J., 1989, A simple illustrative model of open flux motion over the dayside magnetopause, *Plan. Space Sci.* **37**(11):1461–1475.

Konik, R. M., Lanzerotti, L. J., Wolfe, A., Maclennan, C. G., and Venkatesan, D., 1994, Cusp latitude magnetic impulsive events, 2, Interplanetary magnetic field and solar wind contitions, *J. Geophys. Res.* **99**(A8):14831–14853.

Korotova, G. I., and Sibeck, D. G. , 1995, A case study of transient event motion in the magnetosphere and ionosphere, *J. Geophys. Res.* **100**(A1):35–46.

Korotova, G. I., Sibeck, D. G., Singer, H. J., and Rosenberg, T. J., 2002, Tracking transient events through geosynchronous orbit and in the high-latitude ionosphere, *J. Geophys. Res.* **107**(A11):1345, doi: 10.1029/2002JA009477.

Lanzerotti, L. J., Lee, L.C., Maclennan, C. G., Wolfe, A., and Medford, L. V., 1986, Possible evidence of flux transfer events in the polar ionosphere, *Geophys. Res. Lett.* **13**(11):1089–1092.

McHenry, M. A., Clauer, R. C., Friis-Christensen, E., and Kelly, J. D., 1988, Observations of ionospheric convection vortices: Signatures of momentum transfer, *Adv. Space Res.* **8**(9-10):315–320.

McHenry, M. A., Clauer, R. C., and Friis-Christensen, E., 1990, Relationship of solar wind parameters to continuous, dayside, high-latitude traveling ionospheric convection vortices, *J. Geophys. Res.* **95**(A9):15007–15022.

Sibeck, D. G., Baumjohann, W., Elphic, R. C., Fairfield, D. H., Fennell, J. F., Gail, W. B., Lanzerotti, L. J., Lopez, R. E., Luehr, H., Lui, A. T. Y, Maclennan, C. G., Mcentire, R. W., Potemra, T. A., Rosenberg, T. J., Takahashi, K., 1989, The magnetospheric response to 8-minute period strong-amplitude upstream pressure variations, *J. Geophys. Res.* **94**(A3):2505–2519.

Sibeck, D. G., 1990, A model for the transient magnetospheric response to sudden solar wind dynamic pressure variations, *J. Geophys. Res.* **95**(A4):3755–3771.

Sibeck, D. G., 1993, Transient magnetic field signatures at high altitudes, *J. Geophys. Res.* **98**(A1):243–256.

A MODEL FOR THE MHD TURBULENCE IN THE EARTH'S PLASMA SHEET: BUILDING COMPUTER SIMULATIONS

Joseph E. Borovsky

Space and Atmospheric Science Group, Los Alamos National Laboratory, Los Alamos, New Mexico 87545 USA

Abstract: The MHD turbulence of the Earth's plasma sheet in the magnetotail has been examined by satellite measurements of magnetic fields and plasma flows; the measured properties of this turbulence are reviewed. A theoretical analysis indicates that the MHD turbulence in the plasma sheet is a very unusual turbulence because of (1) the very limited range of spatial scales available for MHD flows and (2) the dissipation of vorticity by magnetosphere-ionosphere coupling, which introduces (a) a time dependence to the rate-of-dissipation of a flow, (b) dissipation at all spatial scales, and (c) dissipation rates that depend on the sign of the vorticity. Using a theoretical analysis of flows in the magnetotail and using some transmission-line experiments, two computational models of the plasma-sheet turbulence are being constructed to study the basic properties of this unconventional turbulence. These computational models are discussed extensively. New aspects of the study of plasma-sheet turbulence that are contained in this report are (a) corrected estimates of the Alfvenicity of the turbulence, (b) a strengthened argument that Alfven waves are not important for the dynamics of the turbulence (i.e. that it is a 2D turbulence), (c) an extended discussion about the time dependence of magnetosphere-ionosphere coupling, (d) a description of transmission-line experiments performed to clarify some properties of magnetosphere-ionosphere coupling, and (e) a discussion of numerics for building computer simulations of the magnetotail turbulence.

Key words: MHD turbulence; plasma sheet; magnetosphere; viscoelasticity; non-Newtonian; magnetosphere-ionosphere coupling; simulations; GOY; transmission line.

217

J. –A. Sauvaud and Z. Němeček (eds.),
Multiscale Processes in the Earth's Magnetosphere:From Interball to Cluster, 217-253.
© 2004 *Kluwer Academic Publishers. Printed in the Netherlands.*

1.   INTRODUCTION

MHD turbulence was theorized to exist in the high-Reynolds-number flow of the Earth's magnetotail (Antonova, 1985, 1987, 2000, 2003; Montgomery, 1987; Antonova and Ovchinnikov, 1996a,b, 1997, 1998, 1999a,b, 2000; Antonova et al., 1996; Borovsky et al., 1998; Veltri et al., 1998). Satellite observations of the magnetic field in the magnetotail plasma sheet show that the field has large fluctuations in the MHD range of frequencies (frequencies below the ion-cyclotron frequency) (Hruska and Hruskova, 1970; Hruska, 1973; Bowling, 1975; Coroniti et al., 1977,1980; Tsurutani et al., 1984; Hoshino et al., 1994, 1996; Bauer et al., 1995; Milovanov et al., 1996; Zelenyi et al., 1998; Kabin and Papitashvili, 1998; Troshichev et al., 1999). Subsequently, satellite observations have shown that the flows of the Earth's plasma sheet in the magnetotail are turbulent (Hones and Schindler, 1979; Hayakawa et al., 1982; Sergeev and Lennartsson, 1988; Angelopoulos et al., 1993; Nakamura et al., 1994; Borovsky et al., 1997; Yermolaev et al., 2000; Ovchinnikov et al., 2000, 2002; Neagu et al., 2002; Troshichev et al., 2002; Petrukovich and Yermolaev, 2002; Yermolaev et al., 2002; Pisarenko et al., 2003; Antonova et al., 2002). At present, our understanding of this MHD turbulence is at a rudimentary stage without an understanding of how the turbulence works, or what drives the turbulence, or what effect the turbulence has on magnetospheric dynamics.

In fluid dynamics, where experiments are much more advanced and turbulence is better understood, a lesson was learned long ago: If turbulence is present, it cannot be ignored or "averaged over". Turbulence has consequences for the large-scale behavior of a fluid. Besides producing an enhanced mixing, turbulence alters large-scale flow patterns, it strengthens the coupling between flows and obstacles, and it changes the interaction between flows and boundaries. In fluid-dynamics research a great deal of effort is spent building models for turbulence that can be incorporated into computer codes that are used to simulate flows around bodies. This effort is justified because for engineering problems, getting the right answer matters, and without properly accounting for the effects of turbulence the right answer will not be attained.

And so this lesson is our motivation to study the turbulence in the Earth's magnetotail. We can speculate on consequences, but without a model of the turbulence that will provide an understanding of the turbulence, we cannot assess or quantify the consequences of the plasma-sheet turbulence on magnetospheric dynamics and on the way the magnetosphere operates.

This paper is organized as follows. In section 2 the rudimentary properties of the MHD turbulence in the plasma sheet are reviewed. In section 3 the motivation for building a model of the turbulence in the plasma sheet is discussed. In section 4 the major outstanding questions about the turbulence are investigated. In section 5 the model of the plasma-sheet turbulence is put together. In section 6 the time dependence of magnetosphere-ionosphere coupling is elaborated upon and some simple transmission-line experiments are analyzed. In section 7 an ongoing program to implement computer simulations of the turbulence model is described. Section 8 contains a brief discssion of future work.

2.   RUDIMENTARY PROPERTIES OF THE TURBULENCE

In the Earth's plasma sheet, turbulent flow-velocity fluctuations have been seen by several satellites (e.g. ISEE-2, IMP-6, AMPTE/IRM, Geotail, and INTERBALL) and turbulent magnetic-field fluctuations have been seen by several satellites (e.g. Explorer-33, IMP-3, ISEE-1,-2,-3, AMPTE/IRM, Geotail, and INTERBALL). The references for these satellite observations can be found in section 1.

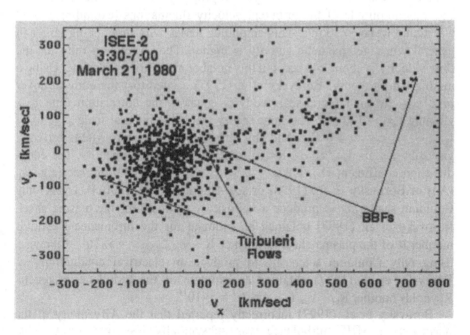

Figure 1. 979 flow measurements from the ISEE-2 FPE instrument (Bame et al., 1978) are displayed. Each flow measurement is a 3-second snapshot.

The turbulent fluctuations in the flow velocity of the plasma sheet measurements can be visualized by the scatter plot of Figure 1. Here, the v_x-versus-v_y flow velocity measured 979 times during 3.5 hours on March 21, 1980 when the ISEE-2 satellite was in the magnetotail plasma sheet is plotted. As can be seen, the flow is not steady. Two populations of flows are discernable in Figure 1: a population of fluctuaing flows with a standard deviation of less than 100 km/sec centered around zero velocity and a population of very fast flows predominantly in the X direction. These two populations are the turbulent flows and the bursty bulk flows (BBFs), respectively.

The MHD turbulent fluctuations of the magnetic field in the plasma sheet can be visualized in Figure 2, where B_x-versus-B_y and B_y-versus-B_z hodograms are constructed from 2 hours of magnetic-field measurements by the ISEE-2 satellite in the plasma sheet on March 21, 1980. As can be seen, there is a mean magnetic field, $<(B_x,B_y,B_z)> = (+21.0,+6.8,+3.1)$ nT, but with very large fluctuations about the mean.

The rudimentary properties of the MHD turbulence of the Earth's plasma sheet in the vicinity of 20 R_E downtail were obtained by Borovsky et al. (1997) based on ISEE-2 plasma-flow-velocity and magnetic-field measurements. The numerical properties are listed in Table 1. The turbulence was found to have the general property $\Delta v/v \gg 1$ and $\Delta B/B \sim 0.5$. The rms amplitude of the turbulent velocity fluctuations (excluding BBFs) was found to be ~75 km/s. This amplitude varies with time, being larger in general when geomagnetic activity is higher. The correlation time for the flow velocity is about 140 sec, but this number also varies with time. Using a mixing-length theory, Borovsky et al (1997) argued that the eddy-turnover time τ_{eddy} for an integral-scale eddy is equal to the correlation time τ_{corr}, yielding $\tau_{eddy} \sim 140$ sec. The same theory yields the integral scale L_{eddy} to be $L_{eddy} \sim v_{rms}\tau_{corr} = 1.6$ R_E. And this mixing-length picture yields an eddy viscosity $v_{eddy} \sim v_{rms}^2\tau_{corr} = 8 \times 10^{15}$ cm^2/sec. A related quantity, the eddy-diffusion coefficient D_{xx}, is $D_{xx} = 0.5$ $v_{rms}^2\tau_{corr} = 2.6 \times 10^{15}$ cm^2/sec (e.g. eq. (A3) of Borovsky et al., (1998) or see Antonova et al. (2002)). Using only Coulomb scattering to produce a viscosity v estimate for the plasma sheet, Borovsky et al. (1997) obtained an estimate for the turbulence Reynolds number R of the plasma-sheet turbulence $R \sim v_{rms}L_{eddy}/v = 5 \times 10^{11}$. Likewise, using only Coulomb scattering to produce an electrical conductivity σ, Borovsky et al. (1997) obtained an estimate of the turbulence magnetic Reynolds number $R_M \sim v_{rms}L_{eddy}4\pi\sigma/c^2 = 5 \times 10^{11}$.

Borovsky et al. (1997) incorrectly reported that the Alfvenicity of the plasma-sheet MHD turbulence was substantially less than unity. That conclusion is corrected here as follows. The Alfvenicity is measured by the Alfven ratio $\varepsilon_v/\varepsilon_B$, where ε_v is the kinetic-energy density of the turbulent

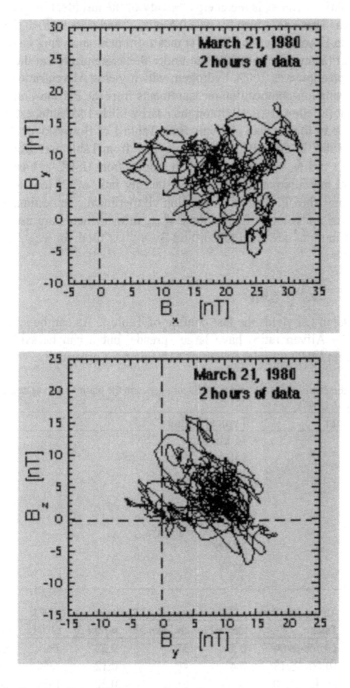

Figure 2. B_x-versus-B_y and B_y-versus-B_z hodograms constructed from 2 hours of measurements of the magnetic field by the Fluxgate Magnetometer (*Russell*, 1978) on the ISEE-2 spacecraft in the plasma sheet on March 21, 1980 during a steady magnetospheric convection event.

flow fluctuations and ε_B is the energy density of the turbulent magnetic-field fluctuations. These are given by $\varepsilon_v = 0.5\ \rho_i\ v_{rms}^2$ and $\varepsilon_B = <(\vec{B}-<\vec{B}>)^2>/8\pi$. The error in Borovsky et al. (1997) comes from their analyzing Fast Plasma Experiment (Bame et al., 1978) data under the assumption that the plasma-sheet composition was 100% hydrogen, which yields Alfven ratios that are too low. Using ion-composition measurements from the Plasma Composition Experiment (Shelley et al., 1978) on the nearby ISEE-1 satellite, estimates of the Alfvenicity for six data intervals from Table 1 of Borovsky et al. (1997) appear in Table 2. The ΔB information comes from Table 1 of Borovsky et al. (1997), and $v_{y\ rms}$ and density n come from ISEE-2 Fast Plasma Experiment measurements. The number-density ratio of O^+ ions to H^+ ions comes from the Plasma Composition Experiment. In estimating the amplitude of the velocity fluctuations, $v_{y\ rms}$ measurements are used so that BBF contributions are minimized, and $3v_{y\ rms}^2$ is used for v_{rms}^2. Thus the Alfven ratio is

$$\varepsilon_v/\varepsilon_B \approx 4\pi\rho 3 v_{y\ rms}^2/B^2 . \tag{1}$$

Expression (1) yields the last column of Table 2. As can be seen in that column, the Alfven ratios have large spreads, but it can be said that the plasma-sheet MHD turbulence has an Alfvenicity of order unity.

Table 1. Rudimentary properties of the MHD turbulence in the plasma sheet as determined by *Borovsky et al. (1997).*

Parameter	Typical Value
v_{rms}	75 km/sec
τ_{corr}	140 sec
L_{eddy}	1.6 R_E
τ_{eddy}	140 sec
v_{eddy}	8×10^{15} cm^2/sec
D_{xx}	2.6×10^{15} cm^2/sec
R	5×10^{11}
R_M	5×10^{11}
$\Delta B/B$	0.5

Table 2. The Alfven ratio for the ISEE data intervals of Borovsky et al. (1997).

Date	Time	ΔB [nT]	Δv_y [km/s]	n [cm-3]	O^+/H^+	$\varepsilon_v/\varepsilon_B$
3/2/79	13:10–16:00	3.5	57	9.22	7%	0.76
03/7/79	10:00–15:13	2.4	42	0.12	13%	0.71
3/9/79	15:40–17:30	6.7	61	0.28	15%	0.47
3/19/79	11:00–12:20	7.0	66	0.15	23%	0.38
3/26/79	12:50–15:10	9.8	90	0.18	15%	0.31
4/2/79	13:40–17:15	8.3	126	0.10	51%	1.23

3. MOTIVATION FOR BUILDING A MODEL OF THE TURBULENCE

The presence of MHD turbulence in the Earth's plasma sheet almost certainly has consequences for the plasma sheet, and perhaps has consequences for the large-scale magnetosphere. Two examples of such consequences are explored in Figures 3 and 4.

In Figure 3 six "convection hodograms" are consructed from single-satellite flow-velocity measurements. A satellite measures the Eulerian flow velocity. In ordinary fluid turbulence, the statistics of Lagrangain-velocity fluctuations are the same as the statistics of Eulerian-velocity fluctuations (Corrsin, 1963; sect. 7.1 of Tennekes and Lumley, 1972). Pretending that the measured turbulent flow velocities are Lagrangian, the flow velocity can be integrated in time to get the path that a Lagrangian fluid element takes through space in the turbulence. In Figure 3, six 2-hour-long intervals of flow-velocity measurements are integrated to get the six paths that are shown. Under the assumption that Eulerian statistics are the same as Lagrangian statistics, these paths would be typical of actual fluid-element paths. As can be seen in the figure, fluid elements in the turbulence tend to make random walks in the plasma sheet and the turbulence should lead to an eddy transport (eddy diffusion) of material and of quantities associated with the material.

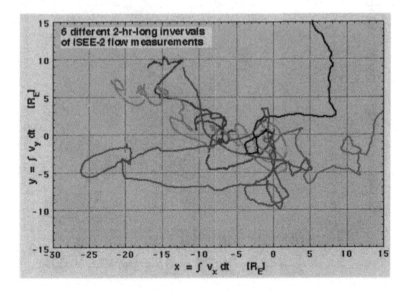

Figure 3. "Convection" hodograms are produced by time integrating the flow-velocity measurements from the ISEE-2 spacecraft in the plasma sheet. Six different two-hour-long intervals of data are used to produce the six different-colored hodograms.

In Figure 4 the magnetic-field measurements that were used to produce the hodograms of Figure 2 are used to produce a hodogram of the magnetic-field direction in the plasma sheet in azimuth-elevation coordinates. The view of the Earth (to scale) from the ISEE-2 spacecraft during the measurements is also depicted. As indicated by the hodogram of Figure 4, the large fluctuations of $\vec{B}$ in the turbulence represent large fluctuations in the direction of $\vec{B}$ It makes sense that these direction changes are local rather than global, otherwise the entire magnetotail would be swinging wildly. (In fact, Borovsky and Funsten (2003) used special "sweeping intervals" to measure the correlation length of such magnetic-field fluctuations and found that their scalesize is about $L_{eddy} \sim 1.6$ R_E.) These large, localized direction changes of the magnetic field indicate that the magnetic field in the plasma sheet is highly disordered. Hence the sketches of the plasma sheet with "spaghetti" magnetic fields (Fig. 4 of Hruska (1973), Fig. 1 of Borovsky et al. (1997), or Fig. 1 of Borovsky and Funsten (2003)).

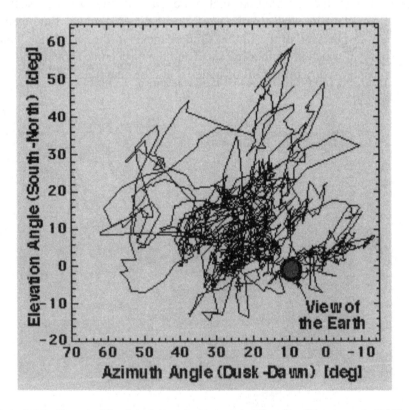

Figure 4. A hodogram of the elevation angle-versus-azimuth angle of the magnetic field in the magnetotail is constructed from 2 hours of magnetic-field measurements taken by the ISEE-2 spacecraft in the plasma sheet on March 21, 1980.

The presence of MHD turbulence can have several consequences. These include mixing, eddy viscosity, eddy transport, eddy resistivity, and internal heating. These also include enhanced localized reconnection. The presence of the turbulence can disorder the magnetic field, change the large-scale flow pattern of the plasma sheet, enhance the momentum coupling between the magnetosheath and the magnetotail, and perhaps alter the large-scale stability and dynamics of the magnetotail.

To be able to quantify these consequences, a model of this MHD turbulence in the plasma sheet is needed. Once such model is obtained, an understanding of how the turbulence works can be developed and realistic calculations of eddy viscosity, Reynolds numbers, heating, etc. can subsequently be performed. Deriving the necessary facts to assemble such a model was the goal of Borovsky and Bonnell (2001) and Borovsky and Funsten (2003). The major parts of the model will be discussed in section 4, the assembly of the model will be discussed in section 5, and the beginnings of numerical simulations of the turbulence based on the model will be discussed in section 7.

4.   PARTS OF THE MODEL

For any turbulence, three major questions that can be asked are (1) what is the dynamics of the turbulence? (i.e. what is the nature of the turbulent fluctuations and how do they interact with each other?), (2) how is the turbulence driven?, and (3) how is the turbulence dissipated? For reference, a spectral sketch of a classic homogeneous turbulence (e.g. Navier-Stokes) appears in Figure 5. For this classic or textbook turbulence, how the answers to the three questions come into play is described. The dynamics of this turbulence is eddy-eddy interactions: two eddies of similar sizes (i.e. two regions of vorticity with similar spatial scales) interact to feed energy into an eddy of a not-too-different size (e.g. sect. 8.2 of Tennekes and Lumley (1972) or sect. 7.3 of Frisch (1995)). By means of such eddy-eddy interactions, there is a net flow of energy from the larger scale sizes (lower frequencies) to the smaller scale sizes (higher frequencies). The turbulence is driven, typically, by large-scale shear flows. The energy in the turbulence is dissipated by ordinary viscosity acting to damp the small-scale eddies. Thus, in this classic turbulence the turbulence is driven at large scales, dissipated at small scales, and there is a region in between where driving and dissipation are unimportant where energy transport dominates. As noted in Figure 5, these three regions are denoted as the energy subrange (driving), the inertial subrange (transport without driving and without dissipation), and the dissipa-

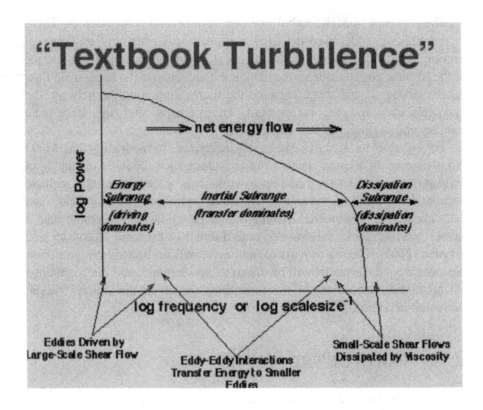

Figure 5. A sketch of the spectral description of classic homogeneous turbulence.

tion subrange. In this classic, familiar turbulence, driving and dissipation are well separated in frequency/scalesize space.

Concerning the dynamics of the turbulence in the Earth's plasma sheet, for fluctuations in the MHD range of periods (many minutes to 10's of seconds in the plasma sheet), the fluctuations could be Alfven waves, magnetosonic (fast) waves, slow-mode waves, mirror-mode waves, or eddies (which are not waves). Borovsky and Funsten (2003) presented arguments that magnetosonic waves, slow-mode waves, and mirror-mode waves are not important for the MHD turbulence in the Earth's plasma sheet. That leaves eddies and Alfven waves. In the MHD range of frequencies and scalesizes, every eddy with a finite extent along the mean magnetic field B, has an Alfven-wave nature: an eddy produces a torsion that propagates along the magnetic field at the Alfven velocity (cf. sect. 3.6 of Alfven and Falthammar (1963)). Eddy-eddy interactions make MHD turbulence go, but Alfven-wave-propagation effects can reduce the coupling between eddies, changing the nature of the turbulence dynamics (Kraichnan, 1965; Hossain et al., 1996). The importance of the Alfven-wave decorrelation of the MHD turbulence in the Earth's plasma sheet is low. Borovsky and Funsten (2003) presented two arguments favoring the eddy nature of the turbulence: one argument, based on the Alfvenicity of the turbulence, is incorrect because of the incorrect estimates of the Alfven ratio that Borovsky and Funsten (2003) utilized (see section 2). The other, based on the scalesize of the plasma sheet and the allowed spectra of Alfven waves is valid. But here, this second argument is improved upon.

Alfven-wave propagation effects do not play a strong role in the dynamics of the MHD turbulence in the plasma sheet, i.e. Alfven-wave propagation effects do not lead to decorrealtion of the nonlinear interactions that drive the turbulent cascade. For decorrelation to occur, the Alfven-wave transit time τ_A must be much shorter than the nonlinear transfer time τ_{nl}, i.e. $\tau_A \ll \tau_{nl}$ (Kraichnan, 1965; Hossain et al., 1996). The nonlinear transfer time is taken to be the eddy-turnover time τ_{eddy} (e.g. sect. 7.2 of Frisch (1995)). For integral-scale eddies in the plasma sheet, $\tau_{eddy} \cong 140$ sec (see Table 1), and for all other eddies in the turbulence $\tau_{eddy} < 140$ sec. So for decorrelation to occur, $\tau_A \ll 140$ sec. In a high-β plasma such as the plasma sheet, there is a significant difference in the properties of left-hand circularly polarized Alfven waves and right-hand circularly polarized Alfven waves (Gary, 1986). Left-hand circularly polarized Alfven waves in a high-β plasma are highly dispersive, with phase and group velocities that can be well below the Alfven velocity. Owing to dispersion and to kinetic damping, in a high-β plasma the left-hand waves have frequencies ω that are limited to $\omega < 0.15$ ω_{ci} (see Fig. 6.4(b) of Gary (1993)), where ω_{ci} is the ion-cyclotron frequency. This means the periods $\tau_A = 2\pi/\omega$ of the left-hand polarized Alfven waves are limited to $\tau_A > 42/\omega_{ci}$. For the plasma sheet with B ~ 10 nT, this becomes $\tau_A > 44$ sec. For integral-scale eddies with teddy $\cong 140$ sec the condition $\tau_A \ll \tau_{nl}$ is marginally met, for other eddies in the turbulence the condition is not met. Hence, decorrelation by the Alfven-wave propagation effect is not important for left-hand polarized Alfven waves. Right-hand circularly polarized Alfven waves with propagation angles (from the magnetic field direction) $\theta > 0°$ are strongly damped in a high-β plasma (see Fig. 6.3(b) of Gary (1993)) owing to proton Landau damping (Gary, private communication, 2003). For any eddy, $k_\perp$ is nonzero, so the condition $\theta > 0°$ will apply. This means that right-circularly polarized Alfven waves in the high-β plasma sheet will be strongly damped and will not act to decorrelate the nonlinear transfers in the plasma-sheet turbulence. With Alfven-wave decorrelations not playing a strong role in the dynamics of the MHD turbulence, the MHD turbulence of the plasma sheet is arguably a 2D turbulence (Matthaeus et al., 1990; Bieber et al, 1996) describable by a reduced set of MHD equations (RMHD) (Montgomery, 1989; Dahlburg et al., 1985).

Concerning the dynamics of the turbulence in the Earth's plasma sheet, another issue must be stressed: the importance of boundaries. This issue will be stressed in two ways. First, the MHD turbulence of the plasma sheet has severely restricted spatial scales. The largest scalesize L_{max} for MHD turbulence is the thickness of the plasma sheet, which is about 6 R_E. The smallest MHD spatial scale L_{min} for a high-β plasma such as the magnetotail plasma sheet is the ion gyroradius, which is about 700 km. Thus the dynamic range of MHD in the plasma sheet is $L_{max}/L_{min} \sim 50$, which is only one and a half decades of wavenumber space. (For the solar-wind turbulence, $L_{max}/L_{min} \sim 10^7$.) This is a severe restriction for the plasma-sheet turbulence. Second, a typical eddy in the plasma sheet is always close to a boundary. The size of an integral-scale eddy is $L_{eddy} \sim 1.5$ R_E. The thickness of the plasma sheet is ~ 6 R_E. So, an integral-scale eddy is always within about 1 eddy diameter of a boundary. The plasma-sheet turbulence is certainly not a "homogeneous turbulence", which is a turbulence that is free from the effects of boundaries.

The driving of the MHD turbulence in the plasma sheet is not understood: i.e. the source of power for the turbulence is not known and there is no understanding of where and when the turbulence will be robust. That the plasma sheet is turbulent is no surprise, since its Reynolds number is so high and, at least in fluid dynamics, all high-Reynolds-number flows are turbulent. Borovsky and Funsten (2003) examined several possibilities for the source of the turbulence and argued that two are likely sources: shear in the large-scale flows of the magnetotail (see also Antonova and Ovchinnikov (1999a)) and stirring by BBFs (see also Angelopoulos (1999) or Hoshino (2000)). The answer to the driving question awaits further analysis of satellite data and future computer simulations of the magnetotail with higher-Reynolds-number MHD computer codes (see section 7.1).

Dissipation of the MHD turbulence in the plasma sheet occurs by two mechanisms: (1) internal dissipation at small spatial scales and (2) external dissipation via magnetosphere-ionosphere coupling. For ordinary fluids, internal dissipation at small spatial scales occurs because viscosity dissipates small-scale vorticity in the turbulence, with the kinetic energy of the turbulence going into heating the fluid. For MHD turbulence, how dissipation at small scales works is a mystery and a topic of current research (Leamon et al., 1998a,b, 1999, 2000; Gary, 1999; Hollweg, 1999; Cranmer and Ballegooijen, 2003; Gary and Borovsky, 2004). The various eddy and Alfven-wave natures of the fluctuations might put energy into ions via cyclotron resoncances, into electrons or ions via Landau resonances, into flows and particle energization via localized reconnections, or into other non-MHD flows and wavemodes via mode couplings.

The second source of dissipation of the MHD turbulence of the plasma sheet is external dissipation by means of magnetosphere-ionosphere coupling. Here the energy of the turbulent fluctuations in the plasma sheet goes into heating the ionosphere. A shear flow (vortex) in the magnetosphere will drive field-aligned currents into the resistive ionosphere, converting flow kinetic energy into heat (Borovsky and Bonnell, 2001). This dissipation of vorticity looks like a "viscosity", but with three complications (see also Borovsky and Funsten, 2003). (1) The magnetosphere-ionosphere coupling evolves temporally via multiple transits of Alfven waves between the magnetosphere and ionosphere (Goertz and Boswell, 1979; Goertz et al., 1993), as examined in detail in section 6. This produces a viscosity that grows with time as the age of a vortex in the plasma sheet increases. This "viscosity" then looks like a "viscoelasticity" (a viscosity that acts with a time delay) (Harris, 1977; Joseph, 1990). (2) The efficiency of the magnetosphere-ionosphere coupling is scalesize dependent, with larger vortices coupling to the ionosphere more efficiently than smaller vortices do (Borovsky and Bonnell, 2001). This "viscosity" then looks like a "hypoviscosity" (a viscosity that is weakened on small spatial scales and strengthened on large scales). (3) The efficiency of the magnetosphere-ionosphere coupling depends on the sign of the vorticity $\vec{\omega}$, i.e. whether $\vec{\omega} \bullet B$, is positive or negative (Borovsky and Bonnell, 2001). This introduces a "sign-vorticity effect" to the "viscosity".

Besides the complications introduced to the turbulence by the viscoelasticity and sign-vorticity effect, which may or may not cause the plasma-sheet turbulence to depart from the classical-turbulence picture of Figure 5, the hypoviscosity introduces a fundamental change to the turbulence. Unlike the classical turbulence with an inertial subrange free from driving and dissipation, with hypoviscosity we now have a turbulence with dissipation at all spatial scales. Driving and dissipation are no longer separated.

5. THE MODEL OF THE MHD TURBULENCE IN THE PLASMA SHEET

The model of the MHD turbulence in the Earth's plasma sheet is described with the use of the spectral sketch in Figure 6. Note in the bottom of the sketch the periods of the fluctuations are indicated (see, e.g. Fig. 2a of Borovsky and Funsten (2003)).

The dynamics of the turbulence contains two regimes, an MHD regime at lower frequencies (larger spatial scales) and a kinetic (non-MHD) regime at higher frequencies (shorter spatial scales). The kinetic regime might involve EMHD (whistler) fluctuations.

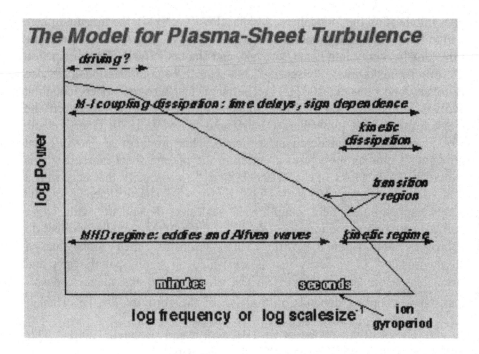

Figure 6. A sketch of the spectral description of the model of MHD turbulence in the Earth's plasma sheet.

The dissipation of the turbulence involves two mechanisms: the first is internal dissipation in the kinetic regime and the second is external dissipation at all MHD spatial scales via magnetosphere-ionosphere coupling. The dissipation via magnetosphere-ionosphere coupling involves the complications of viscoelasticity (time-dependent coupling) and sign vorticity (different dissipation for right-hand and left-hand vortices). Plus, it introduces dissipation at all spatial scales. These aspects are noted in Figure 6.

The driving of the turbulence is a mystery. If shear in the large-scale flows is a source, the driving probably occurs at the larger scales of the turbulence. If stirring by BBFs dominates the driving, then driving at the scale of the BBF width probably dominates, which corresponds to driving at larger scales and at meso scales in the turbulence.

Many details of the model of MHD turbulence in the plasma sheet are not known. (1) As was mentioned above, the manner in which the turbulence is driven is not known. (2) What happens to the turbulence at high k (high frequency) is a mystery: mode conversion and/or dissipation are possibilities. Fortunately, the behavior of turbulence in the larger-spatial-scale regime is not very sensitive to the details of how the turbulence is dissipated (or otherwise destroyed) in the high-k regime (cf. Fig. 5.9 and sect. 5.2 of Frisch (1995)).(3) The sign of the shear is accounted for in the model, but the sign of the polarization of the Alfven-wave nature of the fluctuations is not. For instance, left-hand wave information may dissipate via cyclotron interactions with ions, whereas right-hand wave information may cascade to higher k to become whistler (EMHD) waves. And in the high-β plasma sheet, left-hand polarizations are highly dispersive and travel toward the ionosphere at less than the Alfven speed while right-hand polarizations travel at faster than the Alfven speed but are heavily damped.

Before discussing the construction of numerical codes to simulate the turbulence in the Earth's plasma sheet, the viscoelastic nature of the dissipation via magnetosphere-ionosphere coupling is elaborated upon.

6. VISCOELASTICITY IN THE PLASMA SHEET OWING TO MAGNETOSPHERE—IONOSPHERE COUPLING

To handle the viscoelasticity introduced by the magnetosphere-ionosphere coupling, major modifications turbulence-simulation techniques is needed. To motivate and clarify the purpose of these modifications, in this section a theoretical analysis of viscoelasticity and then an experimental analysis of magnetosphere-ionosphere coupling are performed.

Viscoelasticity is a time-delayed viscous stress on a shear, plus it is a persistence of the stress after a shear has ceased. For ordinary fluids Newton's model holds:

$$S = \eta_o \, \partial u/\partial x \tag{2}$$

(e.g. Feynman et al., 1964) where $\partial u/\partial x$ is the shear in the fluid (where u is a velocity not in the x direction), $\eta_o = v_o\rho$ is the coefficient of viscosity of the fluid (where ρ is the fluid's mass density and v_o is the fluid's kinematic viscosity), and S is the shear stress (shearing force per unit area) in the fluid. In Newton's model the stress S is linearly proportional to the instantaneous value of the shear $\partial u/\partial x$. For viscoelastic fluids, Maxwell's model holds:

$$S + \tau_{relax} \, dS/dt = \eta_{ve} \, \partial u/\partial x \tag{3}$$

(e.g. eq. (10.8) of Faber (1995)) where τ_{relax} is a relaxation time and η_{ve} ($=v_{ve}\rho$) is the viscoelastic viscosity. In Maxwell's model the stress S depends on the time history of the shear $\partial u/\partial x$. These properties are depicted via the example in Figure 7. In the top panel the amplitude of a hypothetical shear at one point in a fluid is plotted; at time t=0 the shear is turned on and at time t=5 the shear is turned off. In the second panel the viscous stress at that point in the fluid is plotted for ordinary viscosity (Newton's model, expression (2)); as can be noticed, the stress is linearly proportional to the instantaneous shear. In the third panel of Figure 7 the stress at that point in the fluid is plotted for viscoelasticity operating (Maxwell's model, expression (3)), where the stress depends on the time history of the flow shear. As seen in the third panel, when the shear turns on at time t=0 the stress begins to turn on but does not reach its full value for some time (depending on the choice of the relaxation time τ_{relax}). And when the shear is abruptly shut off at time t=5, the stress does not suddenly go to zero, rather there is a persistence to the stress (an elasticity) with the stress going to zero over a time that depends again on the choice of τ_{relax}. In Newton's model, vorticity decays as

$$\partial\omega/\partial t = v_o \, \nabla^2 \omega \tag{4}$$

(e.g. eq. (41.17) of Feynman et al. (1964)). In the Maxwell model, every vortex ω dissipates as it ages according to

$$\partial\omega/\partial t = v_{ve} \, \nabla^2 \omega \, [1 - \exp(-\tau_{age}/\tau_{relax})] \,, \tag{5}$$

where τ_{age} is the age of the vortex , v_{ve} is the viscoelastic viscosity, and τ_{relax} is the viscoelastic relaxation time. And according to the Maxwell model, every time an amount of vorticity $\Delta\omega$ disappears, the persistence of stress acts to recreate an opposite vorticity according to

$$\partial\omega/\partial t = v_{ve} \, \nabla^2 (\Delta\omega) \, [1 - \exp(-\tau_{elapsed}/\tau_{relax})] \tag{6}$$

where $\tau_{elapsed}$ is the elapsed time since the change $\Delta\omega$ occurred. In expression (6), for vorticity decreasing, $\Delta\omega$ is a negative quantity. When ordinary viscosity acts along with viscoelasticity (such as when the high-k kinetic dissipation of MHD acts along with the time-delayed magnetosphere-ionosphere coupling), then Zener's model for the shear stress holds:

$$S + \tau_{relax} \, dS/dt = (\eta_{ve} + \eta_o) \, \partial u/\partial x + \eta_o \, \tau_{relax} \, d(\partial u/\partial x)/dt \tag{7}$$

(e.g. Hernandez-Jimenez et al., 2002) (see also eqs (17.19) and (18.13) of Reiner (1958), eq. (1.32) of Harris (1977), and eq. (2.2.7) of Joseph (1990)). In expression (7), η_o is the fluid's ordinary viscosity and η_{ve} is the fluid's viscoelasticity. In expression (7), if $\eta_{ve} \to 0$ or if $\tau_{relax} \to 0$, then Newton's model (expression (2)) is recovered. Or if $\eta_o \to 0$, Maxwell's model (expression (3)) is recovered. In the bottom panel of Figure 7 the shear stress for the Zener model (for η_o that is 40% as large as η_{ve}) is plotted in response to the hypothetical flow shear in the top panel. As can be seen, there is both a visccoelastic response plus a viscous response, a sum of the Maxwell's model response (third panel) and the Newton's molel response (second panel).

The plasma-sheet/ionosphere system has behaviors very simlar to these. When a shear is initiated in the plasma sheet a plasma polarization occurs (Borovsky, 1987), and electrical information about the shear is communicated along the magnetic-field lines to the ionosphere as a transmission-line signal propagating approximately at the Alfven speed (Drell et al., 1965; Scholer, 1970; Goertz and Boswell, 1979). The shear will act as a generator and the ionosphere as a load. When the transmission-line signal reaches the ionosphere, there is a partial reflection off of the ionosphere owing to an impedance mismatch between the transmission line (which has an impedance that is the "Alfven admittance" (Maltsev et al., 1977)) and the ionosphere (which has an impedance that is the height-integrated Pedersen conductivity). The reflected signal heads back toward the velocity shear in the magnetosphere, carrying information about the nature of the load. When the reflected signal reaches the velocity shear (generator region) in the plasma sheet carrying information about the ionospheric load, the generator region adjusts its current output and this sends another transmission-line signal toward the ionosphere, which will partially reflect back. Only after many transit times at the Alfven speed is the generator fully coupled to the resistive load. Throughout this process, there are currents flowing which (1) produce J×B forces that tend to brake the flow shear in the generator and (2) produce Ohmic dissipation in the ionosphere. This temporal development is demonstrated with measurements from the transmission-line experiment of Figure 8. The experimental setup is depicted in the top panel. A generator (a 4-volt battery with a 2.5-Ω external resistance added in series) is connected to a transmission line (a 50-Ω-impedance coaxial cable with a length L = 100 foot with a signal-propagation velocity of 1.98×10^{10} cm/s) which is connected to a resistive load (a 10-Ω resistor). Using a digitizing oscilloscope, the currents flowing through the generator and through the load are measured. These measured currents are displayed in the bottom panel of Figure 8; the current through the generator is in black and the current through the load is in gray. Prior to t=0, all currents are zero. At t=0 the switch is closed turning on the generator; at this time the generator current feeding the 50-Ω transmission line is about 20 mA (2 V into 50 Ω). Now an electrical signal is propagating down the line toward the load: this signal carries the voltage of the generator and carries the current of the generator in each of the two legs of the line

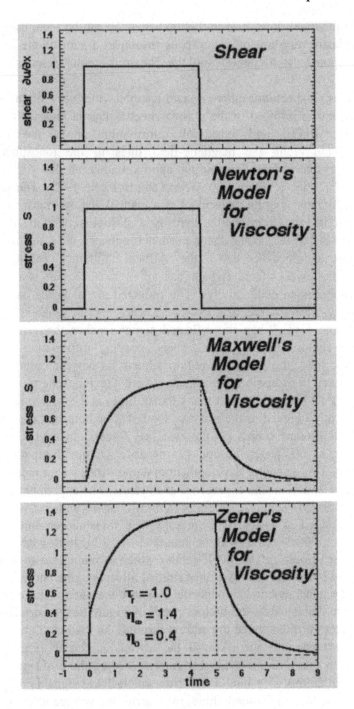

Figure 7. A hypothetical time-dependent shear in a fluid (top panel) and the resulting time-dependent stress in the fluid according to three different models for viscosity: Newton's model (second panel), Maxwell's model (third panel), and Zener's model (bottom panel).

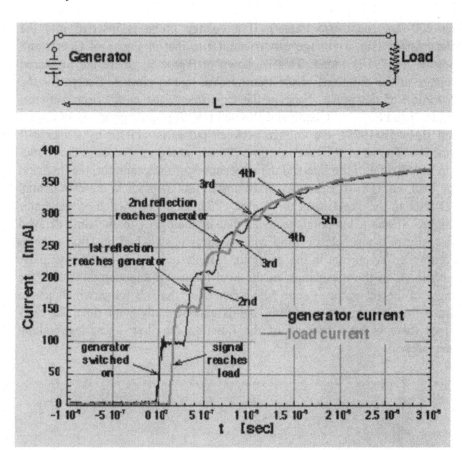

Figure 8. In the top panel a sketch of the transmission-line circuit is shown. In the bottom panel the measured current through the generator (black curve) and through the load (gray curve) are plotted. The data is taken with a digitization rate of 1 nsec run through a 50-nsec boxcar average. At time t=0 the switch is closed.

(negative charge flowing away from the negative terminal of the generator and positive charge flowing away from the positive terminal, each being 20 mA). At time t = 1.4×10^{-7} sec the transmission-line signal reaches the load and current begins flowing through the load. The impedance of the load (10 Ω) is lower than the impedance of the line (50 Ω), so the load will take more current than flows through either leg of the line; hence the current of the load temporarily exceeds the current of the generator. Because of the impedance mismatch, there is a partial reflection of the signal. At time t= 3×10^{-7} sec the reflected signal reaches the generator, and the generator's current output increases. This goes on for multiple transits of the signal. About 7 bounces are discernable in Figure 8 before resistivity in the coaxial transmission line attenuates the signal (for a 200-nsec square pulse, about 4 dB of attenuation per 200-foot round-trip transit). If a voltage probe is inserted into the transmission line, a voltage pattern similar to that of Fig. 3 of Goertz and Boswell (1979) is found. This is shown in Figure 9, where the measured voltage at the midpoint of the transmission line is plotted. Except for the distortion of the pulse shape in the experiment, the comparison between Figure 9 and Fig. 3 of Goertz and Boswell (1979) is quite good. As in Goertz and Boswell (1979), the voltage between pulses is well fit by a 1-exp(-t) function. The fit to the data is exp(-t/726ns); taking into account the round-trip transit time of the line and the impedances of the generator, the line, and the load, eq. (18) of Borovsky and Funsten (2003) predicts the exponential behavior in the experiment to be exp(-t/772nsec). The switch-on generator (Figure 8) has a temporal behavior that is analogous to the time-delayed viscosity of a viscoelastic fluid when a shear is switched on (first edge in Figure 7). The current in the generator-load system builds up exponentially with time, so the dissipation in the load builds up exponentially with the "age" of the generator . For an MHD generator; the J×B force, which acts to brake the flow in the generator, builds exponentially with time. At early times there is little braking of the flow; as the flow ages, the braking increases.

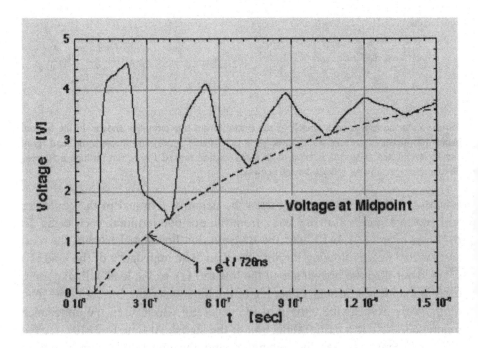

Figure 9. The voltage measured midway along the transmission line (50 feet from both the generator and the load) in the circuit drawn in Figure 8. The data is taken with a digitization rate of 0.4 nsec run through a 40-nsec boxcar average. At time t=0 the switch is closed.

There is a similar, less discussed, behavior when a velocity shear is halted in the plasma sheet. Suddenly stopping a shear will cause a sudden reduction in the cross-field voltage of the MHD generator to zero, which will tend to shut down the field-aligned currents that close in the ionosphere. The sudden decrease in the plasma-sheet voltage will be communicated along the transmission line toward the ionosphere at the Alfven speed, and the transmission-line signal will partially reflect back toward the plasma sheet. Upon arriving at the plasma-sheet generator, the generator will readjust its current (still with voltage = 0) and send this current readjustment as another transmission-line signal, which will be partially reflected. And so forth with multiple reflections. Even though the plasma-sheet voltage is suddenly reduced, the currents flowing through the plasma sheet and ionosphere will be reduced gradually over many Alfven-speed transit times. While the current is being reduced, the J×B force that was acting to brake the flow is still acting, which produces an acceleration in the plasma-sheet that produces a flow in the direction opposite to the original flow. This persistence of force tends to produce a vorticity opposite to the vorticity that was halted -- an elasticity. This electrical property is demonstrated with the transmission-line experiment in Figure 10. The experimental setup is sketched in the top panel: a variable-voltage generator is connected to a load (again, a 10-Ω resistor) via a transmission line (again a 50-Ω coaxial cable with L = 100 feet). The variable voltage of the generator is set to 1.86 volts for a long time (until equilibrium) and then at time t=0 the generator voltage is suddenly reduced to 0.17 volts. As can be seen in Figure 10, before time t=0 the current flowing through the generator (black curve) and the load (gray curve) is about 190 mA. At t=0 the generator voltage is reduced to 0.17 V. At this time the generator current drops, but not to zero. Rather it drops by about 35 mA, which is the current carried by a transmission-line signal of -1.7 V into a 50-Ω line (I = V/R = 1.7V/50Ω = 34 mA). At time t = 1.5×10^{-7} sec the transmission-line signal reaches the load, whereupon the load current is recuced as the current carried by the transmission-line signal passes through the load, with partial reflection. At t = 3×10^{-7} sec the reflected signal reaches the generator whereupon the generator current is readjusted, which sends another signal down the transmission line. This goes on for many bounces; about 6 bounces are discernable in Figure 10 before attenuation in the line smears the signal. As can be seen when the generator voltage is switched off, the currents decay with an exponential time profile. For an MHD generator, while the generator is on the current in the generator represents a J×B force acting to oppose the plasma flow. When the MHD generator is switched off (by flow stoppage), the current still flows in the same direction through the generator, and there is a persistent J×B force in the same direction. This persistent J×B force tends to produce a flow in the direction opposite to the original flow. As can be seen by comparing Figure 10 with the second edge in Figure 7, the persistent currents (and the persistent forces it produces) when a generator is turned off are analogous to the persistent stress in a viscoelastic fluid when shear is stopped.

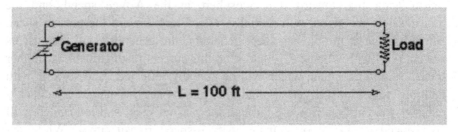

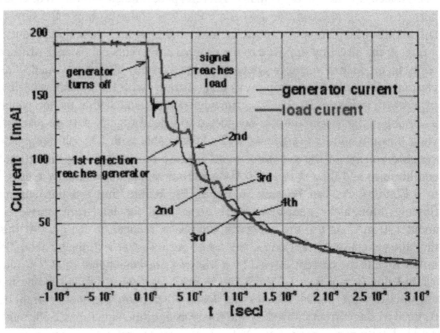

Figure 10. In the top panel a sketch of the transmission-line circuit is shown. In the bottom panel the measured current through the generator (black curve) and through the load (gray curve) are plotted. The data is taken with a digitization rate of 1 nsec run through a 50-nsec boxcar average. At time t=0 the voltage of the generator is reduced from 1.86 V to 0.17 V.

7. CODING THE MODEL FOR COMPUTER SIMULATIONS

Two numerical methods will be used to explore the MHD turbulence of the Earth's plasma sheet. Both methods will be based on various aspects of the turbulence model described in section 5. The two methods and their goals are (1) 3-dimensional MHD simulations in real space to determine how the turblence is driven and to discern the effects of boundaries on the turbulence and (2) GOY (Gledzer-Ohkitani-Yamada) simulations in wavenumber space to discern the effects of dissipation on the turbulence, particularly the effects of hypoviscosity, viscoelasticity, and various possible high-k dissipation mechanisms. The logic going into the computational schemes are discussed in this section.

7.1 3-D MHD simulations

The 3-D MHD codes that are presently used to simulate the dynamics of the magnetosphere operate at Reynolds numbers that are too low and gridspacings that are too coarse to allow turbulence to occur either in the solar wind, the magnetosheath, or the plasma sheet. (The one exception is the ISM code, wherein large-scale vortices of the plasma-sheet turbulence are seen in the magnetotail during simulations (White et al., 2001).) To study MHD turbulence, a simulation code has to be pushed to higher Reynolds numbers. Two questions are: "What should the target Reynolds number R be?" and "Is it possible for a simulation to be run with the target R?" The goal will be to do a sufficient job of representing the turbulence and its effects.

For turbulent flow problems in ordinary (Navier-Stokes) fluids, there are three major types of simulations (Pope, 2000; Matthieu and Scott, 2000): (1) direct numerical simulation (DNS), (2) large-eddy simulations (LES), and (3) Reynolds-averaged Navier-Stokes (RANS). In a DNS simulation the computational grid resolves the large-scale geometry of the fluid flow and resolves all spatial scales in the turbulence down to the turbulence dissipation scale. In general, DNS simulations are computer intensive and so are limited to very simple flow problems at modest Reynolds numbers. In a LES simulation the computational grid resolves the large-scale geometry of the flow and resolves down to mesoscale fluctuations in the turbulence, but does not resolve the smaller-scale fluctuations of the turbulence and does not resolve the dissipation range. Instead LES uses a "subgrid model" to account for the effects that the small-scale turbulent fluctuations have on the resolved mesoscale fluctuations. LES is a technique used for simple fluid-engineering simulations such as channel flow and atmospheric boundary layers. In a

RANS simulation the computational grid resolves only the large-scale fluid flow; it resolves none of the turbulence scales. Instead it incorporates models for the temporal evolution of the amplitude of turbulence based on large-scale shear in the flow and so forth, and it incorporates models for the amplitude-dependent turbulent eddy viscosity into the Navier-Stokes equations (Reynolds equations). The RANS technique is the one most-commonly used for engineering problems in fluid flow, including heat transfer and combustion. A considerable research effort goes into building turbulence models to incorporate into RANS computer codes. For ordinary fluids this modeling effort has been greatly aided by having wind-tunnel experiments to test turbulence models against measurements. The majority of MHD simulations of the magnetosphere are running in the RANS (or "RAMHD") regime, but without the inclusion of a turbulence model. The ISM code may be approaching the LES regime, but without the inclusion of a proper subgrid model.

For simulating turbulent MHD flows (as opposed to Navier-Stokes flows) there are two dilemmas: (1) we do not have any models for determining MHD-turbulence amplitudes, eddy viscosities, eddy resistivities, etc. from large-scale MHD flow properties (in part because we do not have MHD wind-tunnel experiments) and (2) we do not know how the dissipation of the MHD turbulence works. Because of dilemma (1) the RANS technique cannot yet be used for putting together an engineering code in MHD, and such a code would not be of use for studying the physics of turbulence. The LES technique could be of use for putting together an engineering code in MHD, and it would be of some use for studying the physics of MHD turbulence in the plasma sheet, but dilemma (2) interferes with our ability to build a subgrid model for MHD LES. Additionally, there is only a very limited dynamic range of scalesizes in the turbulence of the plasma sheet and to try to resolve only the larger scales of the turbulence and then model the effects of the smaller scales would prove impractical. This leaves the DNS technique, which resolves down into the dissipation scales. Unfortunately we do not know how to implement dissipation into MHD turbulence in collisionless plasams: it is likely that it involes some combination of (a) electron and ion Landau damping, (b) ion-cyclotron damping, (c) mode conversion to EMHD fluctuations, and (d) magnetic-field-line reconnection, all at spatial scales comparable to the ion gyroradius $r_{gi} = v_{Ti}/\omega_{ci}$ and/or the ion skin depth (ion inertial length) $\delta_i = c/\omega_{pi}$. Luckily, in Navier-Stokes fluids the properties of the turbulence (amplitudes, correlation times, large-eddy sizes, etc.) and consequences of the turbulence (drag, etc.) are largely insensitive to the details of the dissipation (cf. Fig. 5.9 and sect. 5.2 of Frisch (1995)). As a preliminary step we will hope this is also true for turbulence in MHD.

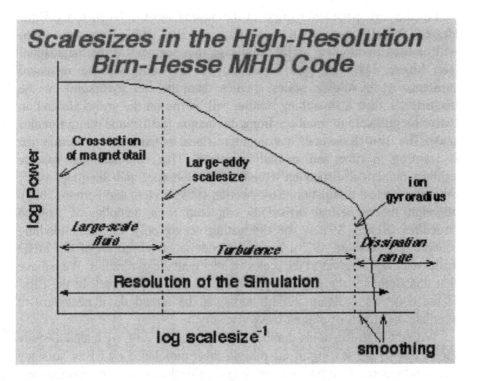

Figure 11. A conceptual sketch of the turbulence in the plasma sheet and the scalesizes that will be resolved by the modification of the Birn-Hesse 3-D MHD code to high Reynolds numbers.

Table 3. A comparison of the two simulation methods for studying plasma-sheet turbulence.

Issue to Be Studied	Code 1 3-D MHD	Code 2 Viscoelastic GOY
Physics of the Plasma-Sheet Turbulence		
effects of underlying tail geometry	good	bad
effects of boundaries	good	bad
effects of ionospheric dissipation	bad	good
spatial dynamic range	good	good
physics of driving	good	poor
effects of driving on the turbulence	moderate	good
Consequences of the Plasma-Sheet Turbulence		
quantifying eddy transport	good	bad
enabling of small-scale reconnection	good	bad
tangling of magnetotail field lines	good	bad
consequences for tail stability	good	bad

For the real-space simulations of the plasma-sheet turbulence, the Birn-Hesse 3-D MHD magnetotail code (Hesse and Birn, 1991; Birn et al., 1996) will be used to simulate MHD turbulence in a segment of the magnetotail (see Figure 11). The grid resolution of the code will be increased substantially to resolve scales smaller than the ion gyroradius in the magnetotail, then a smoothing routine will be run on the grid scale and on twice the gridscale to produce strong dissipation just beyond the gyroradius scale. The Birn-Hesse code has a computational scheme for convection that is leapfrog in time and spatially centered. This computational scheme exhibits numerical dispersion (Rood, 1987; Babarsky and Sharpley, 1997) but no numerical dissipation (Strickwerda, 1989; Periera and Periera, 2001), although the numerical dispersion can lead to a damping of isolated structures (Roach, 1972). The dissipation occurs only via the smoothing scheme (cf. Forester, 1977), hence the code is invicid at all of the MHD scales and hyperviscous at the scales smaller than the gyroradius. We believe that this simulates the physical properties of the plasma sheet. In the Birn-Hesse code, an ordinary viscous term can be turned on if necessary to simulate the effects of plasma-wave viscosity.

With the high-Reynolds-number simulation capability, we intend to study (see Table 3) the stirring of the plasma-sheet turbulence by BBFs, to study the production of turbulence by large-scale shears in the plasma-sheet convection, to study the effects of the magnetotail magnetic-field geometry on the MHD turbulence, and to study the effects of boundaries (the lobes and the LLBLs) on the properties of the turbulence, to obtain a picture of the magnetic-field-line tangling, and to discern the consequences of the turbulence for the enabling of small-scale reconnection events and on the stability of the magnetotail.

7.2   Viscoelastic GOY simulations

The second method of simulation is the viscoelastic GOY (Gledzer-Ohkitani-Yamada) technique. In standard GOY simulations (Schorghofer et al., 1995; Kadanoff et al., 1995; Ditlevsen and Mogensen, 1996), the turbulence spectra is represented on a one-dimensional grid in wavenumber k ($= |\vec{k}|$) space, which is used to keep track of the amount of turbulent energy residing at each wavenumber k. A numerical recipe is implemented which, at each timestep of the simulation, (a) shuffles the amount of energy around in k space, (b) dissipates some energy, and (c) pumps in some energy. The shuffling scheme is written to redistribute the energy on the grid in k space with the restriction that the scheme conserves quantities (such as energy, enstrophy, mean-squared vorticity....) that are conserved in the turbulence that is being simulated (e.g. 2D Navier-Stokes, 3D Navier-Stokes, etc.). Computational GOY schemes have been developed to simulate MHD turbulence (e.g. Gloaguen et al., 1985; Biskamp, 1995; Giuliani and Carbone, 1998; Jensen and Olesen, 1998; Frick et al., 2000; Ching et al., 2003). In GOY simulations, the temporal evolution of the turbulence spectra and the spectral properties of equilibrium can be studied. Two examples of GOY simulations are shown in Figures 12 and 13. In Figure 12 a series of 4 GOY simulations are run until they are in equilibrium. In each simulation, energy is pumped in at low k (large scales) and it cascades toward higher k where it is eventually dissipated at small scales. In each of the four simulations, the magnitude of the fluid viscosity v is changed; as labeled the values are v_o, $10^1 v_o$, $10^2 v_o$, and $10^3 v_o$. As can be seen, as v is increased the point where dissipation dominates over transport moves to lower k. This breakpoint in the spectrum, which separates the inertial subrange from the dissipation subrange, is known as the Kolmogorov scale or inner scale. Note in Figure 12 that the spectrum of turbulence below the Kolmogorov scale is little affected by the value of v.

In Figure 13 a series of 5 GOY simulations is run in which the functional form of the viscosity v is varied. Again, each simulation is run until equilibrium is reached. In the 5 simulations, the functional forms $v = v_o k^\alpha$ vary by the value of the exponent α: $\alpha = 0$ is normal viscosity, exponents $\alpha > 0$ are hyperviscosities (viscosity made stronger on small scales and weakened on large scales), and exponents $\alpha < 0$ are hypoviscosities (weakened on small scales and strengthened on large scales). Recall that magnetosphere-ionosphere coupling produces a hypoviscosity on the plasma sheet (Borovsky and Funsten, 2003), cyclotron-resonance damping produces a hyperviscosity (Stawicki et al., 2001). As can be seen in Figure 13, the shape of the spectrum is similar for the 3 different values of hyperviscosity, and these spectra differ from both the ordinary-viscosity spectrum and the hypoviscosity spectrum. The breakpoint in the spectrum is sharper for the hyperviscoisty cases that it is for the ordinary viscosity case and the breakpoint is more gradual for the hypoviscosity case than it is for the ordinary viscosity case. It is important to notice that hypoviscosity produces a spectrum that is affected for about 2 decades of k below the Kolmogorov scale.

To handle time-dependent viscosity and the persistence of the shearing stress, the GOY method must be generalized from a single 1-dimensional (wavenumber) grid to two 2-dimensional (wavenumber versus age) grids, one grid to handle time-delayed viscosity and the other grid to handle the persistence of stress, plus one 1-dimensional grid to handle the energy-shuffling calculations. The 3 grids are sketched in Figure 14. On the grids, a quantity U that is related to vorticity is kept track of. The primary grid is the 2-D "viscosity grid", where the information about how much energy at each

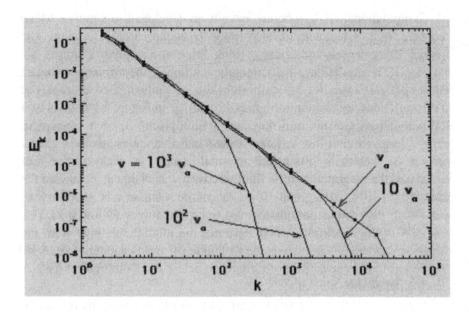

Figure 12. Four GOY simulations in which the magnitude of the viscosity is varied between simulations.

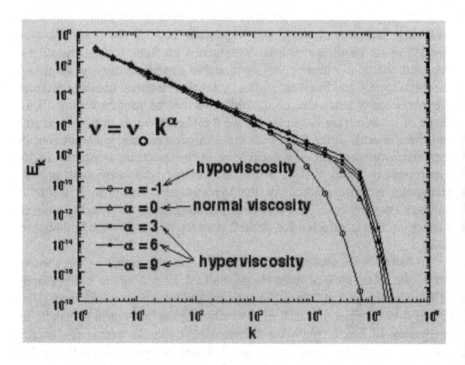

Figure 13. Five GOY simulations in which the functional form of the viscosity is varied between simulations.

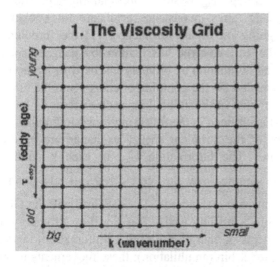

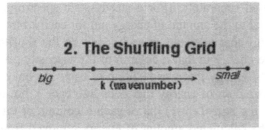

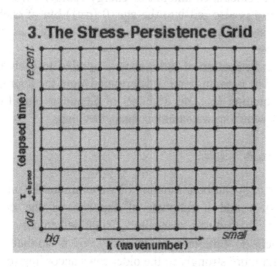

Figure 14. A sketch of the three different computational grids needed for viscoelastic GOY simulations. The indices for the grids are k,n, where the k index increases toward the right (toward higher k) and the n index increases downward (toward greater age or greater elapsed time).

wavenumber k and age τ_{age} is stored and manipulated. The 1-D "shuffling grid" and the 2-D "stress-persistence grid" are for bookkeeping. Several processes are going on in the turbulence: (1) the turbulent exchange of energy, (2a) the time-delayed dissipation by magnetosphere-ionosphere coupling, (2b) the dissipation by ordinary viscosity or hyperviscosity, (3) the stress persistence, and (4) the pumping of energy into the turbulence. During each timestep of the GOY simulation, each process is handled using the various grids as is outlined in the following four paragraphs.

Calculations of the turbulent energy exchange are handled by a GOY shuffling scheme. The information $U_{k,n}$ in each k column of the 2-D viscosity grid is summed over the age $U_k = \sum_{n=1}^{n\,max} U_{k,n}$ and then all of the U_k values are put onto a standard 1-D GOY grid and the GOY shuffling is calculated on this "shuffling grid". The shuffling of energy amongst the k bins has movements of energy into each k bin(creation) and movements of energy out of each k bin (annihilation); these movements in and movements out are calculated separately. (For a steady-state spectrum, the amount of energy in is equal to the amount of energy out for each k bin.) The calculated amount of energy that goes out of each k bin on the shuffling grid is then taken out of the energy in that k column of the viscosity grid, in proportion to the amount of energy that was residing in each age bin of the k column. The calculated amount of energy that goes into each k bin on the shuffling grid is put into the age=0 (n=1) bin of each k column of the viscosity grid. Additionally, the calculated amount of energy removed from each k bin of the shuffling grid is also put into the age=0 (n=1) bin of each k column of the stress-persistence grid, where it will be utilized as described two paragraphs below.

The viscous dissipation of turbulence is applied to each age row of the viscosity grid. The damping rate is determined from the appropriate terms in the vorticity equation (e.g. eq. (3.3.29) of Tennekes and Lumley (1972)) $\partial U/\partial t = \nu \nabla^2 U$, which in wavenumber space becomes

$$\partial U_{k,n}/\partial t = -\nu_{k,n} k^2 U_{k,n} ,\qquad(8)$$

where $\nu_{k,n} = \nu(k,\tau_{age})$ can be a function of wavenumber k (for hyperviscosity or hypoviscosity) and a function of the age of the energy τ_{age} (for viscoelasticity). The ordinary viscosity ν_0 is applied equally to each age row, and the viscoelastic time-delayed viscosity ν_{ve} is applied weakly to the young rows and more strongly to the older rows according to expression (5), which yields

$$\nu_{ve}(\tau_{age}) = \nu_{ve}(0) \left[1 - \exp(-\tau_{age}/\tau_{relax})\right]\qquad(9)$$

for viscoelastic damping in expression (8). Note in expression (9) that v_{ve} can have a k dependence (cf. eqs. (31a) and (31b) of Borovsky and Funsten (2003)). Upon completion of the dissipation, information is propagated downward on the viscous grid toward older age with the timestep Δt.

The stress persistence is calculated according to expression (6), which in k space is written

$$\partial U_{k,n}/\partial t = +v_{ve}(0)\, k^2\, (\Delta U)_{k,n}\, \exp\{-\tau_{elapsed}/\tau_{relax}\}\, . \tag{10}$$

In expression (10) the time dependence of v_{ve} has been taken out and written explicitly in the exponential term, but $v_{ve}(0)$ may still have a k dependence (cf. eqns. (31a) and (31b) of Borovsky and Funsten (2003)). When vorticity $(\Delta U)_k$ is removed from a k bin of the 1-D shuffling grid during a shuffle, that loss $|(\Delta U)_k|$ is added as a positive quantity into the 2-D stress-persistence grid in the k,1 location. With time the $(\Delta U)_{k,n}$ information is moved downward on the stress-persistence grid in the direction of increasing elapsed time. In expression (10) every location k,n on the grid has a $(\Delta U)_{k,n}$ value, a k value associated with the k location, and a τ_{elaped} associated with the n value. At each timestep Δt, each location on the grid produces a $\delta U_{k,n} = (\partial U_{k,n}/\partial t)\Delta t$ from expression (10). These $\delta U_{k,n}$ values represent vorticity that is restored by the stress-persistence property of viscoelasticity. These $\delta U_{k,n}$ values are added as positive quantities to the n=1 row of the viscosity grid.

The pumping of energy into the turbulence (usually at low k) is accomplished by adding random numbers to the age=0 (n=1) row of the 2D viscosity grid.

8.   FUTURE WORK

Our understanding of the MHD turbulence in the Earth's plasma sheet continues to increase. It is becoming clearer that this turbulence has very unusual properties.

Numerical simulations of the plasma-sheet turbulence are coming online. As the 3-D MHD simulations are pushed to higher resolution toward the goal outlined in section 7.1 of resolving scales smaller than the ion gyroradius some aspects of the turbulence are already being seen. One such aspect is the formation of vortices on the edges of bursty bulk flows. Viscoelastic GOY simulations with 2-D grids have commenced (as prescribed in section 7.2) and numerical-stability issues are being worked out.

Further exploration of the kinetic nature of Alfven waves in the magnetotail plasma sheet will be pursued. The plasma sheet, owing to its high-β and the restriction of scalesizes, is not a hospitable environment for Alfven waves: they tend to be highly dispersive and strongly damped. Not only are we finding that Alfven waves are not important for the dynamics of the turbulence, we are also finding that the transmission lines for magnetosphere-ionosphere coupling are resistive and very dispersive, adding a further complication to the picture of MHD turbulence in the plasma sheet.

ACKNOWLEDGMENTS

The author wishes to thank Elizabeth Antonova, Joachim Birn, Herb Funsten, Peter Gary, Bill Matthaeus, Anatoli Petrukovich, and Dave Suszcynsky for their help and to thank Sam Bame, Jack Gosling, Walter Lennartsson, and Chris Russell for satellite data. This work was supported by the NASA SR&T Program and the NSF GEM Program and was conducted with oversight from the U. S. Department of Energy.

REFERENCES

Alfven, H., and Falthammar, C.-G., 1963, *Cosmical Electrodynamics*, second edition, Oxford University Press, Clarendon.

Angelopoulos, V., Kennel, C. F., Coroniti, F. V., Pellat, R., Spence, H. E., Kivelson, M. G., Walker, R. J., Baumjohann, W., Feldman, W. C., Gosling, J. T., and Russell, C. T., 1993, Characteristics of ion flow in the quiet state of the inner plasma sheet, *Geophys. Res. Lett.* 20(16):1711–1714.

Angelopoulos, V., Mukai, T., and Kokubun, S., 1999, Evidence for intermittency in Earth's plasma sheet and implications for self-organized criticality, *Phys. Plasmas* 6(11):4161–4168.

Antonova, E. E., 1985, Nonadiabatic diffusion and equalization of concentration and temperature in the plasma layer of the magnetosphere of the Earth, *Geomag. Aeron.* 25:517.

Antonova, E. E., 1987, On the problem of fundamental harmonics in the magnetospheric turbulence spectrum, *Physica Scripta* 35(6):880–882.

Antonova, E. E., 2000, Large scale magnetospheric turbulence and the topology of magnetospheric currents, *Adv. Space Res.* 25(7/8):1567–1570.

Antonova, E. E., and Ovchinnikov, I. L., 1996a, Equilibrium of the turbulent current and the current layer of the Earth magnetosphere tail, *Geomag. Aeron.* 36(5):7–14.

Antonova, E. E., and Ovchinnikov, I. L., 1996b, Turbulent current sheets and magnetospheric substorms, in: *Proc. International Conference on Substorms (ICS-3)*, E. J. Rolfe and B. Kaldeich, eds., ESA SP-389, pp. 255–258.

Antonova, E. E., and Ovchinnikov, I. L., 1997, Current sheet with medium scale developed turbulence and the formation of the plasma sheet of Earth's magnetosphere and solar prominences, *Adv. Space Res.* 19(12):1919–1922.

Antonova, E. E., and Ovchinnikov, I. L., 1998, Quasi-three-dimensional model of an equilibrium turbulent layer in the tail of the Earth magnetosphere and its substorm dynamics, *Geomag. Aeron.* **38**:14.

Antonova, E. E., and Ovchinnikov, I. L., 1999a, Magnetostatically equilibrated plasma sheet with developed medium-scale turbulence: Structure and implications for substorm dynamics, *J. Geophys. Res.* **104**(A8):17289–17297.

Antonova, E. E., and Ovchinnikov, I. L., 1999b, Quasi-three dimensional modelling of the plasma sheet including turbulence on medium scales, *Adv. Space Sci.* **24**(1):121–124.

Antonova, E. E., and Ovchinnikov, I. L., 2000, Medium scale magnetospheric turbulence and quasi three-dimensional plasma sheet modeling, *Phys. Chem. Earth C.* **25**(1-2):35–38.

Antonova, E. E., Ganushkina, N. Y., and Ovchinnikov, I. L., 1996, The magnetostatic equilibrium in the magnetosphere and substorm plasma properties, in: *Physics of Auroral Phenomena, Proc. XIX Annual Seminar, Apatity*, Kola Science Center, Russian Academy of Science, pp. 7.

Antonova, E. E., Ovchinnikov, I. L., and Yermolaev, Y. I., 2002, Plasma sheet coefficient of diffusion: Predictions and observations, *Adv. Space Res.* **30**(12):2689–2694.

Antonova, E. E., 2003, Investigations of the hot plasma pressure gradients and the configuration of magnetospheric currents from INTERBALL, *Adv. Space Res.* **31**(5):1157–1166.

Babarsky, R. J., and Sharpley, R., 1997, Expanded stability through higher temporal accuracy for time-centered advection schemes, *Monthly Weather Rev.* **125**(6):1277–1295.

Bame, S. J., Asbridge, J. R., Felthauser, H. E., Glore, J. P., Paschmann, G., Hemmerlich, P., Lehmann, K., and Rosenbauer, H., 1978, ISEE-1 and ISEE-2 Fast Plasma Experiment and the ISEE-1 Solar Wind Experiment, *IEEE Trans. Geosci. Electron.* **16**:216.

Bauer, T. M., Baumjohann, W., Treumann, R. A., Sckopke, N., and Luhr, H., 1995, Low-frequency waves in the near-Earth plasma sheet, *J. Geophys. Res.* **100**(A6):9605–9617.

Bieber, J. W., Wanner, W., and Matthaeus, W. H., 1996, Dominant two-dimensional solar wind turbulence with implications for cosmic ray transport, *J. Geophys. Res.* **101**(A2):2511–2522.

Birn, J., Iinoya, F., Brackbill, J. U., and Hesse, M., 1996, A comparison of MHD simulations of magnetotail dynamics, *Geophys. Res. Lett.* **23**(4):323–326.

Biskamp, D., 1995, Scaling properties in MHD turbulence, Chaos *Solitions Fractals* **5**(10):1779–1793.

Borovsky, J. E., 1987, Limits on the cross-field propagation of streams of cold plasma, *Phys. Fluids* **30**(8):2518–2526.

Borovsky, J. E., and Bonnell, J., 2001, The dc electrical coupling of flow vortices and flow channels in the magnetosphere to the resistive ionosphere, *J. Geophys. Res.* **106**(A12):28967–28994.

Borovsky, J. E., and Funsten, H. O., 2003, The MHD turbulence in the Earth's plasma sheet: Dynamics, dissipation, and driving, *J. Geophys. Res.,* **108**(A7):1284, doi: 10.1029/2002JA009625.

Borovsky, J. E., Elphic, R. C., Funsten, H. O., and Thomsen, M. F., 1997, The Earth's plasma sheet as a laboratory for flow turbulence in high-beta MHD, *J. Plasma Phys.* **57**:1–34.

Borovsky, J. E., Thomsen, M. F., and Elphic, R. C., 1998, The driving of the plasma sheet by the solar wind, *J. Geophys. Res.* **103**(A8):17617–17639.

Bowling, S. B., 1975, Transient occurrence of magnetic loops in the magnetotail, *J. Geophys. Res.* **80**:4741–4745.

Ching, E. S. C., Cohen, Y., Gilbert, T., and Procaccia, I., 2003, Active and passive fields in turbulent transport: The role of statistically preserved structures, *Phys. Rev. E.* **67**(1):016304.

Coroniti, F. V., Scarf, F. L., Frank, L. A., and Lepping, R. P., 1977, Microstructure of a magnetotail fireball, *Geophys. Res. Lett.* 4:219–222.

Coroniti, F. V., Frank, L. A., Williams, D. J., Lepping, R. P., Scarf, F. L., Krimigis, S. M., and Gloeckler, G., 1980, Variability of plasma sheet dynamics, *J. Geophys. Res.* 85(NA6):2957–2977.

Corrsin, S., 1963, Estimates of the relations between Eulerian and Lagrangian scales in large Reynolds number turbulence, *J. Atmos. Sci.* 20:115–119.

Cranmer, S. R., and van Ballegooijen, A. A., 2003, Alfvenic turbulence in the extended solar corona: Kinetic effects and proton heating, *Astrophys. J.* 594(1):573–591.

Dahlburg, J. P., Montgomery, D., and Matthaeus, W. H., 1985, Turbulent disruptions from the Strauss equations, *J. Plasma Phys.* 34:1–46.

Ditlevsen, P. D, and Mogensen, I. A., 1996, Cascades and statistical equilibrium in shell models of turbulence, *Phys. Rev. E* 53(5):4785–4793.

Drell, S. D., Foley, H. M., and Ruderman, M. A., 1965, Drag and propulsion of large satellites in the ionosphere: An Alfven propulsion engine in space, *J. Geophys. Res.* 70:3131–3145.

Faber, T. E., 1995, *Fluid Dynamics for Physicists*, Cambridge University Press, New York.

Feynman, R. P., Leighton, R. B., and Sands, M., 1964, *The Feynman Lectures on Physics, Vol. II*, sect. 41-1, Addison-Wesley.

Forester, C. K., 1977, Higher order monotonic convective difference schemes, *J. Comp. Phys.* 23:1–22.

Frick, P., Boffetta, G., Giuliani, P., Lozhkin, S., and Sokoloff, D., 2000, Long-time behavior of MHD shell models , *Europhys. Lett.* 52(5):539–544.

Frisch, U., 1995, *Turbulence*, Cambridge University Press, New York.

Gary, S. P., 1986, Low-frequency waves in a high-beta collisionless plasma: Polarization, compressibility and helicity, *J. Plasma Phys.* 35:431–447.

Gary, S. P., 1993, *Theory of Space Plasma Instabilities*, Cambridge University Press, New York.

Gary, S. P., 1999, Collisionless dissipation wavenumber: Linear theory, *J. Geophys. Res.* 104(A4):6759–6762.

Gary, S. P., and Borovsky, J. E., 2004, Alfven-cyclotron fluctuations: Linear Vlasov theory, to appear in *J. Geophys. Res.* in press.

Giuliani, P., and Carbone, V., 1998, A note on shell models for MHD turbulence, *Eurphys. Lett.* 43(5):527–532.

Gloaguen, C., Leorat, J., Pouquet, A., and Grappin, R., 1985, A scaler model for MHD turbulence, *Physica D* 17:154–182.

Goertz, C. K., and Boswell, R. W., 1979, Magnetosphere-ionosphere coupling, *J. Geophys. Res.* 84(NA12):7239–7246.

Goertz, C. K., Shan, L.-H., and Smith, R. A., 1993, Prediction of geomagnetic activity, *J. Geophys. Res.* 98(A5):7673–7684.

Harris, J., 1977, *Rheology and Non-Newtonian Flow*, Longman, London.

Hasegawa, A., 1975, *Plasma Instabilities and Nonlinear Effects*, Springer-Verlag, New York.

Hayakawa, H., Nishida, A., Hones, E. W., and Bame, S. J., 1982, Statistical characteristics of plasma flow in the magnetotail, *J. Geophys. Res.* 87(NA1):277–283.

Hernandez-Jimenez, A., Hernandez-Santiago, J., Marias-Garcia, A., and Sanchez-Gonzalez, J., 2002, Relaxation modulus in PMMA and PTFE fitting by fractional Maxwell model, *Polymer Test.* 21(3):325–331.

Hesse, M., and Birn, J., 1991, Magnetosphere-ionosphere coupling during plasmoid evolution--1st results, *J. Geophys. Res.* 96(A7):11513–11522.

Hollweg, J. V., 1999, Kinetic Alfven wave revisited, *J. Geophys. Res.* 104(A7):14811–14819.

Hones, E. W., and Schindler, K., 1979, Magnetotail plasma flow during substorms: A survey with IMP 6 and IMP 8 satellites, *J. Geophys. Res.* **84**(NA12):7155–7169.

Hoshino, M., 2000, Small scale plasmoids in the post-plasmoid plasma sheet: Origin of MHD turbulence?, *Adv. Space Res.* **25**(7/8):1685–1688.

Hoshino, M., Nishida, A., Yamamoto, T., and Kokubun, S., 1994, Turbulent magnetic field in the distant magnetotail: Bottom-up process of plasmoid formation?, *Geophys. Res. Lett.* **21**(25):2935–2938.

Hoshino, M., Nishida, A., Mukai, T., Saito, Y., Yamamoto, T., and Kokubun, S., 1996, Structure of plasma sheet in magnetotail: Double-peaked electric current structure, *J. Geophys. Res.* **101**(A11):24775–24786.

Hossain, M., Gray, P. C., Pontius, D. H., Matthaeus, W. H., and Oughton, S., 1996, Is the Alfven-wave propagation effect important for energy decay in homogeneous MHD turbulence?, in: *Proceedings of the 8th International Solar Wind Conference*, AIP Conf. Proc. 382, 358–361.

Hruska, A., 1973, Structure of high-latitude irregular electron fluxes and acceleration of particles in the magnetotail, *J. Geophys. Res.* **78**:7509–7514.

Hruska, A., and Hruskova, J., 1970, Transverse structure of the Earth's magnetotail and fluctuations of the tail magnetic field, *J. Geophys. Res.* **75**:2449–2457.

Jensen, M. H., and Olesen, P., 1998, Turbulent binary fluids: A shell model study, *Physica D* **111**(1-4):243–264.

Joseph, D. D., 1990, *Fluid Dynamics of Viscoelastic Liquids*, sect. 7.6, Springer-Verlag, New York.

Kabin, K., and Papitashvili, V. O., 1998, Fractal properties of the IMF and the Earth's magnetotail field, *Earth Planets Space* **50**(1):87–90.

Kadanoff, L., Lohse, D., Wang, J., and Benzi, R., 1995, Scaling and dissipation in the GOY shell model, *Phys. Fluids* **7**(3):617–629.

Kraichnan, R. H., 1965, Inertial-range spectrum of hydromagnetic turbulence, *Phys. Fluids* **8**:1385–1387.

Leamon, R. J., Smith, C. W., Ness, N. F., Matthaeus, W. H., and Wong, H. K., 1998a, Observational constraints on the dynamics of the interplanetary magnetic field dissipation range, *J. Geophys. Res.* **103**(A3):4775–4787.

Leamon, R. J., Matthaeus, W. H., Smith, C. W., and Wong, H. K., 1998b, Contribution of cyclotron-resonant damping to kinetic dissipation of interplanetary turbulence, *Astrophys. J.* **507**(2):L181–L184.

Leamon, R. J., Smith, C. W., Ness, N. F., and Wong, H. K., 1999, Dissipation range dynamics: Kinetic Alfven waves and the importance of β_e, *J. Geophys. Res.* **104**(A10):22331–22344.

Leamon, R. J., Matthaeus, W. H., Smith, C. W., Zank, G. P., Mullan, D. J., and Oughton, S., 2000, MHD-driven kinetic dissipation in the solar wind and corona, *Astrophys. J.* **537**(2):1054–1062.

Maltsev, Y. P., Lyatsky, W. B., and Latskaya, A. M., 1977, Currents over the auroral arc, *Planet. Space Sci.* **25**:53–57.

Matthieu, J., and Scott, J., 2000, *An Introduction to Turbulent Flow*, ch. 8, Cambridge University Press, New York.

Matthaeus, W. H., Goldstein, M. L., and Roberts, D. A., 1990, Evidence for the presence of quasi-two-dimensional nearly incompressible fluctuations in the solar wind, *J. Geophys. Res.* **95**(A12):20673–20683.

Milovanov, A. V., Zelenyi, L. M., and Zimbardo, G., 1996, Fractal structures and power law spectra in the distant Earth's magnetotail, *J. Geophys. Res.* **101**(A9):19903–19910.

Montgomery, D., 1987, Remarks on the mhd problem of generic magnetospheres and magnetotails, in: *Magnetotail Physics*, A. T. Y. Lui, ed., Johns Hopkins University Press, Baltimore, pp. 204–204.

Montgomery, D., 1989, Magnetohydrodynamic turbulence, in: *Lecture Notes on Turbulence*, World Scientific, Singapore, pp. 75–170.

Nakamura, R., Baker, D. N., Fairfield, D. H., Mitchell, D. G., McPherron, R. L., and Hones, E. W., 1994, Plasma flow and magnetic field characteristics near the midtail neutral sheet, *J. Geophys. Res.* **99**(A12):23591–23601.

Neagu, E., Borovsky, J. E., Thomsen, M. F., Gary, S. P., Baumjohann, W., and Treumann, R. A., 2002, Statistical survey of magnetic field and ion velocity fluctuations in the near-Earth plasma sheet: Active Magnetospheric Particle Trace Explorers/Ion Release Module (AMPTE/IRM) measurements, *J Geophys. Res.* **107**(A7):1098.

Ovchinnikov, I. L., Antonova, E. E., and Yermolaev, Y. I., 2000, Determination of the turbulent diffusion coefficient in the plasma sheet using the Project INTERBALL data, *Comic Res.* **38**(6):557–561.

Ovchinnikov, I. L., Antonova, E. E., and Yermolaev, Y. I., 2002, Turbulence in the plasma sheet during substorms: A case study for three events observed by the INTERBALL Tail Probe , *Cosmic Res.* **40**(6):521–528.

Periera, J. M. C., and Periera, J. C. F., 2001, Fourier analysis of several finite difference schemes for the one-dimensional unsteady convection-diffusion scheme, *Int. J. Numer. Meth. Fluids* **36**:417–439.

Petrukovich, A. A., and Yermolaev, Y. I., 2002, Interball-Tail observations of vertical plasma motions in the magnetotail, *Ann. Geophys.* **20**(3):321–327.

Pisarenko, N. F., Budnik, E. Y., Ermolaev, Y. U. I., Kirpichev, I. P., Lutsenko, V. N., Morozova, E. I., and Antonova, E. E., 2003, The main features of the ion spectra variations in the transition region from dipole to tailward streched field lines, *Adv. Space Res.* **31**(5):1347–1352.

Pope, S. B., 2000, *Turbulent Flows, Part Two: Modelling and Simulation*, Cambridge University Press, New York.

Reiner, M., 1958, Rheology, in: *Handbuch der Physik, Vol. VI*, S. Flugge, ed., Springer-Verlag, Berlin, pp. 434.

Roach, P. J., 1972, *Computational Fluid Dynamics*, sect. III-A-6, Hermosa Publishers, Albuquerque.

Rood, R. B., 1987, Numerical advection algorithms and their role in atmospheric transport and chemistry models, *Rev. Geophys.* **25**(1):71–100.

Russell, C. T., 1978, The ISEE 1 and 2 fluxgate magnetometers, *IEEE Trans. Geosci. Electron.* **16**:239–242.

Scholer, M., 1970, On the motion of artificial ion clouds in the magnetosphere, *Planet. Space Sci.* **18**:977–1004.

Schorghofer, N., Kadanoff, L., and Lohse, D., 1995, How the viscous subrange determines inertial range properties in turbulence shell models, *Physica D* **88**(1):40–54.

Sergeev, V. A., and Lennartsson, W., 1988, Plasma sheet at X ~ 20 R_E during steady magnetospheric convection, *Planet. Space Sci.* **36**(4):353–370.

Shelley, E. G., Sharp, R. D., Johnson, R. G., Geiss, J., Eberhardt, P., Balsiger, H., Haerendel, G., and Rosenbauer, H., 1978, Plasma Composition Experiment on ISEE-A, *IEEE Trans. Geosci. Remote Sensing* **16**:266–270.

Stawicki, O., Gary, S. P., and Li, H., 2001, Solar wind magnetic fluctuation spectra: Dispersion versus damping, *J. Geophys. Res.* **106**(A5):8273–8281.

Strikwerda, J. C., 1989, *Finite Difference Schemes and Partial Differential Equations*, ch. 5, Wadsworth & Brooks, Pacific Grove, California.

Tennekes, H., and Lumley, J. L., 1972, *A First Course in Turbulence*, MIT Press, Cambridge.

Troshichev, O., Kokubun, S., Kamide, Y., Nishida, A., Mukai, T., and Yamamoto, T., 1999, Convection in the distant magnetotail under extremely quiet and weakly disturbed conditions, *J. Geophys. Res.* **104**(A5):10249–10263.

Troshichev, O. A., Antonova, E. E., and Kamide, Y., 2002, Inconsistency of magnetic field and plasma velocity variations in the distant plasma sheet: Violation of the "frozen-in" criterion?, *Adv. Space Res.* **30**(12):2683–2687.

Tsurutani, B. T., Jones, D. E., Slavin, J. A., Sibeck, D. G., and Smith, E. J., 1984, Plasmasheet magnetic fields in the distant tail, *Geophys. Res. Lett.* **11**(10):1062–1065.

Veltri, P., Zimbardo, G., Taktakishvili, A. L., and Zelenyi, L. M., 1998, Effect of magnetic turbulence on the ion dynamics in the distant magnetotail, *J. Geophys. Res.* **103**(A7):14897–14910.

White, W. W., Schoendorf, J. A., Siebert, K. D., Maynard, N. C., Weimer, D. R., Wilson, G. L., Sonnerup, B. U. O., Siscoe, G. L., and Erickson, G. M., 2001, MHD simulation of magnetospheric transport at the mesoscale, in: *Space Weather*, P. Song, H. Singer, and G. Siscoe, eds., American Geophysical Union, Washington.

Yermolaev, Y. I., Petrukovich, A. A., Lelenyi, L. M., Antonova, E. E., Ovchinnikov, I. L., and Sergeev, V. A., 2000, Investigation of the structure and dynamics of the plasma sheet: The CORALL experiment of the INTERBALL project, *Cosmic Res.* **38**(1):13–19.

Yermolaev, Y. I., Petrukovich, A. A., and Zelenyi, L. M., 2002, INTERBALL statistical study of ion flow fluctuations in the plasma sheet , *Adv. Space Res.* **30**(12):2695–2700.

Zelenyi, L. M., Milovanov, A. V., and Zimbardo, G., 1998, Multiscale magnetic structure of the distant tail: Self-consistent fractal approach, in: *New Perspectives on the Earth's Magnetotail*, A. Nishida, D. N. Baker, and S. W. H. Cowley, eds., American Geophysical Union, Washington, pp. 321.

COLD IONOSPHERIC IONS IN THE EXTERNAL DAYSIDE MAGNETOSPHERE

Jean-André Sauvaud and Pierrette Décréau

CESR/CNRS, Toulouse, France and LPCE/CNRS, Orléans, France

Abstract: During periods of quiet magnetic activity, a cold plasma layer with densities reaching 1-3 cm^{-3} is encountered on the magnetospheric side of the dayside magnetopause. Direct density measurements from the plasma frequency indicate that this layer can have a width exceeding 1 R_E in the direction normal to the magnetopause. Plasma composition measurements indicate that the major detected ions are H^+, He^+ and O^+. These cold ionospheric ions show a repetitive pattern of energy changes. While the magnetopause is approaching the satellites, their energy increases from below the detector low-energy threshold up to about 100 eV for protons. After the passage of the satellites into the magnetosheath and just following their re-entry into the magnetosphere, the ion energy decreases from about 100 eV for protons down to the lowest detectable energy. This behavior is interpreted as the effect of the electric field associated with the magnetopause motion. The ion motion is set up when the magnetopause is compressed and relaxed when the boundary is going out. Altogether the measurements clearly show that there are hidden plasma populations inside the dayside magnetosphere, at least during quiet geomagnetic conditions. This paper emphasizes the importance to use the determination of the plasma frequency to probe the magnetospheric density. The use of biased low-energy particle detectors located far enough from the satellite body should allow to probe the distribution function of these low-energy ions in future missions.

Key words: thermal plasma; magnetopause; pressure pulses.

1.   INTRODUCTION

After the discovery of the polar wind by Brinton et al. (1971) and Hoffman et al. (1974) and of the escape of keV ionospheric ions from the auroral region by Shelley et al. (1972, 1976), the consensus view was that the ionosphere is an important source of magnetospheric plasma. The fact

J. –A. Sauvaud and Z. Němeček (eds.),
Multiscale Processes in the Earth's Magnetosphere:From Interball to Cluster, 255-273.
© 2004 *Kluwer Academic Publishers. Printed in the Netherlands.*

that the ionosphere can be the dominant source in both quiet and active magnetic conditions has been first proposed by Chappel et al. (1987) using the available data at that time. However, many questions were and are still open. Primarily in the areas of measuring the transport of plasma through the magnetosphere with a suitably designed set of spacecraft and tracing the steps of particle energization in the ionosphere and the external magnetosphere. Until recently, mass spectrometer flown in the outer magnetosphere have generally only been able to measure a limited portion of the total solid angle. These limitations have been overcome in the frame of the Fast, Polar, Equator-S and Cluster projects; the ion mass and energy distribution functions being measured over $\sim 4\pi$ (see Moore et al., 1995; Shelley et al., 1995; Rème et al. 1997, 2001; Möbius et al., 1998, Carlson and McFadden, 1998, McFadden and Carlson, 1998). Another limitation was the uncontrolled positive satellite potential. A spacecraft in sunlight usually undergoes positive charging owing to photo-electron emission from its sunlit surfaces; this charging prevent positive ions with energies below the spacecraft potential from reaching the detectors. Furthermore photo-electrons with energies lower than the satellite potential return to the satellite body, preventing the electron detectors to measure the natural cold electron population. For the Polar and Cluster projects; the satellite electric potential was reduced using respectively a plasma source (Moore et al., 1995) and an ion emitter (Riedler et al., 1997). However, the spacecraft potential still reaches several volts in the outer magnetosphere which generally precludes the direct measurement of very cold plasma using particle detectors. The satellite electric potential limitation can however be overcome for special events. The difficulty of measuring cold ions disappears if a spacecraft enters the shadow of the Earth, because the vehicle undergoes negative charging in darkness and will detect ambient cold positive ions attracted by the spacecraft's negative charge. Using ion observation during Geotail passages through the solar umbra in 1995-1998, Seki et al. (2003) were able to show the unexpected existence of a cold-ion population in the hot plasma sheet. This cold plasma population, with a density comparable to that of the hot population carries a significant fraction of the mass density, and the effects of its inertia are not negligible when considering the dynamics of large scale phenomena in the Earth's magnetosphere. The Geotail ion spectrometer being not able to separate the ions species, this cold population can consist of H^+ and/or O^+. As stated by Seki et al. (2003), if a non negligible part of the high-density cold population consists of oxygen, this suggests an ionospheric source. On the other hand, if the cold population consists of H^+, the source can be either the ionosphere or the solar wind; the low temperature of the population makes an ionospheric source more likely.

Another way to detect very low-energy ions from a charged spacecraft is presented in this paper. Here we take advantage of boundary motions allowing the low-energy population to reach drift velocities/energies higher than the satellite potential to:

- Present evidence for the existence of a cold ion population in a region adjacent to the dayside magnetopause with densities much higher (up to a factor 10) than that of the hot dayside plasma population. In complement to a previous work by Sauvaud et al. (2001), we will show that this layer can have a width of the order of 1 R_E and that the appearance/disappearance of measurable fluxes of cold ions is due to magnetopause motions; plasma frequency measurements showing that the cold population is present even when not detected by particle spectrometers.
- Establish the ionospheric origin of these ions relying on the energy separation of different cold ion species drifting with the same velocity, and using the Cluster mass spectrometer.

We experimentally show that this layer of cold H^+, He^+ and O^+ ionospheric ions is detected during periods of low magnetic activity, which imply a weak convection electric field, and use our observations made inside the magnetosphere to study the magnetopause motions giving the ions enough energy to become detectable by the Cluster CIS experiment. We show in particular that inside the magnetosphere, the cold ions are responding to magnetopause local outward acceleration or to inward/outward motions linked to magnetosheath pressure increases/decreases in the minute range.

2.   DATA SOURCES

The particle data used in this paper come from the Cluster Ion Spectrometer (CIS) experiment which comprises (Rème et al., 1997; 2001), (i) a Hot Ion Analyser, CIS-2, measuring the ion distributions from 5 eV to 26 keV by combining a classical symmetrical quadri-spherical analyser with a fast particle imaging system based on micro-channel plate electron multipliers and position encoding discrete anodes (Carlson et al., 1982); (ii) a time of flight mass spectrometer, CIS-1, which combines a top-hat analyser with an instantaneous 360° x 8° field of view with a time of flight section to measure the complete 3D distribution functions of the major ion species. Typically these include H^+, He^{++}, He^+ and O^+. The sensor primarily covers the energy range between 0.02 and 38 keV/q. The plasma density is deduced from the WHISPER experiment. The Whisper technique is based on the identification of the electron plasma frequency by analyzing the pattern of resonances triggered in the medium by a pulse transmitter. The central frequency of the pulse, a short sinusoidal wave-train, steps in the frequency range of 4–80 kHz. The total electron density can be derived unambiguously by the sounder in most magnetospheric regions, provided it is in the range of 0.25 to 80 cm^{-3} (see Décréau et al., 1997; 2001, for a complete description of the experiment). The magnetic field data used in this paper come from the two fluxgate magnetometers (FGM) installed onboard each Cluster spacecraft which measure the three components of the magnetic field with an accuracy approaching 0.1 nT (Balogh et al., 2001). These data are averaged over 4 seconds. The electric potential of each satellite is controlled by the ASPOC instrument that lowers the satellite potential by emitting a current of Indium ions (Riedler et al., 1997; Torkar et al., 2001). However, for the cases presented in this paper, ASPOC was off; the satellite potential varied between +10 and +15 volts inside the magnetosphere and reached about +4-6 volts in the magnetosheath.

3.   OBSERVATIONS

3.1   Evidence for cold ionospheric ions in a layer adjacent to the magnetopause

During geomagnetically very quiet times, in association with magnetopause motion, the Cluster spacecraft unexpectedly encounters at high latitudes cold ionospheric plasma (H^+, He^+, O^+) in an extended layer adjacent to the magnetopause. We will first characterize this layer and then use these observations to study the magnetopause small-scale motions at the origin of the detection of the cold plasma. All the observations presented in this paper were performed in the high-latitude regions of the post-noon magnetosphere, southward of the cusp. The local Bz component of the magnetic field was always positive. Moreover, the IMF Bz was positive. Figure 1 give the variations of the Auroral Electrojet indices corresponding to the three cases presented here. For all three days, the auroral activity was weak. According to Kamide and Baumjohann (1985) and Richmond (1990), the corresponding cross-tail potential is small, on the order of 10-20 kV. A first example of the encounter of enhanced plasma densities is given in Figure 2 which pertains to an outbound pass of Cluster from the dayside high-latitude boundary layer up to the magnetosheath in the post noon sector on January 30, 2002.

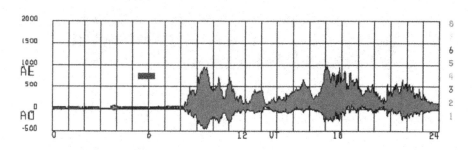

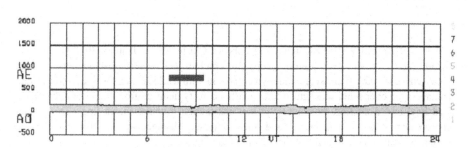

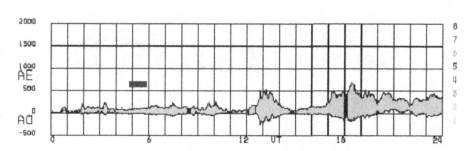

Figure 1. Variations of the preliminary AE and AO auroral indices during three days when Cluster measured cold ionospheric ions in a layer adjacent to the magnetopause. The measurement periods are indicated by an horizontal red bar.

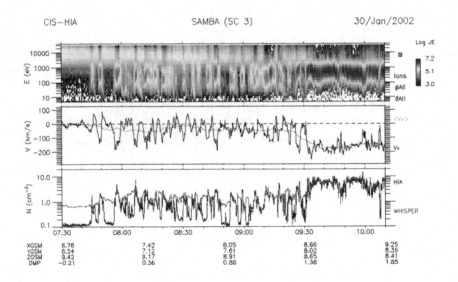

Figure 2. HIA and WHISPER data taken onboard Cluster #3 on January 30, 2002 between 07:30 and 10:15 UT. From top to bottom: ion energy-time spectrogram (keV/(cm^2·s·ster·keV)) antisolar velocity component with a 12 seconds resolution (black line) and averaged over 6 minutes (green line), plasma density deduced from the plasma frequency (red line) and from first moment of the ion distribution function (black line). The satellite coordinates X$_{GSM}$, Y$_{GSM}$, Z$_{GSM}$ and the computed distance to an average Shue-97 magnetopause model are given at the bottom of the Figure.

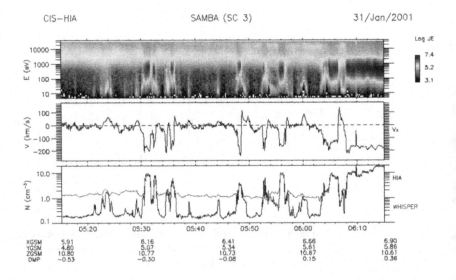

Figure 3. January 31, 2001 from 05:15 to 06:15 UT. Same presentation as Figure 2, except for the averaged antisolar velocity component of the ions which is not shown.

Figure 2 displays from top to bottom: (i) the spin averaged ion energy time spectrogram, (ii) the plasma antisolar velocity component and, (iii) the plasma densities obtained from different means; from the computation of the first moment of the ion distribution function (HIA) and from the electron plasma frequency (WHISPER). The ion energy-time spectrograms shows successive encounters of the dayside plasma sheet and of boundary layer plasma. The plasma sheet is characterized by high-energy fluxes at energies of about 8-10 keV. At the beginning of the time period, between 07:30 and 07:37 UT, only plasma sheet ions are detected. At the end of the time period, after 09:30 UT, the satellite is inside the magnetosheath. The anti sunward velocity reaches here values of the order of -150 km/s and the plasma density is of the order of 5-10 cm^{-3}. Note that inside he magnetosheath HIA and WHISPER give very similar densities. During the period from 07:37 to ~ 09:30 UT, the satellite passes alternatively inside the dayside plasma sheet and inside the high-latitude boundary layer, probably due to magnetosheath plasma pressure variations. In average, inside the boundary layer, the maximum density given by the Whisper experiment is lower than inside the magnetosheath, as is the plasma antisolar velocity given by HIA. A striking feature is that the densities deduced from the HIA ion spectrometer and from the electron plasma frequency are similar every time the satellite is inside the dayside high-latitude boundary layer (plasma temperature about 2 keV) and strongly differ every time the satellite is inside the dayside plasma sheet. Here, their ratio can reach values as high as 10 (09:02 UT). The Whisper experiment alone indicate that over a distance range extending from –0.21 to 1.38 R_E to the model magnetopause of Shue et al. (1997), i.e., over more than 10,000 km, the plasma density is high, varying around 1-2 cm^{-3}. HIA data on the contrary indicate that the density is strongly varying, over one order of magnitude, with its lower values, measured inside the dayside plasma sheet, of the order of 0.1 cm^{-3}. A simple interpretation of this discrepancy could be the existence of a hidden cold plasma population in the dayside plasma sheet which energies lower than the satellite potential. More evidence for this interpretation is given in Figure 3 which provides another example, on January 31, 2001, of such discrepancies between the plasma density computed from the HIA ion spectrometer and directly deduced from the plasma frequency. The presentation of the data is similar to that of Figure 2. On January 31, 2001, Cluster encountered the magnetopause several time in the post noon sector of the high-latitude magnetosphere during the time period 05:22 to 06:02 UT. The Cluster satellites #1 and 3 were providing data on that day. Because their separation distance was less than 600 km close to the magnetopause and the time resolution to obtain a proton 3D distribution was 12 seconds, no direct estimation of time delays for boundary encounters can be obtained. Note that the satellite #3 definitively enters inside the magnetosheath around 06:04 UT. Here the plasma velocity is in the anti solar direction and reaches ~ -170 km/s and the plasma densities given by both the plasma frequency and the HIA spectrometer reach identical values of about 10 cm^{-3}. Starting around 05:30 UT, several encounters with the magnetosheath/boundary layer can be identified. Every one corresponds to high anti-solar velocities and plasma densities of about 10 cm^{-3}. The value of these parameters being very close to those measured later in the magnetosheath, this tend to indicate that the satellite sporadically enters inside the magnetosheath. During these events, both HIA and Whisper give very similar values of the density. This is not the case between two magnetosheath/boundary layer encounters were they generally differ by a factor ~5, the HIA density being systematically lower than that deduced from the plasma frequency. There are however several short time periods when the satellite is inside the magnetosphere and the two densities are close to each other. This is true around 05:23, 05:44, 05:49, 05:54, 05:57, 06:00, 06:04 UT, when HIA detects a low-energy plasma which either appear as an "inverted V" in the energy time spectrogram (05:23 and 06:00 UT), or as a "foot", either following the retreat of the magnetosheath/boundary layer (05:49, 05:54 UT) or preceding its encounter with the satellite (06:04 UT). Such small structures are difficult to distinguish in Figure 2, but are clearly seen on a detailed energy time spectrogram. This is made clear by the data presented in Figure 4 which, like Figure 2, pertains to January 30, 2002. However, here the data are shown with a better resolution, between 07:44 and 07:52 U. The top panel presents the spin averaged energy time spectrogram. The two following panels respectively show the variation of the perpendicular velocity of the ions and their antisolar parallel velocity. The bottom panel gives the variations of the density deduced from HIA and Whisper. The satellite is located inside the dayside plasma sheet (E > 3 keV) and encounters two time the high-latitude boundary layer/magnetosheath. During the first event, around 07:45 UT, the satellite clear stays inside the magnetosphere as indicated by the very small value of the plasma antisolar parallel velocity. During the second event, around 07:50 UT, the plasma velocity becomes much higher, indicating that the satellite could have reach the magnetosheath for a while. Just before (~07:48:12 UT) and just after (~ 07:50:30 UT) this second encounter with solar plasma, the energy-time spectrogram shows foots of low-energy plasma (~10 eV to 100 eV) whose energy increases towards the center of the main structure. Inside these foots, the densities measured by the ion spectrometer and by the Whisper experiment are nearly the same (~1-2 cm^{-3}). A similar foot is easily distinguishable after the first event, around 07:46:30 UT. Note that during the appearance/disappearance of these foots, the plasma perpendicular velocity is highly increasing/decreasing.

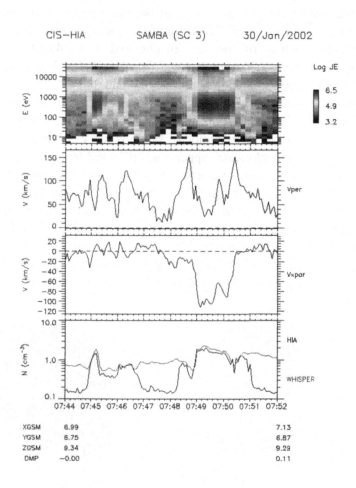

Figure 4. A close-up of the HIA and WHISPER measurements performed between 07:44 and 07:52 UT on January 30, 2002 (see Figure 2). From top to bottom: the ion energy time spectrogram, the ion perpendicular velocity, the ion antisolar parallel velocity, the plasma densities as deduced from the plasma frequency (red line-WHISPER) and from the ion 3D distribution (black line-HIA).

To summarize these observations, a low-energy ion population becomes detectable just before the arrival at the satellite of the boundary layer/magnetosheath solar plasma and just after the disappearance of this boundary layer/magnetosheath plasma from the satellite location; disappearance probably due to an outward motion of the magnetopause. The density of this low-energy plasma component is high and reliably measured by both HIA and Whisper. The motion of the low-energy component which is only detected during a brief period (~1 minute) preceding/following the motion of the outer boundary of the magnetosphere, is mainly perpendicular to the magnetic field (second panel of Figure 4).

The characteristics of the motion of the ion forming these "foots" have been determined using their 3D distribution function. Figure 5 displays the distribution function of the ion encountered at 07:46:24 UT in two perpendicular planes, ($V_{//}$ and $V_\perp$) and ($V_{\perp 1}$, $V_{\perp 2}$). The left part is for the complete distribution function between 5 eV and 38 keV. The plasma sheet appears as an isotropic high-energy component. Close to the center of the distribution, bright small spots refer to the low-energy component. The distribution given on the right side of figure is for V < 400 km/s. Here the low-energy component clearly appears as cold ions drifting perpendicularly to the magnetic field with a velocity lower than 200 km/s. Since the ions are cold (small ion Larmor radii) and the **E×B** velocity is large compared to their thermal velocity, they appear at the same phase in the spin, i.e., for a given value of $V_{\perp 1}$. The distribution is taken in the middle of the foot displayed in Figure 5. The observed dynamical behavior of the foot, i.e., the decrease in energy due to a slow shift in perpendicular energy/velocity towards very low values, is strongly indicative of the adiabaticity of the motion of the ions.

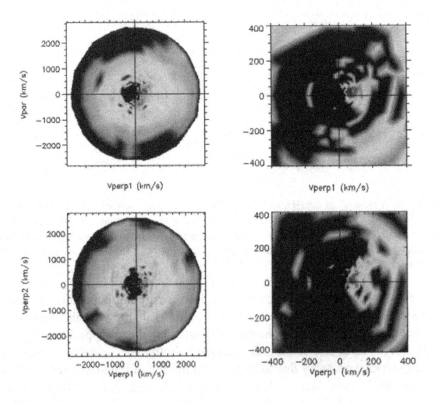

Figure 5. Left: Ion distribution function ($V_{//}$, $V_\perp$) and ($V_{\perp 1}$, $V_{\perp 2}$) for the total velocity range measured by the HIA ion spectrometer. Right: A close-up for velocities lower than 400 km/s. The ($V_{//}$, $V_\perp$) plane contains the onboard computed bulk velocity. The data are taken at 07:46:24 on January 30, 2002 (see Figure 4).

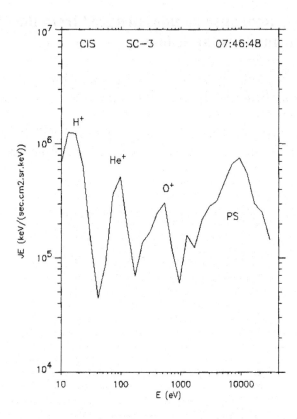

Figure 6. Ion energy spectra taken in the plasma flow direction by the non-mass resolving HIA spectrometer at 07:46:48 on January 30, 2002 (see Figure 4). Note the appearance of peaks corresponding to H^+, He^+ and O^+ with a velocity of ~ 60 km/s, i.e., an electric field of ~ 1.2 mV/m. The H^+ temperature estimated from a maxwellian fit is close to 10 eV.

The important question pertaining to the origin of the very low-energy component still remain unanswered at this stage.

Figure 6, showing the ion energy spectrum taken at 07:46:48 UT inside the foot following the first structure displayed in Figure 4, gives the answer. The ion energy spectrum clearly indicates that the cold ion population is a mixture of H^+, He^+ and O^+ ions. The masses are here translated in energies owing to the common drift velocity of the ions. This result, sustained by direct O^+ measurements by the mass spectrometer, demonstrates the ionospheric origin of the cold ion layer adjacent to the magnetopause. We thus arrive to the first conclusion that the low-energy plasma gains a drift velocity under the influence of the magnetopause motion and is originating from the ionosphere. The disappearance of this plasma from the ion spectrometer energy range corresponds to periods when the magnetopause is steady, or too far from the satellite to induce a drift high enough to allow the ions to overcome the satellite potential.

3.2   Magnetopause motions deduced from the dynamics of cold ionospheric ions

The ion drift velocity associated with inward/outward magnetopause motions is inward/outward directed, as evidenced by the low-energy plasma acceleration displayed in Figure 4.

For weak separation distances of the Cluster spacecraft in 2001 and 2002 (Figures 2 and 3), almost no time delay can be found from plasma measure-

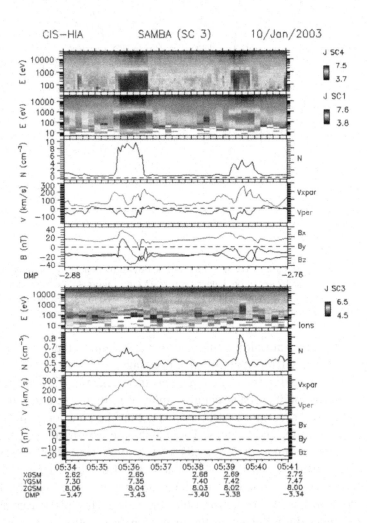

Figure 7. Ion measurements made onboard satellites # 4, #1 and #3 during briefs encounters of the high-latitude boundary layer on January 10, 2003 between 05:34 and 05:41 UT. From top to bottom: SC-4, energy time spectrogram; SC-1, energy time spectrogram, ion density, perpendicular velocity (red line) and antisolar parallel velocity (black line), GSM magnetic field components (Bx-black line, By blue line, Bz-red line); SC-3, same presentation as for SC-1.

ments. This is due to the CIS time resolution which is at best equal to the satellite spin period, i.e., 4 seconds.

In 2003, the separations between the Cluster satellites were higher and it happened that one satellite was still in the magnetosphere while the others were sporadically inside the dayside boundary layers/magnetosheath. This is exemplified in Figure 7 which displays ion measurements performed simultaneously onboard satellites 4, 1 (top) and 3 (bottom) during a brief time period, from 05:34 to 05:31 UT on January, 10, 2003. Satellite 4 was the closest to the magnetopause. According to the Shue-97 model, it was located 2.80 R_E inside the magnetosphere. Satellite 3 was the deepest inside the magnetosphere, 3.40 R_E from the magnetopause. Thus satellite 1 and 3 were separated by ~ 3800 km along the magnetopause normal. Satellite 4 is in between. Satellites 1 and 4 encountered two times the dayside high-latitude boundary layer/magnetosheath. These regions are characterized by a plasma with a high density, respectively ~10 and 3 cm^{-3}, a high plasma velocity and quite strong changes in the magnetic field components. During the first boundary layer encounter, the B_Y component of the magnetic field reverses. Located 3800 km deeper inside the magnetosphere, satellite 3 did not encounter the high-latitude boundary layer/magnetosheath for more than 4 seconds. This satellite measures a low energy near mono-energetic population showing clear energy enhancements related to the motion of the outer boundary of the dayside magnetosphere as probed by satellite 1 and 4. Only during a very brief period, at 05:39:37 UT, satellite 3 encounter a plasma with the properties of the boundary layer. The low-energy plasma composition obtained from the energy separation of the ions with different masses indicates that the ions forming the low energy, quasi mono-energetic population are predominantly H^+ with a tenuous O^+ component (not shown). Thus, satellite 3 measures ionospheric cold plasma put in drift by the magnetopause motions. Note that the parallel velocity of this low-energy component is almost zero. Onboard satellite 3, during the first event, the ion perpendicular velocity component increases up to 300 km/s, while during the second event the perpendicular velocity stays below 200 km/s. In order to examine the plasma flow in a physical coordinate system, we used the magnetic field variations from the magnetosphere to the magnetosheath (encountered after 06:30 UT), to perform a minimum variance analysis and to compute the magnetopause normal. The vector normal to the magnetopause (X_{GSE}= -0.57, Y_{GSE}= -0.57, Z_{GSE}= -0.57) is used together with the vector L, directed along the magnetospheric B field, and the vector M located, inside the magnetopause plane, to display the plasma velocity components in the L, M, N system for the event registered between 05:34:00 and 05:37:30 UT (Figure 8).

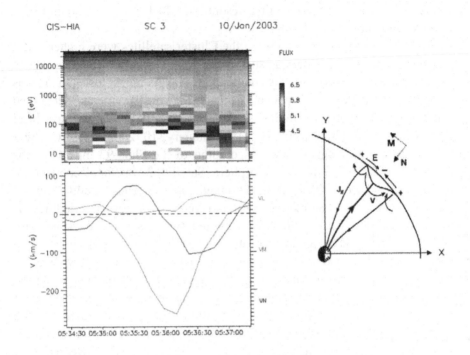

Figure 8. Left: Ion energy-time spectrogram between 05:34:30 and 05:37:30 on January 10, 2003 and the V_L, V_M, V_N ion velocity components. N is normal to the magnetopause, L is along the magnetospheric B field and M is in the magnetopause plane. Right: Deformation of the magnetopause by a pressure pulse, associated electric field, plasma flows and currents (adapted from Lysak et al.(1994)).

The V_L, V_M, V_N components of the velocity show clear variations. V_L stays about constant. V_N shows a bipolar variation, first positive and then negative. V_M has a large negative variation. As indicated in the right panel Figure 8, adapted from Lysak et al. (1994), the plasma velocity along the normal, V_N, first inward directed and then outward directed, varies as expected from a local compression of the magnetopause skimming anti sunward. Note that the inward velocity region is on the antisunward side of the pulse, i.e., at earlier local times. From the model, we expect the flow between the inward and outward velocities regions to be driven azimuthally towards noon. The measured large positive V_M velocity correspond to this flow. As stated by Lysak (1994), from an electrodynamic point of view, the inward and outward flows are accompanied by dawnward and duskward directed azimuthal electric fields. Thus the electric field has a negative divergence at dusk. Therefore, the region between the inward and outward flows has a negative charge at dusk. The charges should discharge along the magnetic field lines, producing upward currents at dusk, preceded and followed by currents of opposite polarities. Although, it is not possible to compute the current using the four Cluster spacecraft as there are not located in the same current region during the event (SC-1 and 4 are inside the magnetosheath, while SC-3 is inside the magnetosphere), we can assess that the plasma flow measurements made inside the magnetosphere during a pressure pulse characterized from three Cluster spacecraft well fit the model of pressure pulse proposed by Lysak et al. (1994).

There are however numerous observations of low-energy plasma "inverted V" which do not fit this model. Figure 9 provide such an example. This event pertains to the cold plasma encountered on January 31, 2001 (see Figure 2). The same procedure as before was here used to compute the L, M, N system. The normal to the magnetopause, N, has the following

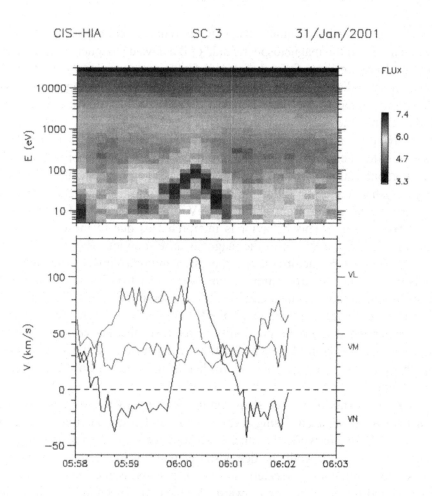

Figure 9. Same as the right part of Figure 8, for January 31, 2001 between 05:58 and 06:03 UT (see also Figure 3).

components, $X_{GSE}= 0.65$, $Y_{GSE} = 0.32$ and $Z_{GSE} = 0.69$. As illustrated in Figure 9, the variations of the V_M and V_L components of the velocity are weak compared to that of the V_N component. Before the event and after it, V_N was slightly negative. During the event, this component shows a positive burst corresponding to a strong outward motion of the plasma. Thus in this case, the cold plasma acceleration corresponds to an outward motion of the magnetopause, i.e., to a magnetosheath pressure decrease. The cold plasma is accelerated outward when the magnetopause is moving outward and is then loosing its drift energy when the magnetopause stops moving and stays far from the satellite. This kind of motion is quite common in the Cluster data base.

3.3   Discussion and conclusion

During periods of quiet magnetic activity, a cold plasma layer is encountered on the magnetospheric side of the dayside magnetopause. Direct density measurements from the plasma frequency given by the WHISPER experiment indicate that this layer can have a width exceeding 1 R_E in the direction normal to the magnetopause. Plasma composition measurements indicate that the major ions are H^+, He^+ and O^+. The cold ions show a repetitive pattern of energy changes. While the magnetopause is approaching the satellite, their energy increases from the detector low-energy threshold up to about 100 eV for protons. After the passage of the satellites into the magnetosheath and just following their re-entry into the magnetosphere, the ion energy decreases from about 100 eV for protons down to the lowest detectable energy. This behavior is interpreted as due to the effect of the electric field associated with the magnetopause motions. The ion motion is set up when the magnetopause is compressed and relaxed when the boundary is going out. This interpretation is substantiated by the change of the proton distribution function, which suggests the conservation of the first adiabatic invariant in the frame moving with $\mathbf{E \times B}$. In the opposite case, the ion would be accelerated without the possibility to return to thermal energy when the magnetopause is going far away. That the energetization of the ions is linked to the magnetopause motion can further be deduced from their velocity distribution.

Characteristic Inverted-V shapes in the time-energy spectrograms of cold ions are often detected. Using a case when one Cluster satellite was inside the magnetosphere while the other spacecraft were for a brief time interval inside the magnetosheath, we demonstrated that this ion structure is due to a bulk acceleration then deceleration of the plasma as the magnetopause is accelerated inward and then outward. We also been able to show that inverted V structure can also be produced when the magnetopause is going quickly outward and then slows down its motion. This drift of the thermal plasma caused by magnetopause motion allows us to determine the composition of the initially cold plasma even with the non mass resolving instrument. Knowing the dynamics of the cold ions allow to reconcile the plasma density measurements made from the plasma frequency and the particle spectrometers located on the satellite body and to conclude that the density of ionospheric ions can be 5 to 10 times higher than the plasma sheet density, and can reach up to 2-3 cm^{-3}. Altogether these observations clearly show that there are hidden plasma populations inside the dayside magnetosphere. A similar population has already been found in the magnetospheric tail by Seki et al. (2001) and Sauvaud et al. (2004) and in the dayside by Sauvaud et al. (2001). This paper emphasizes the importance to use the determination of the plasma frequency to probe the magnetospheric density. The use of biased low-energy particle detectors located far enough from the satellite body should allow to probe the distribution function of this low energy plasma in future missions.

ACKNOWLEDGMENTS

The CESR data decommutation and display softwares have been performed by E. Penou and A. Barthe. The work at CESR and LPCE is supported by French Space Agency (CNES). We thank Elena Budnik for helpful discussions and for computations. The auroral preliminary indices have been provided by WDC-2, Kyoto.

REFERENCES

Balogh, A., Dunlop, M. W., Cowley, S. W. H., Southwood, D. J., Thomlinson, J. G., Glassmeier, K. H., Musmann, G., Luhr, H., Buchert, S., Acuna, M. H., Fairfield, D. H., Slavin, J. A., Riedler, W., Schwingenschuh, K., and Kivelson, M. G., 1997, The Cluster magnetic field investigation, *Space Science Reviews* **79**(1-2):65–91.

Brinton, H. C., Grebowski, J. M., and Mayr, H. G., 1971, Altitude variation of ion composition in the mid-latitude through region: Evidence for upward plasma flow, *J. Geophys. Res.* **76**:3738–3745.

Carlson, C., Curtis, D. W., Paschmann, G., and Michael, W., 1982, An instrument for rapidly measuring plasma distribution functions with high resolution, *Adv. Space Res.* **2**:67.

Carlson, C. W., and McFadden, J. P., 1998, Design and application of imaging plasma instruments, in: *Measurements techniques in space plasma*, R. F. Pfaff, J. E. Borovsky, and D. S. Young, ed., *AGU Geophysical monograph* 102, pp 125–140.

Chappell, C. R., Moore, T. E., and Waite Jr., J. H., 1987, The ionosphere as a fully adequate source of plasma for the Earth's magnetosphere, *J. Geophys. Res.* **92**(A6):5896–5910.

Decreau, P. M. E., Fergeau, P., Krasnnoselskikh, V., Leveque, M., Martin, P., Randriamboarison, O., Sene, F. X., Trotignon, J. G., Canu, P., Mogensen, P. B., Vasiljevic, C., Guyot, E., Launay, L., CornilleauWehrlin, N., deFeraudy, H., Iversen, I., Gustafsson, G., Gurnett, D., and Woolliscroft, L., 1997, Whisper, a resonance sounder and wave analyzer: Performances and perspectives for the Cluster mission, *Space Science Review* **79**(1-2):157–193.

Décréau, P. E. M., Fergeau, P., Krasnoselskikh, V., Le Guirriec, E., Leveque, M., Martin, P., Randriamboarison, O., Rauch, J. L., Sene, F. X., Seran, H. C., Trotignon, J. G., Canu, P., Cornilleau, N., de Feraudy, H., Alleyne, H., Yearby, K., Mogensen, P. B., Gustafsson, G., Andre, M., Gurnett, D. C., Darrouzet, F., Lemaire, J., Harvey, C. C., and Travnicek, P., 2001, Early results from the Whisper instrument on Cluster: an overview, *Ann. Geophys.* **19**(10-12):1241–1258.

Hoffman, J. H., Dodson, W. H., Lippincott, C. R., and Hammack, H. D., 1974, Initial ion composition results from the ISIS-2 satellite, *J. Geophys. Res.* **79**:4246–4251.

Kamide, Y., and Baumjohann, W., 1985, Estimation of electrid fields and currents from international magnetospheric study magnetometer data for the CDAW 6 intervals: Implications for substorms dynamics, *J. Geophys. Res.* **90**(NA2):1305–1317.

Lysak, R. L., Song, Y., and Lee, D.-H., 1994, Generation of ULF waves by fluctuations in the magnetopause position, in: *Solar Wind Sources of Magnetospheric Utra-Low-Frequency Waves*, M. Engebretson, K. Takahashi, and M. Scholer, eds., *AGU Geophysical Monograph* 81, pp. 273–281.

McFadden, J. P., and Carlson, C. W., 1998, Computer simulation in designing electrostatic optics for space plasma experiments, in: *Measurements techniques in space plasma*, R. F. Pfaff, J. E. Borovsky, and D. S. Young, eds., *AGU Geophysical monograph* 102, pp. 249–256.

Möbius, E, Kistler, L. M., Popecki, M. A., et al., 1998, The 3-D plasma distribution function analyzers with time-of-flight mass discrimination for Cluster, Fast, and Equator-S, in: *Measurements techniques in space plasma*, R. F. Pfaff, J. E. Borovsky, and D. S. Young, eds., *AGU Geophysical monograph* 102, pp. 243–248.

Moore, T. E., Chappell, C. R., Chandler, M. O., Fields, S. A., Pollock, C. J., Reasoner, D. L., Young, D. T., Burch, J. L., Eaker, N., Waite, J. H., Mccomas, D. J., Nordholdt, J. E., Thomsen, M. F., Berthelier, J. J., Robson, R., 1995, The thermal ion dynamics experiment and plasma source instrument, *Space Science Reviews* **71**(1-4):409–458.

Rème, H., Bosqued, J.-M., Sauvaud, J.-A., Cros, A., Dandouras, J., Aoustin, C., Bouyssou, J., Camus, T., Cuvilo, J., Martz, C., et al., 1997, The Cluster Ion Spectrometry (CIS) Experiment, *Space Sci. Rev.* **79**(1-2)303–350.

Rème, H., Aoustin, C., Bosqued, J. M., Dandouras, I., Lavraud, B., Sauvaud, J.-A., Barthe, A., et al., 2001, First multispacecraft ion measurements in and near the Earth's magnetosphere with the identical Cluster ion spectrometry (CIS) experiment, *Ann. Geophysicae* **19**(10-12):1303–1354.

Richmond, A. D., Kamide, Y., Akasofu, S. I., Alcayde, D., Blanc, M., Delabeaujardiere, O., Evans, D. S., Foster, J. C., Friischristensen, E., Holt, J. M., Pellinen, R. J., Senior, C., Zaitzev, A. N., 1990, Global measures of ionospheric electrodynamic activity inferred from combined incoherent scatter data and ground magnetometer observations, *J. Geophys. Res.* **95**(A2):1061–1071.

Riedler, W., Torkar, K., Rüdenauer, F., Fehringer, M., Pedersen, A., Schmidt, R., Grard, R. J. L. , Arends, H., Narheim, B. T., Troim, J., Torbert, R., Olsen, R. C., Whipple, E., Goldstein, R., Valavanoglou, N., Zhao, H., 1997, Active spacecraft potential control, *Space Science Reviews* **79**(1-2): 271–302.

Sauvaud, J.-A., Louarn, P., Fruit, G., Stenuit, H., Vallat, C., Dandouras, J., Rème, H., André, M., Balogh, A., Dunlop, M., Kistler, L., Möbius, E., Mouikis, C., Klecker, B., Parks, G. K., McFadden, J., Carlson, C., Marcucci, F., Pallocchia, G., Lundin, R., Korth, A., McCarthy, M., 2004, Case studies of the dynamics of ionospheric ions in the Earth's magnetotail, *J. Geophys. Res.* **109**(A1):A01212, doi: 10.1029/2003JA009996.

Sauvaud, J.-A., Lundin, R., Reme, H., McFadden, J., Carlson, C. W., Parks, G. K., Möbius, E., Kistler, L. M., Klecker, B., Amata, E., DiLellis, A. M., Formisano, V., Bosqued, J. M., Dandouras, I., Decreau, P., Dunlop, M., Eliasson, L., Korth, A., Lavraud, B., and McCarthy, M., 2001, Intermittent thermal plasma acceleration linked to sporadic motions of the magnetopause, first Cluster results, *Ann Geophys.* **19**(10-12):1523–1532.

Seki, K., Hirahara, M., Hoshino, M., Terasawa, T., Elphic, R. C., Saito, Y., Mukai, T., Hayakawa, H., Kojima, H., and Matsumoto, H., 2003, Cold ions in the hot plasma sheet of Earth's magnetotail, *Nature* **422**:589–592.

Shelley, E. G., Johnson, R. G., and Sharp, R. D., 1972, Satellite observations of energetic heavy ions during a geomagnetic storm, *J. Geophys. Res.* **77**:6104–6110.

Shelley, E. G., Johnson, R. G., and Sharp, R. D., 1976, Satellite observations of an ionospheric acceleration mechanism, *Geophys. Res. Lett.* **3**:654–656.

Shelley, E. G., Ghielmetti, A. G., Balsiger, H., Black, R. K., Bowles, J. A., Bowman, R. P., Bratschi, O., Burch, J. L., Carlson, C. W., Coker, A. J., Drake, J. F., Fischer, J., Geiss, J., Johnstone, A., Kloza, D. L., Lennartsson, O. W., Magoncelli, A. L., Paschmann, G., Peterson, W. K., Rosenbauer, H., Sanders, T. C., Steinacher, M., Walton, D. M., Whalen, B. A., and Young, D. T., 1995, The toroidal imaging mass-angle spectrograph (TIMAS) for the Polar mission, *Space Science Reviews* **71**(1-4):497–530.

Shue, J-H, Chao, J. C., Russell, C. T., Song, P., Khurara, K. K., and Singer, H. P., 1997, A new functional form to study the solar wind control of the magnetospause size, *J. Geophys. Res.* **102**(A5):9497–9511.

Torkar, K, Riedler, W., Escoubet, C. P., Fehringer, M., Schmidt, R., Grard, R. J. L., Arends, H., Rudenauer, F., Steiger, W., Narheim, B. T., Svenes, K., Torbert, R., Andre, M., Fazakerley, A., Goldstein, R., Olsen, R. C., Pedersen, A., Whipple, E., and Zhao, H., 2001, Active spacecraft potential control for Cluster — implementation and first results, *Ann. Geophys.* **19**(10-12):1289–1302.

ROLE OF ELECTROSTATIC EFFECTS IN THIN CURRENT SHEETS

Lev M. Zelenyi[1], Helmi V. Malova[1, 2], Victor Yu. Popov[3], Dominique C. Delcourt[4], and A. Surjalal Sharma[5]

[1]*Space Research Institute, RAS, 117810, Profsoyusnaya street 84/32, Moscow, Russia;*
[2]*Scobeltsyn Institute of Nuclear Physics of Moscow State University,119992,Moscow,Russia;*
[3]*Faculty of Physics, Moscow State University, Vorobyevy gory, 119899, Moscow, Russia;*
[4]*Centre d'études des Environnements Terrestres et Planétaires-CNRS, Saint-Maur des Faussés, France;* [5]*Department of Astronomy, University of Maryland, College Park, MD 20742, USA*

Abstract: Thin current sheets (TCSs) are sites of energy storage and release in the Earth's magnetosphere. A self-consistent analytical model of 1D TCS is presented in which the tension of the magnetic field lines is balanced by ion inertia rather than plasma pressure. The influence of the electron population and the corresponding electrostatic electric fields required to maintain quasineutrality are taken into account under the realistic assumption that electron motion is fast enough to support quasi-equilibrium Boltzmann distribution along field lines. Electrostatic effects can lead to specific features of local current density profiles inside TCS, for example, to their partial splitting. The dependence of electrostatic effects on the electron temperature, the form of electron distribution function, and the curvature of magnetic field lines are analyzed. Possible implications of these effects on the fine structure of current sheets and some dynamic phenomena in the Earth's magnetotail are discussed.

Key words: thin current sheets; nonlinear particle dynamics; electrostatic effects; self-consistent model.

1.   INTRODUCTION

Recent findings of a very thin structures (in fact "singular" on a scale of magnetosphere) observed in the magnetotail and magnetopause have attracted a lot of attention in space and plasma physics communities (Hoshino,1996; Runov, 2003; Zelenyi, 2003). The thin current sheet (TCS)

J. –A. Sauvaud and Z. Němeček (eds.),
Multiscale Processes in the Earth's Magnetosphere:From Interball to Cluster, 275-288.
© 2004 *Kluwer Academic Publishers. Printed in the Netherlands.*

observed in the Earth's magnetotail with thicknesses of about ion Larmor radius are probably most exciting example of such plasma structures. Many evidences on the existence of thin current sheets have been accumulating from an number of earlier missions (Mitchell et al., 1990; Pulkkinen et al., 1993, 1994; Sergeev et al., 1993, 1998; Hoshino et al., 1996), but dramatic breakthrough has been achieved with the recent Cluster measurements.

The trajectories of ions and electrons are very different within the TCS. While the behavior of the thermal (usually about few keV) ions is principally nonadiabatic near the equatorial plane (Büchner and Zelenyi, 1989), the electrons (having energies 5-6 times lower than ions) are mostly magnetized in the current sheet (except for a very small region of about an electron gyroradius wide, near the X- and O-lines).

Many numerical simulations using full kinetic, hybrid, Hall-MHD codes have given evidence that electrons could carry substantial currents, especially in the vicinity of the very weak magnetic field regions, e.g., X and O-lines (Pritchett and Coroniti, 1994, 1995; Hesse et al., 1996; Birn et al., 1998; Yin and Winske, 2002). Hoshino et al. (1996) and Asano et al. (2003) proposed that the electron current, which dominates at the edges of thin current sheet, might form characteristic double-humped or split structure of current sheet profile. Current sheets with such interesting structure (known also as "bifurcated" current sheets) are shown to exist in the Earth's magnetosphere during substorm onset (Runov et al., 2003a,b; Sergeev et al., 2003). The physical mechanism of current sheet (CS) bifurcation is still unknown. Greco et al. (2002) suggested that current bifurcation might result from chaotic particle scattering away from the tail midplane due to magnetic fluctuations. Zelenyi et al. (2002) developed picture of bifurcation in which the thin current sheet gradually "deteriorates" due to the nonadiabatic scattering of quasi-trapped ions (this process was referred to as CS "aging"). The natural question of the relative contributions of the electron current and the current due to the quasi-trapped ions in the formation of TCS bifurcated structure is the subject of this study.

The aim of this study is to take into account the electrostatic effects in a self-consistent 1D-model of a very thin current sheet (which was elaborated earlier by Sitnov et al., 2000a; and Zelenyi et al., 2000) in both the cases of isotropic and weakly anisotropic electron pressures and to determine the relative roles of the two effects under different parameter regimes of the CS structure.

2. BASICS EQUATIONS AND SOLUTION OF TCS MODEL: ISOTROPIC ELECTRON PRESSURE

2.1 The current of transient ions

We consider a self-consistent thin current sheet model where the magnetic field line tension is balanced by the finite inertia of ions, moving along strongly curved magnetic field lines, instead of plasma pressure gradients typical for a Harris-like equilibria. The TCS is considered homogeneous along the "Sun–Earth" (X-coordinate in GSM system of reference) and the "dawn–dusk" (Y-coordinate) directions (Sitnov et al., 2000a; Zelenyi et al., 2000). The sources of impinging plasma beams supporting TCS formation are located in the northern and southern magnetotail lobes and could be associated with plasma mantle fluxes. We consider a magnetic field that has two components and depends only one spatial coordinate, i.e. $\boldsymbol{B}=\{B_x(z),0,B_n\}$.

Our essential assumptions are: (a) the ion plasma population consists of transient ions with Speiser orbits (moving with average thermal speed v_T and flow speed v_D); (b) the dynamics of ion population is quasi-adiabatic (Büchner and Zelenyi, 1989); (c) the electron component could be described by hydrodynamic equations with isotropic pressure tensor; (d) the plasma is quasineutral ($n_i \approx n_e$); and (e) the electron motion along the field lines is fast enough to support quasi-equilibrium Boltzmann distribution in the presence of electrostatic potential and mirror forces. Last three assumptions (c)-(e) represent the extension of our 1D self-consistent TCS model which have been presented earlier in papers by Sitnov et al. (2000a), Zelenyi et al. (2000), Malova et al. (2000) where the electron component was not taken into account.

The quasiadiabaticity assumption is valid under the condition $\kappa = \sqrt{R_c / \rho_{\max}} \ll 1$ where ρ_{max} is the maximum gyroradius of the ion in the magnetic field with the smallest curvature radius R_c. This condition has been shown to be valid in some of the TCS observations (Pulkkinen et al., 1994). The estimates for the parameter κ (Lui, 1993; Sergeev et al., 1993), have shown that the parameters κ for electrons and ions in the midtail are $\kappa_e \geq 1, \kappa_i \approx 0.1 \cdot \kappa_e < 1$. Therefore, ion parameter κ_i is usually small enough to for the quasi-adiabatic approximation to be valid. The quasiadiabaticity condition allows one to use, in the calculation of ion current $j_i(z)$, an additional integral of motion (although approximate) to avoid explicit solution of Vlasov-Maxwell equations, i.e. the action integral of the "fast" motion along the Z-coordinate: $I_z = (m/2\pi)\oint v_z dz$ (Sonnerup, 1971; Francfort and Pellat, 1976; Büchner and Zelenyi, 1989; Whipple et al.,

1990). The exact integrals of motion are the total particle energy $W = mv^2/2 + e\tilde{\varphi}(z)$ ($\tilde{\varphi}$ is the electrostatic potential) and the canonical momentum $P_y = mv_y - (e/c)A_y(x,z)$. Generally there exist three kinds of magnetotail ion populations in TCS: Speiser's (transient) ions, quasi-trapped population on closed (so-called "cucumber") orbits, and trapped ions with ring-like orbits. These three distinct populations may be classified according to their I_z values (Büchner and Zelenyi, 1989). In the present study we consider only ions with Speiser's orbits, and discard the effects of "cucumber" and "ring" orbits since their effects have been already explored in our previous publications (Zelenyi *et al.*, 2002, 2003). The procedure for the derivation of a set of self-consistent dimensionless TCS equations for a one-component plasma system has been described in details by Sitnov et al. (2000a,b), and Zelenyi et al. (2000). Here we use a similar approach to obtain self-consistent solutions in a two-component system.

Using directly the equations of motion and the conservation of P_y, the quasiadiabatic (approximate) invariant I_z can be expressed in the form

$$I_z(\vec{v},z) = 2m/\pi \int_{z_0}^{z_1} \left(v_y^2 + v_z^2 + 2e\{\tilde{\varphi}(z) - \tilde{\varphi}(z')\} \right.$$

$$\left. - \left\{ v_y + e/mc \int_z^{z'} B(z'')dz'' \right\}^2 \right)^{1/2} dz' \ . \tag{1}$$

Here the limits of integration are obtained from the condition (Sitnov et al., 2000b)

$$v_y \pm \sqrt{v_y^2 + v_z^2 + 2e\{\tilde{\varphi}(z) - \tilde{\varphi}(z')\}/m} = -(e/mc)\int_{z_0}^{z_{0,1}} B(z'')dz'' \ , \tag{2}$$

with the additional condition $z_0 = 0$ if the solution of Eq.(2) is negative.

For the 1D geometry under consideration the self-consistent Vlasov-Maxwell equations acquire the following simple form:

$$\frac{dB_x}{dz} = (4\pi/c)\left(j_{yi}(z) + j_{ye}(z) \right), \tag{3}$$

$$j_{yi} = (4\pi/c)\sum_i e \int v_y f_i(z,\vec{v})d\vec{v} \ . \tag{4}$$

Here j_{yi} is the ion current along the y-direction. The ion distribution function can be obtained by mapping from the original source distribution to a location within the current sheet using the Liouville theorem: $f_i \sim \exp\{-[(v_{||}(\vec{v}, I(z)) - v_D)^2 + v_\perp^2(\vec{v}, I(z))]/v_T^2\}$ (v_T is the thermal velocity, and v_D is the drift velocity). The electron current can be calculated using the Boltzmann approximation and the details are discussed in the next section. We introduce the dimensionless variables $\vec{r} = \vec{R}/(\omega_0 v_D \varepsilon^{4/3})$, $\zeta = z\omega_0/\varepsilon^{4/3}v_D$, $\vec{w} = \vec{v}/(v_D\varepsilon^{2/3})$, $I = I_z\varepsilon^{2/3}\omega_0/(m_i v_T)$, $\varphi = m_i^2\varepsilon^{4/3}\tilde{\varphi}/2e$, $\tilde{j}_e = \tilde{n}w_e$, $n = \tilde{n}/N_0$, $b = \tilde{b}/B_0$ where $\vec{r}$ is a normalized distance (in this case ζ is the dimensionless z coordinate), $\vec{w}$ is the dimensionless particle velocity vector, I is the dimensionless adiabatic invariant, $\varepsilon = v_T/v_D$ is the ion source anisotropy parameter, $\omega_0 = eB_0/mc$ is the gyrofrequency in the magnetic field B_0 at the edges of the sheet, b is the normalized magnetic field, φ is the dimensionless electrostatic potential and $\tilde{j}_e$ is the normalized electron current in y direction : $j_e = eN_e v_D\varepsilon^{2/3}\tilde{j}_e$. Using these variables and the form discussed above, one can now rewrite the ion distribution function in the right hand side of Eq.(3) in the normalized form (Sitnov et al., 2000b)

$$f_i(\vec{w}, I(\zeta)) = \frac{n_0 \exp(-\varepsilon^{-2/3}I) \times \exp\left(\varepsilon^{-2/3}\left(\left[\sqrt{w_0^2 + \varphi - I(\zeta)} - \varepsilon^{-2/3}\right]^2 + I(\zeta)\right)\right)}{\left(\pi^{3/2}v_T^3\left[1 + erf(\varepsilon^{-1})\right]\right)}.$$

(5)

The Maxwell equation, Eq. (3), can then be re-written as

$$\frac{db}{d\zeta} = \frac{4\varepsilon}{\pi^{3/2}}\left(\frac{v_D}{v_A}\right)^2\left\{\int\frac{w_y}{\left[1 + erf(\varepsilon^{-1})\right]}\exp\left(\varepsilon^{-2/3}\left(\left[\sqrt{w_0^2 + \varphi - I} - \varepsilon^{-2/3}\right]^2 + I\right)\right)d^3w\right.$$

$$\left. + \frac{\varepsilon\,\pi^{3/2}}{4}\,\tilde{j}_e\right\}.$$

(6)

Introducing the dimensionless vector-potential $\eta = \varepsilon^{2/3}\int_0^\zeta b(\zeta')d\zeta'$ one can re-write Eq.(3) in the final form:

$$b(\eta) = \frac{2\sqrt{2}\varepsilon^{1/6}}{\pi^{3/4}}\left(\frac{v_D}{v_A}\right)\left\{\int_0^\eta\left\{\int\frac{w_y}{\left[1 + erf(\varepsilon^{-1})\right]}\exp\left(\varepsilon^{-2/3}\left(\left[\sqrt{w_0^2 + \varphi - I} - \varepsilon^{-2/3}\right]^2\right.\right.\right.$$

$$\left.\left.\left. + I\right)\right)d^3w + \frac{\varepsilon\,\pi^{3/2}}{4}\,\tilde{j}_e\right\}d\eta'\right\}^{1/2}.$$

(7)

Eq.(7) may be solved using the usual boundary condition $b(\infty) = 1$.

The important problem here is how the electron currents $\tilde{j}_e$ (the right hand side of Eq.(6)) is taken into account in framework of 1D kinetic TCS model. This problem was not resolved earlier. Sitnov et al. (2000b) made the first attempt to estimate this contribution, but the influence of electrons was introduced only by a redistribution of the ions in the electrostatic potential, arising from the finite temperature of electrons. At the same time, the net electron current was not taken into account. It was shown by Sitnov et al. (2000b) that the contribution of the electrons on the total current is negligibly small. The procedure to include net electron drifts in a similar self-consistent numerical TCS model was described earlier by Peroomian et al. (2002). We extend here their approach and apply it to our self-consistent TCS model. Such an approach works for both the cases of isotropic and anisotropic electron pressures. In this paper, we present in detail only the calculations for isotropic $\hat{p}_e$.

2.2   Electron current contribution: A semi-fluid approach

To calculate electron effects we adopt an electron fluid model in the direction perpendicular to the magnetic field. Further, the electron pressure tensor is assumed to be isotropic so that in this model $\hat{p}_{eij} = p_e \delta_{ik}$ and the more complicated case of anisotropic pressure tensor will be discussed in another publication. Thus the electron motion in the perpendicular direction is

$$m_e \frac{d\vec{v}_{e\perp}}{dt} = -e\left(E + \frac{\mathbf{v}_e \times \mathbf{B}}{c} \right) - \frac{\nabla P_e}{n_e}. \tag{8}$$

In the parallel direction the magnetized electrons with a magnetic moment μ are acted on by the mirror force $-\mu\nabla B$, so that :

$$m_e \frac{d\vec{v}_{e\parallel}}{dt} = -eE_\parallel - \frac{\nabla P_{\parallel e}}{n_e} - \mu\nabla B. \tag{9}$$

Here e is the magnitude of the electron charge. Neglecting the electron inertia one can get from (8)

$$\vec{v}_{e\perp} = c\frac{\mathbf{E} \times \mathbf{B}}{B^2} + c\frac{\nabla P_e \times \mathbf{B}}{en_e B^2}, \tag{10}$$

$$\vec{j}_e = -en_e \vec{v}_{e\perp} . \tag{11}$$

Similarly Eq.(9) gives

$$e\nabla \varphi - \frac{\nabla P_e}{n_e} - \mu \nabla B = 0 , \tag{12}$$

where $E_{||} = -\nabla_{||}\varphi(s)$, $\varphi(s)$ being the electrostatic potential along field lines.

The electron contribution to the perpendicular current corresponds to the electron part of the Hall current in the generalized two-fluid Ohm's law (Braginskii, 1965) with Hall MHD effect included (Hoshino, 1996):

$$\frac{J}{\sigma} = \frac{ne^2}{m_e} \left(E + \frac{V_0 \times B}{c} - \frac{J \times B}{nec} \right) , \tag{13}$$

where V_0 is the bulk plasma velocity.

Eq.(12) could be easily integrated if one assumes the specific form of the equation of state for the electron fluid

$$P_e = n_e T_e \sim c_0 n_e^\gamma , \tag{14}$$

where γ is the polytropic index with $\gamma = 1$ for the isothermal case and $\gamma = 5/3$ for the adiabatic case. For isothermal electron case, Eq.(9) could be rewritten in a form corresponding to the Boltzmann distribution

$$\frac{n_e(s)}{n_0} = \exp\left\{ \frac{e(\varphi(s)-\varphi_0) - \mu(B(s)-B_0)}{T_e} \right\} , \tag{15}$$

with the corresponding boundary condition at the edges of the current sheet:

$$\varphi(L) = \varphi_0 \equiv 0 . \tag{16}$$

Another important equation, which we use in this model, is the quasi-neutrality condition:

$$n_i\left(\vec{r}, \varphi(z)\right) = n_e\left(\vec{r}, \varphi(z)\right) = n , \tag{17}$$

This condition effectively governs the motion of charged particles in plasma, because the electrostatic potential $\varphi(\vec{r})$ acts both at ions and electrons redistributing their densities along field lines to make them approximately equal at each point.

Using the dimensionless variables defined earlier, we obtain from Eqs. (10)–(11) the following normalized equation for electron drift currents in a semi-fluid approach:

$$\tilde{j}_e = -\tilde{n}\frac{\left[\vec{\xi}'_\perp, \vec{b}\right]}{b^2}, \tag{18}$$

where $\vec{\xi}'_\perp = cE'_\perp/(B_0 v_D \varepsilon^{2/3})$, $E'_\perp = E_\perp + T_e/e\nabla \ln n(\mathbf{r})$. To close the system of equations $\{(7), (17), (18)\}$, one can calculate dimensionless electric field $E'_\perp$ using Eq.(12). Using the simple expression for an averaged magnetic moment $\mu = m_e v_{\perp e}^2/2B_0 \approx T_{\perp e}q^2/B_0$ (where $q = \sin\theta$, θ is the pitch angle averaged over electron population) and using ion flow parameter $\varepsilon = v_T/v_D$, we get

$$\vec{\xi}'_\perp = -\vec{\nabla}\varphi, \qquad \varphi(s) = \varphi_0 + \frac{\varepsilon^{2/3}}{\tau}\left[\ln n(s,\vec{r}_\perp) + q^2(b - b_n)\right], \tag{19}$$

where $\tau = T_i/T_e \equiv v_T^2/v_{\perp e}^2$. Contrary to our previous model with one plasma component where all solutions were expressed in terms of a single dimensionless parameter ε, this system of equations depends on five independent parameters: ε, τ, q^2, b_n, and γ. Therefore, Eqs. (7), (17), (18), (19), and the normalized equations (1) and (2) (derived earlier by Sitnov et al. (2000a)), with corresponding boundary conditions (7) and (16) for the magnetic field and the electrostatic potential, represent the closed system of equations for the self-consistent magnetic field and currents for a one-dimensional two component TCS equilibrium.

2.3   Numerical methods of solution

Numerical solutions of the above system of equations have been obtained by using a double iteration technique. At the start of the iteration the electron current is neglected and the ion current, density and magnetic field are computed using a standard iteration process (Eq. (7) with $\tilde{j}_e = 0$). To recalculate the electrostatic potential we substitute the ion density and quasi-neutrality condition (17) into Eq. (19). The resulting electrostatic potential is then used to calculate the electron current (18). This iteration procedure is repeated taking into account the electron current in Eq.(6) until the solution converges (usually it requires 5-7 iteration steps).

2.4   Results of calculations

The system of equations (7), (17), (18) and (19) has been solved numerically and the profiles of the self-consistent current density and the electrostatic potential were obtained. In Fig. 1a we show the partial current densities of ions and electrons for different initial values of the magnetic moments (parameter q). Figure 1b demonstrates the corresponding electrostatic potential within the TCS. One can see that the smaller is the magnetic moment of electrons the lower is the contribution of electrons in comparison with the ion current. The mirror force for larger values of q is larger and therefore stronger values of electrostatic field are required to overcome the mirror force and redistribute the electrons along field lines to reach their balance with ions ($n_i \approx n_e$). Correspondingly, the growing cross-field electron drifts are responsible for the ion currents. The maximum of the electron current for very large averaged magnetic moments ($q=0.9$) is still nearly two times smaller than the ion current.

Fig. 2 demonstrates the dependence of the partial electron currents on the ratio of the ion and electron temperatures τ. For typical values of the parameter τ in the magnetotail ($\tau \sim 5$) the contribution of electron drift currents are not very large, as one can see from Fig. 2. The electron current (and corresponding electrostatic fields) are very sensitive to the value of electron temperature. For hotter electrons, stronger electric fields are required to withstand the electron pressure gradients. Correspondingly, first and second terms in Eq.(10) increase with an increase of T_e (i.e., decrease of τ).

The dependence of the electron current and the corresponding scalar potential on the normal component of the magnetic field b_n (i.e., the effect of the field line curvature) are presented in Fig. 3. This dependence is also strong, although the electron current density for characteristic magnetotail values of $b_n \sim 0.1$ is only about 30% of the ion current density. The mechanism responsible for the b_n influence on electron current are again evident from Eq.(10). A strong dependence of electric and diamagnetic drifts on $|B| \rightarrow B_n|_{z=0}$ arises from B^2 in the denominator of both terms. We also found that the dependence of the current sheet profiles on the polytropic index $\gamma=1$, $5/3$, 2 is negligibly small and do not lead to the effects comparable the ones presented here.

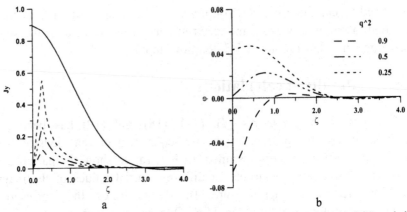

Figure 1. Profiles of the dimensionless electron and ion current densities in TCS and the corresponding electrostatic potential for different values of parameter q^2 as functions of dimensionless z-coordinate ζ. (a) Ion current (solid line) and electron currents are demonstrated at q^2=0.9 (dotted line), 0.5 (dotted-dashed line), 0.25 (dashed line); b_n=0.1, τ=5, γ=1, ε=1

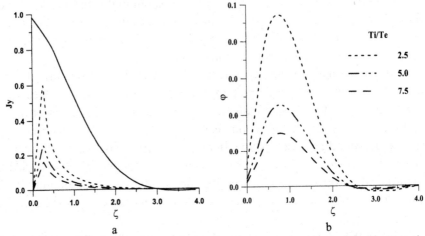

Figure 2. Profiles of the dimensionless electron and ion current densities and the electrostatic potential in TCS for different values of parameter τ as functions of coordinate ζ. (a) Ion current (solid line) and electron currents are shown at τ=2.5 (dotted line), 5.0 (dotted-dashed line), 7.5 (dashed line); b_n=0.1, q^2 =0.5, γ=1, ε=1.

3.   CONCLUSIONS

The 1D model of the current sheet presented here shows that in the case of isotropic pressure the electrons can carry a significant part of the cross-tail current and could produce a weak splitting in the current profile. This is illustrated in Fig. 4a, where the net current density profiles for three values of the parameter τ (the same as in Fig. 2) are shown. Current sheet has a double

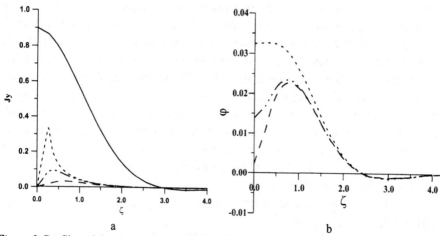

a   b

Figure 3. Profiles of the electron and ion currents and the electrostatic potential for different b_n as functions of coordinate ζ. (a) Ion current is shown by solid line; electron currents are, correspondingly, shown at $b_n = 0.1$ (dotted line), 0.25 (dotted-dashed line), 0.5 (dashed line); $\tau = 5$, $q^2 = 0.5$, $\gamma = 1$, $\varepsilon = 1$.

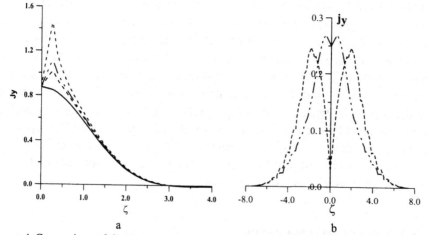

a   b

Figure 4. Comparison of the double-humped current sheet structures; (a) taking into account the electrostatic effects and (b) formed during the accumulation of the quasitrapped plasma in TCS ("aging" process, Zelenyi et al., 2001). (a) The notations are the same as in Fig.2, (b) profiles of ion current density are shown at two different instants before a current sheet disruption   (Zelenyi et al., 2003).

humped (sometimes called "splitted" or "bifurcated") structure due to currents mostly related with $\left[\vec{E} \times \vec{B}\right]$ drifts of magnetized electrons offset from CS central plane ($z=0$). In our case the ion current has "classical" appearance with a peak in the midplane.

One more mechanism of CS splitting, related with particle scattering at magnetic turbulent fluctuations, was recently investigated by Zimbardo et al.

(2004). Few other mechanisms of splitting which have been discussed thus far were related to peculiarities of non-adiabatic ion drifts. "Double humped" structure of TCS have been described, e.g., by Eastwood (1972) and Harold and Chen (1996), who have used particles with a specific non-Maxwellian distribution function as an ion source. It is known that the single ion crossing CS plane carries the local current with "double-peak" shape due to topology of its nonadiabatic meandering motion. Therefore, if the current carrying ions of such type prevail in the system the corresponding double-humped current profile could emerge even for the sheet supported solely by transient (Speiser) orbits. For the "usual" shape of the ion distribution at the source (shifted Maxwellian), bell-shape distribution forms due to superposition of currents carried by particles with various initial pitch-angles (Sitnov et al., 2000; Zelenyi et al., 2000).

As was shown by Zelenyi et al. (2002, 2003), another mechanism of bifurcation due to trapping of ions at closed (so-called "cucumber") orbits gradually become effective after 40-60 minutes since CS formation. As we have shown here, the current carried by electrons (even with isotropic pressure) could also contribute to complexification of CS structure. Our result is in agreement with the Asano (2001) model of the current system near the reconnection X-line, where the TCS current is dominated by ions in the center of the sheet, but is governed by electrons near the edges, where Hall electron drifts might play an important role. Although j_e increases with a decrease of B_n and T_i/T_e, for realistic values of CS parameters, it will hardly reach 40-50% of $j_{\perp i}$. A comparison of the two current density profiles at two instants before CS destruction, shown in Fig. 4b (from Zelenyi et al., 2003) show that the "aging" mechanism is more effective in yielding a splitting of the current profile of TCS compared to the contributions from an electron fluid with an isotropic pressure.

ACKNOWLEDGMENTS

The authors are grateful to V. Sergeev (Institute of Physics, St. Petersburg State University, Russia) and A.Runov (Institut für Weltramforschung der OAW, Graz, Austria) for their continued interest in our work and for very helpful discussions. This work was supported by the Russian Foundation of Basic Research grants 02-02-16003, 03-02-16967, grant of Council of the President of the Russian Federation for Support of Leading Scientific Schools HIII-1739.2003.2, INTAS grant 03-51-3738 and NASA grant NAG510298.

REFERENCES

Asano, Y., 2001, *Configuration of the Thin Current Sheet in Substorms*, Ph.D. thesis, Univ. Tokyo.

Asano, Y., Mukai, T., Hoshino, M., Saito, Y., Hayakawa, H., and Nagai, T., 2003, Evolution of the thin current sheet in a substorm observed by Geotail, *J. Geophys. Res.* **108**(A5): 1189, doi: 10.1029/2002JA009785.

Birn, J., Hesse, M., and Schindler, K., 1996, MHD Simulations of Magnetotail Dynamics, *J.Geophys.Res.* **101**(6):12939–12954.

Braginskii, S I., 1965, Transport processes in a plasma, in: *Rev. Plasma Phys., Vol. 1*, M. A. Leontovich, ed., Consultants Bureau Enterprises, Inc., New York, NY, pp. 256–277.

Büchner, J., and Zelenyi, L. M., 1989, Regular and chaotic charged particle motion in magnetotaillike field reversals: 1. Basic theory of trapped motion, *J. Geophys. Res.* **94**:11821–11842.

Francfort, P., and Pellat, R., 1976, Magnetic merging in collisionless plasmas, *Geophys. Res. Lett.* **3**:433–436.

Greco, A., Taktakishvili, A. L., Zimbardo, G., Veltri, P., and Zelenyi, L. M., 2002, Ion dynamics in the near-Earth magnetotail: magnetic turbulence versus normal component of the average magntic field, *J.Geophys. Res.* **107**(A10):1267, doi: 10.1029/2002JA009270.

Eastwood, J. W., 1972, Consistency of fields and particle motion in the 'Speiser' model of the current sheet, *Planet. Space Sci.*, **20**:1555–1568.

Harold, J. B., and Chen, J., 1996, Kinetic thinning in one- dimensional self-consistent current sheets, *J. Geophys. Res.* **101**(A11):24899–24910.

Hesse, M., Winske, D., Kuznetsova, M. M., Birn, J., and Schindler, K., 1996, Hybrid modeling of the formation of thin current sheets in magnetotail configurations, *J. Geomagn. Geoelectr.* **48**:749–763.

Hoshino, M., Nishida, A., Mukai, T., Saito, Y., and Yamamoto, T., 1996, Structure of plasma sheet in magnetotail: double-peaked electric current sheet, *J. Geophys. Res.* **101**(A11):24775–24786.

Lui, A. T. Y., 1993, Inferring global characteristics of current sheet from local measurements, *J. Geophys. Res.* **98**:13423–13427.

Malova, H. V., Sitnov, M. I., Zelenyi, L. M., and Sharma, A. S.. 2000, Self-consistent model of 1D current sheet: the role of drift, magnetization and diamagnetic currents, in: *Proceedings of Chapman Conference: Magnetospheric Current Systems, Vol. 118*, Ed. by S. Ohtani, R.Fujii, M.Hesse,R.L.Lysak, AGU, Washington, .pp. 313–322.

Mitchell, D. G., Williams, G. J., Huang, C. Y., Frank, L. A., and Russell, C. T.. 1990, Current carriers in the near-Earth cross-tail current sheet during substorm growth phase, *Geophys. Res. Lett.* **17**:583–586.

Peroomyan, V., Zelenyi, L. M., and Schriver, D., 2002, Imprints of small-scale nonadiabatic particle dynamics on large-scale properties of dynamical magnetotail equilibria, 2002 COSPAR, Publ. by Elsevier Science Ltd, Pergamon, Great Britain, *Adv. Sp. Res.* No **12**:2657–2662.

Pritchett, P. L., and Coroniti, F. V., 1992, Formation and stability of the self-consistent one-dimensional tail current sheet, *J. Geophys. Res.* **97**:16773–16787.

Pritchett, P. L., and Coroniti, F. V., 1995, Formation of thin current sheets during plasma sheet convection, *J. Geophys. Res.* **100**:23551–23565.

Pulkkinen, T. I., Baker, D. N., Owen, C. J., Gosling, J. T., and Murthy, N., 1993, Thin current sheets in the Deep Geomagnetotail, *Geophys. Res. Lett.* **20**:2427–2430.

Pulkkinen, T. I., Baker, D. N., Mitchell, D. G., McPherron, R. L., Huang, C. Y., and Frank, L. A., 1994, Thin Current Sheets in the Magnetotail During Substorms: CDAW 6 Revisited, *J. Geophys. Res.* **99**:5793–5804.

Runov, A., Nakamura, R., Baumjohann, W., Zhang, T. I., and Volverk, M., 2003a, Cluster observation of a bifurkated current sheet, *Geophys. Res. Lett.* **30**:1036, doi: 10.1029/2002GL016136.

Runov, A, Nakamura, R., Baumjohann, W., Treumann, R. A., Zhang, T. L., Volwerk, M., Vörös, Z., Balogh, A., Glassmeier, K.-H., Klecker, B., Rème, H., and Kistler, L., 2003b, Current sheet structure near magnetic X-line observed by Cluster, *Geophys. Res. Lett.* **30**:1579, doi: 10.1029/2002GL016730.

Sergeev, V. A., Mitchell, D. G., Russell, C. T., and Williams, D. J., 1993, Structure of the tail plasma/current sheet at 11 Re and its changes in the course of a substorm, *J. Geophys. Res.* **98**:17345–17365.

Sergeev, V. A., Angelopoulos, V., Carlson, C., and Sutcliffe, P., 1998, Current sheet measurements within a flapping plasma sheet, *J. Geophys. Res.* **103**(A5):9177–9188.

Sergeev, V. A., Runov, A., Baumjohann, W., Nakamura, R., Zhang, T. L., Volwerk, M., Balogh, A., Rème, H., Sauvaud, J.-A., André, M., and Klecker, B., 2003, Current sheet flapping 15 motion and structure observed by Cluster, *Geophys. Res. Lett.* **30**:1327, doi: 10.1029/2002GL016500.

Sitnov, M. I., Zelenyi, L. M., Malova, H. V., Sharma, A. S., 2000a, Thin current sheet embedded within a thicker plasma sheet: Self-consistent kinetic theory, *J. Geophys. Res.* **105**(A7):13029–13044.

Sitnov, M. I., Zelenyi, L. M., Sharma, A. S., and Malova, H. V., 2000b, Distinctive features of forced current sheets: Electrostatic effects, in: *Proc. Int. Conf. Substorm-5*, Ed. by A. Wilson, The Netherlands, ESA Publications Division, ESTEC, St.Petersburg, Russia, pp. 197–200.

Sonnerup, B. U. Ö., 1971, Adiabatic particle orbits in a magnetic null sheet, *J. Geophys. Res.* **76**:8211–8222.

Voronov, E. V., and Krinberg, I. A., 1999, The magnetospheric convection as a reason of the formation of very thin plasma sheet, *Geomagn.Aeron.* (transl. Russian journ.) **39**:24–32.

Whipple, E. C., Rosenberg, M., and Brittnacher, M., 1990, Magnetotail acceleration using generalized drift theory: A kinetic merging scenario, *Geophys. Res. Lett.* **17**:1045–1048.

Yin, L., and Winske, D., 2002, Simulations of current sheet thinning and reconnection, *J.Geophys. Res.* **107**:1485, doi: 10.1029/2002JA009507.

Zelenyi, L. M., Sitnov, M. I., Malova, H. V., and Sharma, A. S., 2000, Thin and superthin ion current sheets, Quasiadiabatic and nonadiabatic models, *Nonlinear processes in Geophysics* **7**:127–139.

Zelenyi, L. M., Delcourt, D. C., Malova, H. V., and Sharma, A S., 2002, "Aging" of the magnetotail thin current sheets, *Geophys. Res. Lett.* **29**:1608, doi: 10.1029/2001GL013789.

Zelenyi, L. M., Malova, H. V., and Popov, V. Yu., 2003, Bifurcation of thin current sheets in the Earth's magnetosphere, *JETP Letters* (Transl. from Russian), **78**:296–299.

Zimbardo, G., Greco, A., Veltri, P., Taktakishvili, A. L., and Zelenyi, L. M., 2004, Double peak structure and diamagnetic wings of the magnetotail current sheet, *Planet. Space. Sci.*, in press.

BURSTY BULK FLOWS AND THEIR IONOSPHERIC FOOTPRINTS

Victor A. Sergeev

Institute of Physics, St.Petersburg State University, 198904 St.Petersburg, Russia

Abstract: The bursty bulk flows (BBFs), which provide a major contribution to the Earthward convection in the high-beta plasma sheet region of the magnetotail, are nearly uniformly distributed in distance between 40-50 Re and the inner magnetosphere. Most of them are now confirmed to be plasma bubbles, the underpopulated plasma tubes with a smaller value of plasma tube entropy (pVγ). Many BBFs are visible in the ionosphere due to the associated plasma precipitation and 3d-electric currents, which provides an excellent possibility to study the global dynamics of BBFs by observing their auroral footprints. A number of recent studies, including studies of associated precipitation, convection and field-aligned currents indicate that main mechanism providing a bright optical image of the BBF is the electric discharge (field-aligned electron acceleration) from the dusk flank of the BBF where the intense upward FAC is generated. The auroral signatures have variable forms, with auroral streamers being the most reliable and easily indentified BBF signature. The picture of BBFs emerging from these results corresponds to the powerful (up to several tens kV in one jet) sporadic narrow (2-3 Re) plasma jets propagating in the tail as the plasma bubbles, which are probably born in the impulsive reconnection process but filtered and modified by the interchange process. Penetration of BBFs to less than 6.6 Re distance in the inner magnetosphere was frequently observed, with indications of flow jet diversion and braking (with associated pressure increase and magnetic field compression). Such interaction also creates long-lived drifting plasma structures, particularly those which can be related to torch and omega-type auroras. Role of BBFs in generating other types of auroral structures is briefly discussed.

Key words: magnetotail; plasma sheet; convection; bursty bulk flows; aurora.

J. –A. Sauvaud and Z. Němeček (eds.),
Multiscale Processes in the Earth's Magnetosphere:From Interball to Cluster, 289-306.
© 2004 *Kluwer Academic Publishers. Printed in the Netherlands.*

1.   BURSTY FLOWS IN THE PLASMA SHEET BUBBLE MODEL AND ITS PREDICTIONS

The plasma circulation in the magnetosphere gives birth to a chain of important processes powered by the solar wind energy The most interesting part of the convection circuit is the Earthward convection in the tail plasma sheet, which includes the crucial processes like energy conversion and particle energization, the generation of 3d-current systems and explosive growth of magnetospheric substorms. This is the only part of convection circuit where the plasma pressure is greater/comparable to the magnetic pressure, therefore the plasma behavior (via compression and associated pressure gradients) is able to modify strongly the properties of convection circuit and even change the state of the magnetotail. An example may be a so called Pressure Balance Inconsistency (PBI, e.g. Erickson, 1992) which is considered by many researchers as the most developed (and most theoretically and observationally supported) hypothesis to explain the loading (and subsequent explosive unloading) of magnetic energy during magnetospheric substorms. Its core is a statement that in realistic tail-like configuration the closed plasma sheet tubes can not be in the state of the laminar adiabatic Earthward convection because of excessive plasma compression and too strong plasma pressure gradients which quickly grow and brake such a convection. Instead of laminar convection spectacular fast and short plasma flow bursts are observed (Baumjohann, 1993). These bursty bulk flows (BBFs) were shown to provide the dominant contribution to the total mass, energy and magnetic flux transport in the midtail plasma sheet (Angelopoulos et al.,1992; 1994). A summary of BBFs properties with emphasize laid on the global characteristics inferred from recent studies of their ionospheric images are the main purpose of our brief review.

An easy way to introduce the BBF and differentiate them from the turbulence comes from considering the probability distribution function (PDF) of Vx and Vy plasma flow components — Figure 1. As distinct from isotropic exponential distribution characterizing the turbulence at small velocities, the pronounced high-speed tails appear above roughly 250 km/s, and only in Vx component. These are the BBFs, the anisotropic high-speed flows accelerated in the direction aligned along the action of the Ampere force in the tail. The BBFs were initially defined based on velocity threshold (>400km/s, Angelopoulos et al., 1992), however the PDF looks different at different distances, with a fraction of high-speed flows strongly decreasing with decreasing distance in the tail (possibly, due to the progressive deceleration of Earthward moving plasma tubes). Recent study by Schödel et al (2001) showed that a better definition could be based on the flux transfer threshold, which provides the continuity of the magnetic flux transport contribution from BBFs between 50Re and 15Re, with only ~30% drop below its average value in the nearest to the Earth (10-15 Re) bin. (They used the threshold $|V \times Bz| > 2$ mV/m, which is roughly 10 times the average magnetotail convection Ey (~0.2 mV/m), a so defined objects were called Rapid Flux Transfer events. In the following we shall have in mind this definition while keeping the previous name, the BBF). The average total transport accomplished by such BBFs between X= –50 Re and –10 Re is roughly 50% of total circulated mass, energy and flux (numbers may vary with varying threshold) while being observed only ~5% of time (Schödel et al., 2001a). The BBF is a universal phenomenon observed at any state of the magnetotail, with occurrence increasing from quiet state to very active periods. Their occurrence rate is highest near the midnight and decreases toward the flanks of the tail (Angelopoulos et al., 1994).

The BBF interpretation is much dependent on whether they represent the bulk flows or particle beams. This long-debated issue has been addressed in recent systematic study by Raj et al (2002) who investigated hundreds distribution functions at high time resolution in the plasma sheet at 10–25 Re. They found that most of samples in high-β CPS regions are true convective flows. At the same time the beam distributions (predominantly field-aligned, with frequent low-energy cutoffs) were mostly observed in the outer (low-β) plasma sheet region. Such bimodal appearance is consistent with BBFs being the true convecting plasma tubes if transient character and kinetic effects accompanying the fast Earthward contraction of plasma tubes are taken into account (e.g. Ji and Wolf, 2003).

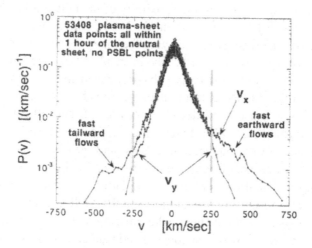

Figure 1. Probability distribution function for plasma flow Vx and Vy-components in the central plasma sheet (after Borovsky et al., 1997).

The burstiness of convection electric field (with variability of both velocity and electric field much greater than their average value, e.g., Borovsky et al., 1997) is well recognized. The shortest series (4s or 12s, dictated by the instrument cycle) are most frequently observed (Baumjohann, 1993), and distributions of bursts durations do not change with distance between 10 and 50 Re in the tail (Schödel et al., 2001a). The typical BBF often appears at r~20 Re as a group of bursts with the ~0.5-1 min rise-and-fall timescale, and the duration roughly 10 min altogether (Angelopoulos et al 1994). With |Vx| ~400 km/s such durations corresponds to the characteristic length about 4 Re (40 Re) along the tail. The complete spatio-temporal description of the BBF is not easy to obtain based on single spacecraft observations alone, it is where the studies of global ionospheric images of BBFs are of great importance. With reference to this information (section 2) we may refer to 3 –10 min as the lower estimate of characteristic BBF lifetime, and to $\Delta Y \sim$ 2-3 Re as the characteristic scale-size across the tail (also confirmed by direct dual spacecraft studies in the plasma sheet, e.g. Sergeev et al., 1996a, Angelopoulos et al., 1997, although in situ estimates of BBF cross-tail are rarely possible and not very accurate). Also with the reference to optical BBF manifestations, the strong BBFs appear as the well-defined mesoscale objects, the narrow plasma streams, rather than a kind of turbulence.

Impulsive magnetic reconnection is generally considered as the basic generation mechanism for the fast flow bursts. Its signatures were firmly established in the midtail including the particle beam structure and quadrupolar Hall magnetic field shear (Nagai et al., 2001) and current sheet structure (Runov et al., 003) in the proximity of the tailward/Earthward flow reversal, the Walen tests (Øieroset, et al., 2000), simultaneous observations of tailward and earthward BBFs carrying the magnetic flux of corresponding polarity (Petrukovich et al., 1998; Slavin et al., 2002) etc. In the midtail region at r ~ 15–30 Re during early substorm expansion phase the BBFs can be observed in the thinned current sheet (presumably near the reconnection region which was most often detected between 20 and 30 Re at substorm onsets, e.g. Nagai et al., 1998). Later on, during substorm recovery phase as well as during steady convection events and nearly quiet periods the Earthward BBFs are most frequently observed in the thick current sheets. In the following we shall concentrate on the latter type, the Earthward BBFs registered in the closed flux tube region at r~10–30 Re, where the flux tubes are closed across the neutral sheet, where the pressure crisis is most sharply pronounced, and which was best covered and studied in observations.

In that region most BBFs show the properties of plasma bubbles, the plasma depleted tubes which have stronger (and more dipolar) magnetic field than that in the plasma sheet proper (Sergeev et al., 1996a). This was observed in those statistical (superimposed epoch) studies in which spacecraft stayed inside the plasma sheet both before and during the event (e.g. Wind observations at X ~ −12 Re in Kauristie et al. 2000, Geotail observations at 14–30 Re in Nakamura et al., 2001b and Schödel et al., 2001b, and IRM observations of magnetic impulse events at 11–18 Re in Sergeev et al., 2001b). According to these studies, inside the BBF proper (1) the plasma pressure is reduced (mostly due to the density decrease); (2) the magnetic field intensity and its latitude angle are increased, (3) the leading front is often well-defined and corresponds to a boundary between two different plasmas.

Being predicted theoretically (Pontius and Wolf, 1990; see also Chen and Wolf, 1993; 1999) the mature bubble is thought to be a kind of transient narrow plasma stream moving Earthward with respect to the surrounding plasma tubes. This relative motion is primarily due to the smaller plasma tube entropy ($S = pV^{5/3}$) in the bubbles, which causes them to be electrically polarized due to the difference of cross-tail currents inside and outside the bubble. The inward bubble motion (under conserved S_B in the bubble) in the ambient tail-like configuration characterized by strong outward S gradient (Figure 2) is expected to continue until $S_B \sim S$ (other parameters like plasma pressure and magnetic field also approach those in the surrounding plasma tubes), unless is diverted around the inner magnetosphere before that time. Until that time the bubble polarization and associated flow vorticity are expected to generate the field-aligned currents of R1 sense, directed downward (upward from ionosphere) at the dawn (dusk) flank of the plasma bubble, as schematically shown in Figure 3a. Such current system together with associated enhanced equatorward convection stream and associated precipitation are the main predictions for bubble manifestations in the ionosphere.

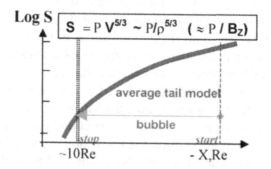

Figure 2. Schematic distribution of plasma tube entropy in the tail and the motion of plasma bubble in such configuration.

Plasma Sheet Bubble and its Ionospheric Mapping

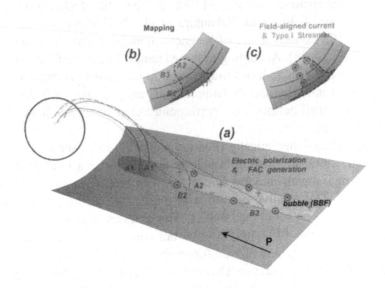

Figure 3. Scheme showing the ionospheric mapping and manifestations of the narrow bursty flow as predicted by the bubble model (from Sergeev et al., 2004a).

The process of the bubble growth has been modeled in the thin filament approximation (Chen and Wolf, 1999; Ji and Wolf, 2003) and, recently, the bubble evolution was simulated numerically within the ideal MHD in realistic 2d magnetic geometry (Birn et al., 2004). These studies confirmed the basic role of entropy difference and electric polarization in the Earthward motion to form the narrow plasma stream, but also showed a more complicated dynamics along the contracted flux tube, which are specific for short-duration transient phenomenon like the BBF is.

Magnetic reconnection is a natural mechanism to form the bubbles at the entry to the closed flux tubes region Earthwards of the reconnection region, since it cuts the long low-density plasma tubes decreasing their volume while modest heating doesn't compensate the loss of entropy (see e.g. simulation results by Birn and Hesse, 1996). It is a combination of localized impulsive magnetic reconnection and interchange instability (providing easy Earthward propagation of the bubbles in the closed tubes of midtail plasma sheet region) which probably form the transient narrow fast plasma streams in the midtail. At times the interchange instability itself may be the additional mechanism creating the bubble. Whereas the configurations with $d(pV^{5/3})/dr < 0$ are known to be interchange-stable, the presence of azimuthal pressure gradients consistent with generation of realistic field-aligned currents (Golovchanskaya and Maltsev, 2003) as well as occasional appearance of local region with dBz / dr > 0 in the neutral sheet (Pritchett and Coronity, 2000) will changes the instability threshold suggesting it can sometimes be excited in the corresponding portions of the plasma sheet.

2. IONOSPHERIC IMAGES OF BURSTY FLOWS

Geometry, large-scale characteristics and global development of the narrow fast plasma streams are difficult to study *in situ*, therefore the auroral imaging could be the best possibility to monitor their development. Fortunately the fast flow bursts are systematically accompanied by auroral effects at the right time and location, e.g. Fairfield et al. (1999), Lyons et al. (1999), Ieda et al. (2001), Sergeev et al. (2000; 2001a), Nakamura et al. (2001a; 2001b), Zesta et al. (2000), Miyashita et al. (2003). The particular form of auroral response may vary depending on the flow burst conditions (tailward or Earthward flow, in thin or expanded plasma sheet etc), they include the localized brightenings and auroral expansions, bright spots and north-south forms or auroral streamers (most studies have been done with global imagers which does not resolve the details below ~50 km). Among them one type, the auroral streamer, has the most obvious connection to the bursty bulk flow. A streamer can be best defined as a transient narrow structure initiated in the poleward oval and propagating towards the equatorial oval boundary — see examples in Figure 4 where the development of three streamers B, C, D is easy to discern in this ~10°-wide auroral oval. (From our experience, the wide oval occurring most often near the maximum substorm expansion and during recovery phase, also during continuously active periods, is the best condition to register the long auroral streamers). Its development includes three stages (Sergeev et al., 2001a): (1) auroral brightening in the poleward part of the oval, (2) propagation from poleward oval to diffuse equatorward oval, (3) after contacting the diffuse equatorward oval the streamer leaves here the bright patch (like A and B), which can be visible for long time (sometimes exceeding an hour). In terms of plasma sheet origin, such streamer dynamics and orientation (not always exactly north-south, e.g., the streamer C as contrasted to streamer D) implies a development of some narrow plasma structure in Earthward direction from the distant tail toward the inner magnetosphere. Not only the general dynamics is similar, a close temporal and spatial association between these two phenomena has been established in case studies (Sergeev et al., 2000; Nakamura et al., 2001a; 2001b).

To exploit this relationship for the BBF studies and monitoring the mechanism generating the precipitation should be established. This is not a

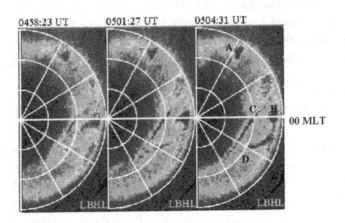

Figure 4. Development of auroral streamers B, C, D (and remnants of streamer A) observed by Polar UVI imager during steady convection event on December 11, 1998 (courtesy by K.Liou).

trivial task as there exist several possibilities of generating the electron precipitation as well as a number of factors which influence the outcome (see e.g. Sergeev 2002). The mechanisms may include (*A*) direct precipitation from the fast plasma stream; (*B*) field-aligned electron acceleration in the upward field-aligned current generated by the plasma stream, and (*C*) acceleration by the field-aligned electric field generated by the difference of ion and electron pitch-angle distributions in the stream (e.g. Serizawa and Sato, 1984). Observationally these three mechanisms should differ in the relationship between the precipitation and field-aligned currents (FACs, which are important in the mechanism *B* but play no role in the mechanisms *A, C*), as well as in the relationship between the electron and proton acceleration and precipitation (here *A* differs from *C*).

Two different streamer-associated precipitation patterns have been established in recent survey of the precipitation, FAC and convection observed during a dozen DMSP traversals over the streamer-conjugate region (Sergeev et al, 2004a). Figure 5 provides an example of observations for the type I streamer precipitation. Here the narrow active streamer near midnight was the brightest aurora seen by Polar UVI along the DMSP trajectory which crossed it from dawn to dusk in equatorward direction (shown by arrow on the UVI image at 1116:49 UT). Accordingly, the DMSP spectrogram shows the strongest intense narrow (L ~ 40 km) electron precipitation with signatures of field-aligned acceleration at ~1118 UT, at the streamer-conjugate location. It has associated a depressed proton precipitation but intense sheet-like upward FAC ($j_\parallel$ > 7 A/km^2, visible as sharp

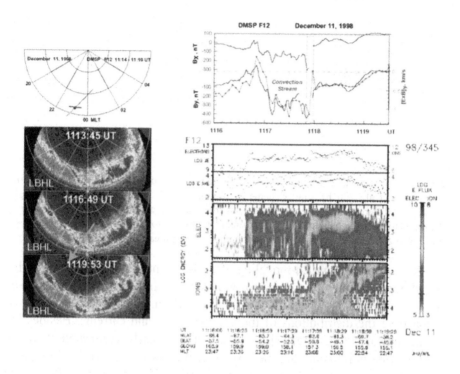

Figure 5. Summary of DMSP F12 observations including the crossing of auroral streamer at 1118 UT (Sergeev et al., 2004a). Left: trajectory of DMSP in the southern hemisphere with indicated position and orientation of the FAC sheet, and Polar UVI images with the arrow showing the approximate mapping of DMSP trajectory. Right: summary of transverse magnetic variations and y-component of convection velocity (dash-dot line), streamer-associated precipitation is marked with a dashed box; bottom — traces of electron/proton energy fluxes and their spectrograms.

positive By variation). This streamer is observed at the dusk flank of a single strong stream of equatorward convection (negative [E×B]y) , which is rather wide (~350 km) and strong (> 1 km/s), resulting in the potential drop across the stream ~13 kV. The downward FAC sheet at the dawnside edge of the convection stream (~1117UT) completes the R1-sense FAC system (together with the upward FAC sheet at ~1118UT, coinciding with the streamer location). Such pattern is consistent with the scheme shown in Figure 3c. Association of field-aligned accelerated electrons with the upward FAC sheet and the large intensity of this sheet current (exceeding a few A/km² normally required for the field-aligned electric field to occur, e.g. Lyons, 1980) are the basic features of type I precipitation.

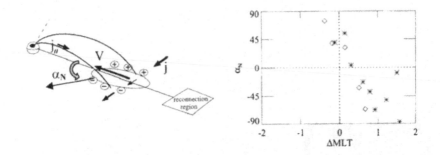

Figure 6. Spatial relationship between BBF footpoint and associated auroral activation (Nakamura et al., 2001b). Left: Scheme of plasma bubble qand auroral activation. Right: Local time differences between the spacecraft footpoint and auroral activation depending on the orientation of the bubble boundary normal determined from Minimal Variance Analysis. Positive/negative α_N correspond to the crossing of westward/eastward edge of the flow burst.

The comprehensive 2d-modeling of a 3d-current system associated with a NS-aligned auroral structure has been made by Amm et al (1999) who used the radar electric field observations and the magnetometer network data to reconstruct the distributions of the conductivities and currents. Besides the strong upward FAC sheet ($j_{max} \sim 25$ A/km^2) nearly collocated with the 100km wide north-south auroral structure, they also inferred the downward FAC sheet (up to 15 A/km^2) at the eastern side of entire 250km-wide structure, as well as the south-west electric field of 20–30 mV/m (possibly the lower limit) corresponding to the > 0.5 km/s equatorward convection along the streamer. This pattern as well as parameter values are consistent with those found in the abovementioned DMSP analysis (Sergeev et al., 2004a) with the exception of a few times larger current density of field-aligned currents, which can be attributed to a very disturbed conditions (>1000 nT in AE index) during the Amm et al. (1999) event.

Important nontrivial result has been obtained by Nakamura et al (2001b), who studied statistically strong short-duration isolated BBFs accompanied by isolated (thus, well identified) auroral signatures. After determining which BBF flank was crossed by the Geotail spacecraft (as given by the angle α_N obtained from minimum variance analysis, see a scheme in Figure 6), and performing the spacecraft mapping into the ionosphere with event-adjusted magnetospheric model, they were able to show that BBF-associated structures (both localized brightnings and auroral streamers) are conjugate to the spacecraft when it occurs at the western flank of the BBF (Fig.6, right panel). The mappings from the dawn-flank were systematically displaced eastward of the auroral structure, suggesting that the auroral streamer shows us a limited part of the entire BBF plasma stream, whose ionospheric projection has a spatial scale about 0.5–1 h MLT (or 300–600 km in E-W direction). These results confirm and complete the picture, in which the auroral streamer is formed due to the field-aligned acceleration in the upward field-aligned current mapped from the dusk edge of the narrow plasma stream. According to this picture, the auroral streamer is the image of upward field-aligned current rather than of the whole BBF stream. These results altogether nicely agree with the predictions of the bubble model (Figure 3).

Whereas type I patterns were observed in the fresh streamers crossed in their poleward or central parts, the type II precipitation pattern consisting of a more broad (a few hundreds km) diffuse precipitation of both electrons and protons (without signatures of field-aligned acceleration) was observed at their latest stage near the equatorward oval: over the bright patch in the equatorward oval (like patches A, B in Figure 4) or over the torch-like structure. This enhanced diffuse-like precipitation is expected to be the mapping of the entire stream, it can be formed either due to the specific mapping distortions (Sergeev et al., 2004a) or/and due to preferential increase of parallel pressure at some stage of the plasma bubble dynamics (Birn et al., 2004). (The mapping distortions due to large, up to 0.5 MA, R1 sense field-aligned currents can be enough strong. As schematically shown in Fig. 3 a, b, the dipolarized magnetic tubes inside the streamer, e.g. A1, are mapped to the inner magnetosphere, therefore it carries much larger pressure and energy flux and provides the stronger precipitation as compared to the tubes outside the streamer, which map to more distant parts of the plasma sheet, like B2 in Fig. 3 a, b).

The auroral observations provide the estimates of two important parameters of the bubbles. The low estimate for the lifetime of bubble propagation in the plasma sheet is given by the duration of auroral streamer propagation (stage (2)), which was 3 to 10 min. Also, the scale-size of the bubble across the plasma jetting was found to be 0.5–1 h MLT or 300–600 km at the ionospheric level. This follows from different kinds of estimates, including (a) total width of R1-sense sheet FAC system (Amm et al., 1999; Sergeev et al., 2004a), (b) the spread of mapped Geotail footpoints around the location of streamer or localized brightening in Figure 6 (Nakamura et al., 2001b). Also, the upper estimate of the bubble width could be given by the minimal longitudinal separation between different streamers simultaneously developing in the auroral zone, which is about ~1 h MLT. If mapped to the plasma sheet, 1h MLT gives 3–4 Re at the distances 10–30 Re, therefore the cross-tail scale of the BBF-bubble from these estimates is 2–3 Re, in perfect agreement with estimates obtained by in situ measurements by two closely-spaced spacecraft (Sergeev et al., 1996a; Angelopoulos et al., 1997).

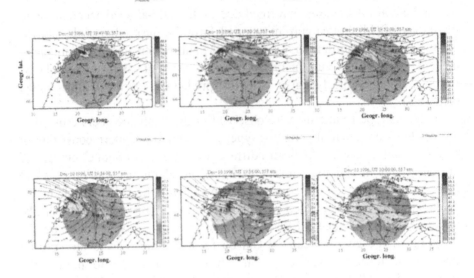

Figure 7. Dynamics of equivalent currents and auroral precipitation during intrusion of the auroral streamer into the diffuse auroral oval (Kauristie et al., 2003).

The final stage of BBF inward motion may include important dynamic processes like the flow braking and/or diversion around the inner magnetosphere (e.g, Shiokawa et al., 1997), whose physics is of great interest to study. The late stage of the streamer life was addressed in the study by Kauristie et al. (2003) based mostly on ground-based observations. Figure 7 allows to see important details of the dynamics of the electron auroras and (upward continued) equivalent currents (expected to flow opposite to the ionospheric convection). Those include, first, a close association of discrete auroras with (equatorward) convective jet, with brightest electron auroras on the dusk side of this jet, which is consistent with the type-I pattern discussed in above. The second feature is the westward deflection of convection jet (and associated discrete auroras) when reaching the equatorward edge of the auroral oval (diffuse auroras) which occurred nearly in the center of all-sky camera field-of-view at ~1952–1954 UT. Besides these clear signatures of azimuthal deflection of the BBF, an interesting dynamics of hydrogen emission was also observed (intensification and poleward shift at the arrival of streamer) which was interpreted as the signature of equatorial magnetic field dipolarization and ~30% plasma pressure increase in the equatorial plane in the interaction region. These observations show a large potential of ground observations to address the late BBF stage of life.

Since the streamer-associated ground magnetic variation can be quite large (~500 nT in the case by Amm et al., 1999, and ~300 nT in Kauristie et al., 2003 event) and sharp, these transient variations can easily be mixed with the substorm onsets or activations if interpreted alone.

3. CONSEQUENCES OF BURSTY FLOWS

Let us briefly summarize the important consequences and conclusions learned from studying the global BBF characteristics, particularly, their auroral manifestations. First is a clear demonstration that fast bursty flows represent well-organised and relatively large-scale and long-lived structures, namely, the narrow fast plasma streams, as distinct from the picture of chaotic plasma turbulence. This probably occurs due to the combination of magnetic reconnection and interchange instability (filtering the bubbles), which creates these individual elongated structures inside the extended closed flux tube region of the plasma sheet. A second important fact is that, when multiplying the cross-tail size 3 Re by 3 mV/m, one gets the total flux transport rate about 60 kV, that is, the individual stream can sometimes support alone the total flux transport in the tail at the required rate. (Much smaller cross-stream potential drops in the ionosphere, ≤ 10 kV, could be due to imperfect mapping of potential due to reflection etc). Many individual BBFs can reach the inner magnetosphere from the far tail, as follows from observations of auroral streamers connecting poleward and equatorward parts of the oval. Also the average flux transport rate provided by the BBFs is continuous from ~ 40–50 Re to 10–15 Re (Schödel et al., 2001a). This implies that the BBFs are really the basic mean the plasma sheet use to solve the pressure crisis (PBI) and complete the convection circle throughout the high-β plasma sheet region. The BBF-related shear instabilities can be the basic driver of the turbulence in the plasma sheet (Neagu et al., 2002).

Important consequences could follow from the penetration of narrow plasma streams (bubbles) into the inner magnetosphere. Extensive evidence exist that during moderately disturbed conditions the auroral streamers intruding into the equatorward oval are accompanied by localized (width about 1–2 h in MLT) and short-duration (minutes) energetic particle injections to the geosynchronous orbit (Henderson et al., 1998; Sergeev et al., 1999; 2000; 2001; 2004b). However, there is no complete understanding of the final stage of the bubble life, neither from theory or simulations, nor from the observations. Particularly, the innermost penetration distance of the BBFs, or the conditions when the BBFs will be diverted or stopped are not elucidated yet. As a consequence the question about the role of BBF-related plasma injections during magnetic storms is still waited to be answered.

The latter question can be related to the possible role of the BBFs in the transport of energetic particles (e.g. those accelerated at the near-Earth reconnection line and by Fermi/betatron acceleration within BBFs) to the radiation belts. Normally this is prohibited by the magnetic drift which deflects the particles azimuthally, thus forming the boundary of Alfven forbidden region at rather large distance from the Earth. However, if there is a region of reversed B-field gradient (which exists at the leading front of the bubble), this would decrease the total azimuthal deflection allowing the particle trajectory to reach a smaller distance. Modeling results (e.g. Li et al., 2003) confirmed this effect is effective only if such 'surfing wave' moves with relatively small velocities (a few tens to a hundred km/s, as opposed to ~ 1000 km/s sound speed in the near-Earth plasma sheet). The mechanism creating such slow-moving dipolarization region was not yet actually identified, we suggest the bubbles could be the vehicle bringing these energetic particles inward to populate the radiation belts and ring current with rather energetic particles, which also could play a role of seed population to get later (during magnetic storms) the significant numbers of MeV energy particles.

Intrusion of narrow plasma streams can considerably modify the structure of the inner magnetosphere by creating a long-lived plasma inhomogeneities. These can be seen as a long-lived (up to > 1.5 hours) bright patches, which appear in the diffuse equatorward oval after the intrusion of auroral streamers (like patches A, B in Figure 4) and then move along the oval with the direction and speeds consistent with plasma convection (Sergeev et al., 2004). Most bright, long-lived and fast-drifting structures are seen in the dawnside oval, some of them could be identified with omega-bands or torch-like structures (see also Henderson, 2002).

Convection streamlines

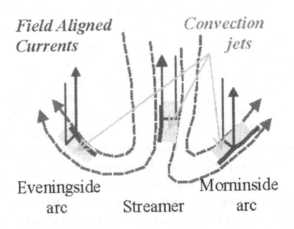

Figure 8. Scheme of convection jets and 3d sheet-like field-aligned currents accompanying both the auroral streamers and premidnight/postmidnight discrete arcs (Sergeev et al. 2002).

There was a little progress in solving the long-standing problem concerning the origin of various auroral structures. It is a one more aspect where we have to look for the possible role of the BBFs which create plasma inhomogeneities in different parts of the plasma sheet. As already discussed in above, the BBFs may give birth to various auroral structures, including the auroral streamers and patches (omega- and torch-structures) in the equatorward oval. Relationship between the auroral arcs and narrow plasma streams (BBFs?) is also an interesting option; particularly, many narrow premidnight and postmidnight arcs are known to be located at one side of relatively wide (about 100 km) convection stream (at poleward/equatorward side from the arc in morning/premidnight MLT sector, e.g. Timofeev and Galperin, 1991). These features are common also for streamers, allowing to suggest the same origin of these discrete structures as a result of plasma stream intrusion and its azimuthal deflection in the inner magnetosphere — see Figure 8. Transient narrow streams of sunward convection of the required scale-size have recently been observed by SuperDARN radars (Senior et al., 2002). This could be an interesting issue to explore to really understand more about magnetospheric mechanisms creating auroral arcs and discrete structures in general.

ACKNOWLEDGEMENTS

The author thank K. Liou, R. Nakamura, W. Baumjohann, J.Birn, M.Kubyshkina, K. Kauristie and O. Amm for their contributions and useful discussions The work was partly supported by INTAS grant 03-51-3738, Leading Schools grant 760.2003.5 and by Intergeophysics Program.

REFERENCES

Amm, O., Pajunpaa, A., and Brandstrom, U., 1999, Spatial distribution of conductances and currents associated with a north-south auroral form during a multiple-substorm period, *Ann. Geophys.* **17**(11):1385–1396.

Angelopoulos, V., Baumjohann, W., Kennel, C. F., Coroniti, F. V., Kivelson, M. G., Pellat, R., Walker, R. J., Luhr, H., and Paschmann, G., 1992, Bursty bulk flows in the inner central plasma sheet, *J. Geophys. Res.* **97**(A4):4027–4039.

Angelopoulos, V., Kennel, C. F., Coroniti, F. V., Pellat, R., Kivelson, M. G., Walker, R. J., Russell, C. T., Baumjohann, W., Feldman, W. C., and Gosling, J. T., 1994, Statistical characteristics of bursty bulk flow events, *J. Geophys. Res.* **99**(A11):21257–21280.

Angelopoulos, V., Phan, T. D., Larson, D. E., Mozer, F. S., Lin, R. P., Tsuruda, K., Hayakawa, H., Mukai, T., Kokubun, S., Yamamoto, T., Williams, D. J., McEntire, R. W., Lepping, R. P., Parks, G. K., Brittnacher, M., Germany, G., Spann, J., Singer, H. J., Yumoto, K., 1997, Magnetotail flow bursts: association to global magnetospheric circulation, relationship to ionospheric activity and direct evidence for localization, *Geophys. Res. Lett.* **24**(18):2271–2274.

Baumjohann, W., 1993, The near-Earth plasma sheet: An AMPTE/IRM perspective, *Space Sci. Rev.* **64**:141.

Birn, J., Raeder, J., Wang, Y. L., Wolf, R. A., and Hesse, M., 2003, On the propagation of bubbles in the geomagnetic tail, *Ann. Geophys.* in press.

Borovsky, J. E., Elphic, R. C., Funsten, H. O., Thomsen, M. F., 1997, The Earth's plasma sheet as a laboratory for flow turbulence in high-β MHD, *J. Plasma Phys.* **57**:1–34.

Chen, C. X., Wolf, R. A., 1993, Interpretation of high-speed flows in the plasma sheet, *J. Geophys. Res.* **98**(A12):21409–21419.

Chen, C. X., and Wolf, R. A., 1999, Theory of thin-filament motion in Earth's magnetotail and its application to bursty flows, *J. Geophys. Res.* **104**:14613–14626.

Erickson, G. M., 1992, A quasi-static magnetospheric convection model in two dimensions, *J. Geophys. Res.* **97**(A5):6505–6522.

Fairfield, D. H., Mukai, T., Brittnacher, M., Reeves, G. D., Kokubun, S., Parks, G. K., Nagai, T., Matsumoto, H., Gurnett, D. A., Yananoto, T., 1999, Earthward flow bursts in the inner magnetotail and their relation to auroral brightenings, AKR intensifications, geosynchronous particle injections and magnetic activity, *J. Geophys. Res.* **104**(A1):355–370.

Golovchanskaya, I. V., Maltsev, Y. P., 2003, Interchange instability in the presence of the field-aligned current: Application to the auroral arc formation, *J. Geophys. Res.* **108**(A3):1106, doi: 10.1029/2002JA009505.

Henderson, M. G., Reeves, G.D., and Murphree, J. S., 1998, Are north-south aligned auroral structures the ionospheric manifestations of bursty bulk flows?, *Geophys. Res. Lett.* **25**(19):3737–3740..

Henderson, M. G., Kepko, L., Spence, H. E., Connors, M., Sigwarth, J. B., Frank, L. A., Singer, H. J., Yumoto, K., 2002, The evolution of north-south aligned auroral forms into auroral torch structures: the generation of omega bands and ps6 pulsations via flow bursts, in: *Proc. 6th Intern. Conf. on Substorms*, R. M. Winglee, ed., Univ. of Washington, Seattle, pp. 169–174.

Ieda, A., Fairfield, D. H., Mukai, T., Saito, Y., Kokubun, S., Liou, K., Meng, C.-I., Parks, G. K., Brittnacher, M. J., 2001, Plasmoid ejection and auroral brightenings, *J. Geophys. Res.* **106**(A3):3845–3857.

Ji, S., Wolf, R. A., 2003, Double-adiabatic MHD theory for motion of a thin magnetic filament and possible implications for bursty bulk flows, *J. Geophys. Res.* **108**(A5):1191, doi: 10.1029/2002JA009655.

Kauristie, K., Sergeev, V. A., Kubyshkina, M., Pulkkinen, T. I., Angelopoulos, V., Phan, T., Lin, R. P., and Slavin, J. A., 2000, Ionospheric current signatures of transient plasma sheet flows, *J. Geophys. Res.* **105**(A5):10677–10690.

Kauristie, K., Sergeev, V. A., Amm, O., Kubyshkina, M. V., Jussila, J., Donovan, E., Liou, K., 2003, Bursty bulk flow intrusion to the inner plasma sheet as inferred from auroral observations, *J. Geophys. Res.* **108**(A1):1040, doi: 10.1029/2002JA009371.

Li, X., Sarris, T. E., Baker, D. N., Peterson, W. K., 2003, Simulation of energetic particle injections associated with a substorm on August 27, 2001, *Geophys. Res. Lett.* **30**(1),:1004, doi: 10.1029/2002GL015967.

Lyons, L. R., 1980, Generation of large-scale regions of auroral currents, electric potentials, and precipitation by the divergence of convection electric field, *J. Geophys. Res.* **85**(NA1):17–24.

Lyons, L. R., Nagai, T., Blanchard, G. T., Samson, J. C., Yamamoto, T., Mukai, T., Nishida, A., and Kokubun, S., 1999, Association between Geotail plasma flows and auroral poleward boundary intensifications observed by CANOPUS photometers, *J. Geophys. Res.* 104(A3):4485–4500.

Miyashita, Y., Machida, S., Liou, K., Mukai, T., Saito, Y., Meng, C.-I., and Parks, G., 2003, Relationship between magnetotail variations and auroral activities during substorms, *J. Geophys. Res.* 108(A1):1022, doi: 10.1029/2001JA009175.

Nagai, T., Fujimoto, M., Saito, Y., Machida, S., Terasawa, T., Nakamura, R., Yamamoto, T., Mukai, T., Nishida, A., and Kokubun, S., 1998, Structure and dynamics of magnetic reconnection for substorm onsets with Geotail observations, *J. Geophys. Res.* 103(A3):4419–4440.

Nagai, T., Shinohara, I., Fujimoto, M., Hoshino, M., Saito, Y., Machida, S., and Mukai, T., 2001, Geotail observations of the Hall current system: Evidence of magnetic reconnection in the magnetotail, *J. Geophys. Res.* 106(A11):25929–25949.

Nakamura, R., Baumjohann, W., Brittnacher, M., Sergeev, V. A., Kubyshkina, M., Mukai, T., Liou, K., 2001a, Flow bursts and auroral activations: Onset timing and foot point location, *J. Geophys. Res.* 106(A6):10777–10789.

Nakamura, R., Baumjohann, W., Schödel, R., Brittnacher, M., Sergeev, V. A., Kubyshkina, M., Mukai, T., Liou, K., 2001b, Earthward flow bursts, auroral streamers, and small expansions, *J. Geophys. Res.* 106(A6):10791–10802.

Neagu, E., Borovsky, J. E., Thomsen, M. F., Gary, S. P., Baumjohann, W., Treumann, R. A., 2002, Statistical study of magnetic field and ion velocity fluctuatrions in the near-Earth plasma sheet: AMPTE/IRM measurements, *J. Geophys. Res.* 107:(A7):1098, doi: 10.1029/2001JA000318.

Øieroset, M., Phan, T. D., Lin, R. P., Sonnerup, B. U. O., 2000, Walen and variance analyses of high-speed flows observed by Wind in the midtail plasma sheet: Evidence for reconnection, *J. Geophys. Res.* 105(A11):25247–25263.

Petrukovich, A. A., Sergeev, V. A., Zelenyi, L. M., Mukai, T., Yamamoto, T., Kokubun, S., Shiokawa, K., Deehr, C. S., Budnick, E. Y., Buchner, J., Fedorov, A. O., Grigorieva, V. P., Hughes, T. J., Pissarenko, N. F., Romanov, S. A., Sandahl, I., 1998, Two spacecraft observations of a reconnection pulse during an auroral breakup, *J. Geophys. Res.* 103(A1):47–59.

Pontius, D. H., Jr., and Wolf, R. A., 1990, Transient flux tubes in the terrestrial magnetosphere, *Geophys. Res. Lett.* 17(1):49–52.

Pritchett, P. L., and Coroniti, F. V., 2000, Localized convection flows and field-aligned current generation in a kinetic model of the near-Earth plasma sheet, *Geophys. Res. Lett.* 27(19):3161–3164.

Raj, A., Phan, T., Lin, R. P., Angelopoulos, V., 2002, Wind survey of high-speed bulk flows and field-aligned beams in the near-Earth plasma sheet, *J. Geophys. Res.* 107(A12):1419, doi: 10.1029/2001JA007547.

Runov, A., Nakamura, R., Baumjohann, W., Treumann, R. A., Zhang, T. L., Volwerk, M., Vörös, Z., Balogh, A., Glaßmeier, K.-H., Klecker, B., Rème, H., Kistler, L., 2003, Current sheet structure near magnetic X-line observed by Cluster, *Geophys. Res. Lett.* 30(N 11):1579, doi: 10.1029/2002GL016730.

Schödel, R., Baumjohann, W., Nakamura, R., Sergeev, V. A., Mukai, T., 2001a, Rapid flux transport in the central plasma sheet, *J. Geophys. Res.* 106(A1):301–313.

Schödel, R., Nakamura, R., Baumjohann, W., Mukai, T., 2001b, Rapid flux transport and plasma sheet reconfiguration, *J. Geophys. Res.* 106(A5):8381–8390.

Senior, C., Cerisier, J.-C., Rich, F., Lester, M., and Parks, G., 2002, Strong sunward propagating flow bursts in the night sector during quiet solar wind conditions: SuperDARN and satellite observations, *Ann. Geophys.* 20:771–779.

Sergeev, V. A., Angelopoulos, V., Cattell, C. A., and Russell, C. T., 1996, Detection of localized, plasma-depleted flux tubes or bubbles in the midtail plasma sheet, *J. Geophys. Res.* **101**(A5):10817–10826.

Sergeev, V. A., Liou, K., Meng. C.-I., Newell, P. T., Brittnacher, M., Parks, G., Reeves, G. D., 1999, Development of auroral streamers in association with localized impulsive injections to the inner magnetotail, *Geophys. Res. Lett.* **26**(3):417–420.

Sergeev, V. A., Sauvaud, J.-A., Popescu, D., Kovrazhkin, R. A., Liou, K., Newell, P. T., Brittnacher, M., Parks, G., Nakamura, R., Mukai, T., Reeves, G. D., 2000, Multiple-spacecraft observation of a narrow transient plasma jet in the Earth's plasma sheet, *Geophys. Res, Lett.* **27**(6):851–854.

Sergeev, V. A., Kubyshkina, M. V., Liou, K., Newell, P. T., Parks, G., Nakamura, R., Mukai, T., 2001a, Substorm and convection bay compared: Auroral and magnetotail dynamics during convection bay, *J. Geophys. Res.* **106**(A9):18843–18855:

Sergeev, V. A., Baumjohann, W., Shiokawa, K., 2001b, Bi-directional electron distributions associated with near-tail flux transport, *Geophys. Res. Lett.* **28**(19):3813–3816.

Sergeev, V. A., 2002, Ionospheric signatures of magnetospheric particle acceleration in substorms — How to decode them?, in: *Proc. 6th Intern. Conf.on Substorms*, R. M. Winglee, ed., Univ. of Washington, Seattle, pp. 39–46.

Sergeev, V. A., Liou, K., Newell, P. T., Ohtani, S.-I., Hairston, M. R., Rich, F., 2004a, Auroral streamers: Characteristics of associated precipitation, convection and field-aligned currents, *Ann. Geophys.* **22**(2):537–548.

Sergeev, V. A., Liou, K., Thomsen, M. F., Yahnin, D. A., 2004b, Narrow plasma streams (plasma bubbles) as a candidate to populate the inner magnetosphere, in: *Physics and Modelling of the inner magnetosphere*, AGU monograph, in press.

Serizawa, Y., and Sato, T., 1984, Generation of large scale potential difference by currentless plasma jets along the mirror field, *Geophys. Res. Lett.* **11**(6):595–598.

Slavin, J. A., Fairfield, D. H., Lepping, R. P., Hesse, M., Ieda, A., Tanskanen, E., Ostgaard, N., Mukai, T., Nagai, T., Singer, H. J., Sutcliffe, P. R., 2002, Simultaneous observations of earthward flow bursts and plasmoid ejection during magnetospheric substorms, *J. Geophys. Res.* **107**(A7):1106, doi: 10.1029/2000JA003501.

Shiokawa, K., Baumjohann, W., and Haerendel, G., 1997, Braking of high-speed flows in the near-Earth tail, *Geophys. Res. Lett.* **24**(10):1179–1182.

Timofeev, E. E., and Galperin, Y. I., 1991, Convection and currents in stable auroral arcs and inverted V's, *J. Geomagn. Geoelectr.* **43**:259–274.

Zesta, E., Lyons, L. R., and Donovan, E., 1997, The auroral signature of Earthward flow bursts observed in the Magnetotail, *Geophys. Res. Lett.* **27**:3241–3244.

MULTI-POINT CLUSTER OBSERVATIONS OF VLF RISERS, FALLERS AND HOOKS AT AND NEAR THE PLASMAPAUSE

J. S. Pickett[1], O. Santolík[2,1], S. W. Kahler[1], A. Masson[3], M. L. Adrian[4], D. A. Gurnett[1], T. F. Bell[5], H. Laakso[3], M. Parrot[6], P. Décréau[7], A. Fazakerley[8], N. Cornilleau-Wehrlin[9], A. Balogh[10], and M. André[11]

[1]*Department of Physics and Astronomy, The University of Iowa, Iowa City, IA, USA;* [2]*Faculty of Mathematics and Physics, Charles University, Prague, Czech Republic;* [3]*RSSD of ESA, ESTEC, Noordwijk, The Netherlands;* [4]*Marshall Space Flight Center, Huntsville, AL, USA;* [5]*Starlab, Stanford University, Stanford, CA, USA;* [6]*LPCE/CNRS, Orleans, France;* [7]*LPCE et Université d'Orléans, Orléans, France;* [8]*Mullard Space Science Laboratory, University College, London, UK;* [9]*CETP/UVSQ, Velizy, France;* [10]*The Blackett Laboratory, Imperial College, London, UK;* [11]*Swedish Institute of Space Physics, Uppsala Division, Uppsala, Sweden*

Abstract: The four Cluster Wideband (WBD) plasma wave receivers occasionally observe electromagnetic triggered wave emissions at and near the plasmapause. We present the remarkable cases of such observations. These triggered emissions consist of very fine structured VLF risers, fallers and hooks in the frequency range of 1.5 to 3.5 kHz with frequency drifts for the risers on the order of 1 kHz/s. They appear to be triggered out of the background whistler mode waves (hiss) that are usually observed in this region, as well as from narrowband, constant frequency emissions. Occasionally, identical, but weaker, emissions are seen to follow the initial triggered emissions. When all the Cluster spacecraft are relatively close (< 800 km, with interspacecraft separations of around 100–200 km), the triggered emissions are correlated across all the spacecraft. The triggered emissions reported here are observed near the perigee of the Cluster spacecraft (around 4–5 R_E) within about 20 degrees, north or south, of the magnetic equator at varying magnetic local times and generally at times of low to moderate Kp. In at least one case they have been observed to be propagating toward the magnetic equator at group velocities on the order of 5–9 × 10^7 m/s. The triggered emissions are observed in the region of steep density gradient either leading up to or away from the plasmasphere where small-scale density cavities are often encountered. Through analysis of images from the EUV instrument onboard the IMAGE spacecraft, we provide evidence that Cluster may sometimes be immersed in a low density channel or other complex

J. –A. Sauvaud and Z. Němeček (eds.),
Multiscale Processes in the Earth's Magnetosphere:From Interball to Cluster, 307-328.
© 2004 *Kluwer Academic Publishers. Printed in the Netherlands.*

structure at the plasmapause when it observes the triggered emissions. Examples of the various types of triggered emissions are provided which show the correlations across spacecraft. Supporting density data are included in order to determine the location of the plasmapause. A nonlinear gyroresonance wave-particle interaction mechanism is discussed as one possible generation mechanism.

Key words: triggered emissions; plasmapause; risers; fallers; hooks; Cluster observations.

1.   INTRODUCTION

Triggered and discrete narrowband emissions have been observed on the ground and in space for decades (Smith and Nunn, 1998). These emissions are generally observed both inside and outside the plasmapause at L values ranging from about 2.5 to 10. They are observed primarily in the form of risers, fallers and upward and downward hooks, or combinations thereof [cf., Helliwell, (1965); Smith and Nunn, (1998)]. Discrete emissions have been defined as emissions having no obvious trigger source, whereas triggered emissions have a clearly recognizable trigger signal (Nunn et al., 1997). A review of the observations and generation theories of triggered emissions was contained in Matsumoto (1979) and Omura et al. (1991) who characterized these emissions as 1) being narrow bandwidth, usually less than 100 Hz; 2) having long durations, sometimes on the order of 1 s; 3) having large amplitudes which saturate at a level as much as 30 dB above that of the triggering signal; 4) having continuously sweeping frequency, typically at a rate of order kHz/s; and 5) taking place almost always within the plasmasphere. Triggered emissions are observed to be produced by constant-frequency wave (CW) transmissions from terrestrial VLF transmitters (Helliwell, 1965; Bell, 1985), by lightning VLF whistlers (Nunn and Smith, 1996), by magnetospheric lines and strong Power-Line Harmonic Radiation (PLHR) induction lines (Helliwell et al., 1975; Luette et al., 1979; Parrot and Zaslavski, 1996; Nunn et al., 1997) and by hiss (Smith and Nunn, 1998).

Discrete emissions have similar characteristics as triggered emissions but appear as isolated events, perhaps arising spontaneously because of high instability conditions in the plasma triggered by random noise, very weak PLHR, or unducted VLF signals (Nunn et al., 1997). They can be either ducted or unducted. VLF chorus is composed of a sequence of closely spaced, discrete emissions, often overlapping in time and most often consisting of rising tones (Helliwell, 1965). Thus chorus shares many of the characteristics of the more isolated discrete emissions. Often a background of hiss is present when chorus is observed. For a review of chorus observations and theory, see Sazhin and Hayakawa (1992).

Several more recent studies have been published that deal with triggered, discrete and chorus emissions. These include, but are not limited to, LeDocq, et al. (1998), Trakhtengerts (1999), Bell et al. (2000), Trakhtengerts and Rycroft (2000), Meredith et al. (2001), Pasmanik, et al. (2002), Lauben et al. (2002), Santolík and Gurnett (2003), and Parrot et al. (2003). Several studies are ongoing at the present time to provide further understanding of the propagation characteristics of these emissions, as well as their generation mechanisms. For example, one surprising aspect of chorus was the discovery by Gurnett et al. (2001) that correlated chorus elements appear at different frequencies on the various Cluster spacecraft separated by distances of a few hundred km or less.

The main aim of this paper is to bring a multi-spacecraft perspective to the observations of triggered emissions in the form of risers, fallers, hooks and combinations thereof that are made on Cluster at and inside the plasmapause. We refer to these emissions as triggered emissions as they appear in many cases to be triggered from observable wave sources. In a few of the cases where the triggering emission is somewhat in doubt, the emissions might better be referred to as discrete since the individual risers, fallers and hooks are well isolated from one another. We begin by describing the suite of instruments on the Cluster and IMAGE spacecraft whose observations are the focus of this study. We show examples of the types of triggered emissions that are observed and their correlation across several spacecraft separated by hundreds of km. This is followed by a discussion of the observations in terms of their significance and of a possible generation mechanism.

2.   INSTRUMENTATION

The primary observations discussed below are from the Cluster Wideband (WBD) plasma wave receiver (Gurnett et al., 1997) which makes a one-axis measurement of the electric or magnetic field using one 88 m dipole antenna or one of three orthogonal magnetic searchcoils. The electric antennas of the four Cluster spacecraft are always oriented close to the ecliptic plane and are spinning with a period of approximately 4 seconds. WBD continuously samples waveforms using a 9.5 kHz bandwidth filter with filter roll-off occurring at about 50 Hz on the low end and 9.5 kHz on the high end. It contains an automatic gain system, implemented in hardware, with gain update rate of 0.1 s, which helps to keep the high intensity triggered emissions within the dynamic range of the instrument without clipping. The data are sampled at high time resolution, 36.5 μs, by transmitting the data directly to the Deep Space Network (DSN) ground stations located in Canberra, Australia and Goldstone, California, USA. WBD data on each spacecraft are obtained over approximately 4% of any one 57-hour Cluster orbit. All of the WBD data shown in this paper were obtained while the spacecraft were near perigee (4.0-4.5 Re). Typically, in this region of space WBD cycles between an electric antenna and a magnetic searchcoil antenna with a 42s/10s duty cycle, respectively. Since the onboard time counter only allows for absolute time accuracy to about 2 ms, WBD makes use of the ground receive time tags provided by DSN, which are accurate to 10 μs, in order to determine time delays between spacecraft.

In situ density measurements are obtained by the Whisper sounder (Décréau et al., 1997) provided the plasma frequency associated with that density falls within the frequency range of the instrument (80 kHz). The Whisper sounder is designed to provide an absolute measurement of the total plasma density by the means of a resonance sounding technique in the range of $0.2 - 80$ cm^{-3}. These measurements are usually made for a period of about 3 s every 52 s during WBD operation. The EFW instrument measures electric fields and waves with two pairs of probes on wire booms in the spin plane of each satellite (the same probes used by WBD to make its wave electric field measurements). Each pair has a probe-to-probe separation of 88 m (Gustafsson et al., 1997). The potential of the probes with respect to the spacecraft is sampled at 5 samples/s, and this measurement can be used to estimate the plasma density. This technique can be used in all but the most tenuous magnetospheric plasmas traversed by Cluster where the spacecraft current balance becomes more sensitive to the energy of the ambient electrons (Pedersen et al., 2001, André et al, 2001). The lower time resolution density measurements from the Whisper Sounder have been used here to better calibrate the density values obtained from the higher time resolution EFW potential measurements. Simultaneous measurements of the waves using the two spin plane electric field antennas and the three magnetic searchcoils at frequencies between 8 Hz and 4 kHz are made by the STAFF-SA instrument (Cornilleau-Wherlin, 1997). STAFF-SA provides the complete auto- and cross-spectra over a frequency range of nine octaves, which are then analyzed to obtain the wave vector, Poynting vector, ellipticity and planarity of polarization. The averaged magnetic field measurements with a time resolution of 4 seconds that are used to calculate the electron cyclotron frequency and to determine each of the spacecraft's locations in a magnetic field aligned coordinate system are provided by the FGM instrument (Balogh et al., 1997). FGM consists of two triaxial fluxgate magnetometers, with one of the sensors located at the end of one of the two 5.2 m radial booms of the spacecraft and the other at 1.5 m inboard from the end of the boom. Remote sensing of the plasmasphere is provided by the Extreme Ultraviolet Imager (EUV) located on the IMAGE spacecraft (Sandel et al., 2000). EUV is designed to image the distribution of plasmaspheric He^+ ions (typically plasmaspheric $He^+/H^+ \sim 0.1$–0.2) via resonant scattering of solar 30.4-nm UV radiation. Tuned specifically to detect the 30.4-nm resonance line of plasmaspheric He^+, EUV consists of three wide–field (30°) cameras such that the field-of-view (FOV) of the three cameras overlap slightly to form a fan-shaped instantaneous FOV of dimensions 84° x 30° (Sandel et al., 2000; 2001). As the IMAGE satellite proceeds through a single spin, the fan–shaped FOV of EUV sweeps across an 84° x 360° swath of sky recording the intensity of detected 30.4-nm radiation in an array of approximately 0.6° x 0.6° pixels. Since the plasmaspheric He^+ scattered 30.4-nm emission is an optically thin medium, the measured intensity is directly proportional to the He^+ column density along the line of sight through the plasmasphere. Each frame of EUV imaged data is produced from a 10-minute accumulation that encompasses 5–spins of the IMAGE satellite.

3.   OBSERVATIONS

We present WBD observations of triggered emissions in the form of VLF risers, fallers and hooks from three different dates, March 11, 2002, June 23, 2003 and August 31, 2003, when Cluster was located at or near the plasmapause. Table 1 shows the separation distances between the four Cluster spacecraft along the direction of the magnetic field and perpendicular to that direction. The Kp index for each of these three dates was moderate to low at 2-, 4, and +1, respectively. The Kp index is probably important as it is based on processes that have been shown to affect the location of the plasmapause (Moldwin et al., 2002).

Table 1. Spacecraft separations along and across B for time periods shown in Figures 1, 4 and 8.

	SC1-SC2 (km)	SC1-SC3 (km)	SC1-SC4 (km)	SC2-SC3 (km)	SC2-SC4 (km)	SC3-SC4 (km)
Mar. 11, 2002						
// to B	105	162	216	57	111	54
ζ to B	151	195	306	71	166	111
Jun. 23, 2003						
// to B					23	
ζ to B					121	
Aug. 31, 2003						
// to B	464	560	825	96	361	265
ζ to B	518	421	747	98	243	333

3.1   March 11, 2002

We begin our survey of the triggered emissions by showing an example from March 11, 2002 when the four spacecraft were relatively close to each other (~100 km separations). Figure 1 is a typical frequency, in kHz, vs. time, in UT hours:minutes:seconds, spectrogram with gray scale indicating electric field power spectral density. The panels from top to bottom show the WBD data from Cluster spacecraft (SC) 1 through 4, respectively. This particular example spans a time period of 4 seconds in which triggered emissions in the form of linked risers and fallers, or upward and downward

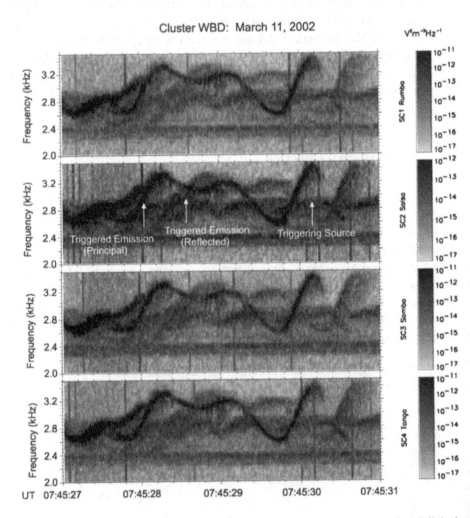

Figure 1. Cluster WBD spectrogram of linked risers and fallers (or upward and linked downward hooks) observed on all four Cluster spacecraft on March 11, 2002. The hooks appear to be triggered out of a constant frequency wave of frequency about 2.8 kHz.

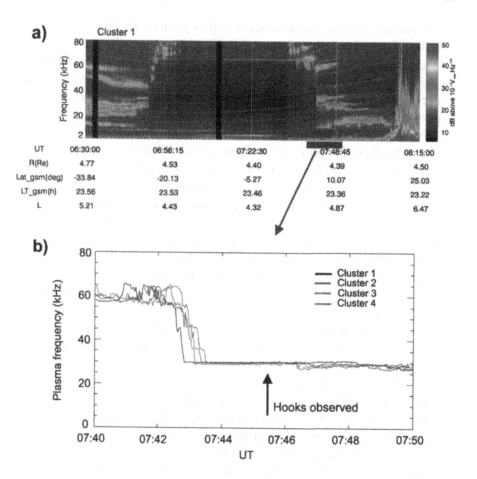

Figure 2. Profile of the electron plasma frequency constructed using the Whisper sounder measurements onboard Cluster 1 on March 11, 2002. a) The power spectrogram showing both the active and the passive spectra. b) Detail of the profile of the electron plasma frequency close to the time interval where the WBD instrument measures triggered emissions (indicated by an arrow).

hooks, are observed to form a nearly sinusoidal signature in the frequency domain. The triggered emissions span the frequency range of about 2.5-3.5 kHz and appear to be triggered out of a narrowband, nearly constant frequency emission observed at about 2.8 kHz. The triggered emissions are well-correlated across all four spacecraft. One particularly interesting feature of these triggered emissions is their reappearance, suggestive of reflection of the principal emission, approximately ½ s after the initial principal intense emissions, but at a much reduced amplitude and slightly more diffuse. These data were obtained while the spacecraft were at 4.4 R_E, 19.3° magnetic latitude, 23.3 hours magnetic local time and 4.9 L-shell. Although not shown in Figure 1, WBD data obtained just prior to 07:45:00, while the instrument was sampling with the magnetic searchcoil as a sensor for 10 s, clearly show that the triggered emissions have magnetic components where the E/B ratio is higher for the waves from which the emissions are triggered than for the triggered emissions themselves. In addition, the overall intensity of the triggered emissions is much higher than the waves from which they are triggered.

The in situ density profile for this March 11, 2002 triggered emission case can be obtained from measurements of the Whisper sounder. Figure 2a shows the power spectrogram from Cluster 1. Both active and passive regimes of the instrument are combined. The slowly varying dotted lines are interpreted as electron cyclotron harmonics appearing when the active sounding mode takes place (once per 52 seconds). Interpretation of these measurements in terms of the local plasma frequency is shown in Figure 2b. We show here a small subinterval of time between 07:40 and 07:50 UT, when the spacecraft exited from the plasmasphere. This time interval is indicated by a bar under the spectrogram in Figure 2a. In Figure 2b we combine results from all four Cluster spacecraft (distinguished per the code in the upper right-hand corner). The time when the triggered emissions were observed on the WBD instrument is shown by an arrow. These emissions were observed outside of the plasmasphere in the plateau region at a density around 10 cm^{-3}. Other observations of triggered emissions during this pass are again localized in the low-density region from 07:47 to 07:49 UT.

An observation of the plasmasphere recorded by the IMAGE EUV instrument post–Cluster measurements is presented in Figure 3. Figure 3(a) shows the EUV image taken at 11:55 UT on 11 March 2002 with the direction to the sun annotated with an arrow. The image indicates that the Cluster measurements presented in Figure 1 and 2 were made in the presence of a complex plasmapause/plasmasphere region characterized by a distinct plasmapause and an embedded low-density channel located radially inward of the plasmapause. The low-density channel is observed to extend from dusk, through Earth's shadow, and on toward dawn. The salient plasmaspheric features of the observation are annotated in Figure 3(b). These features are mapped onto the equatorial plane of the solar magnetic (SM) coordinate system in Figure 3(c) and presented in relation to the orbit of the Cluster tetrahedron from 04:00–10:00 UT projected onto this plane (note that from the perspective of IMAGE EUV at 11:55 UT, the ~100-km separation of the individual Cluster spacecraft is not discernable). The segment of the Cluster orbit spanning the measurements presented in Figures 1 and 2 (07:27–08:03 UT) is highlighted (from triangle to square). Had the Cluster/IMAGE observations been concurrent, Figure 3(c) would indicate that Cluster encountered the plasmasphere within the region of Earth's

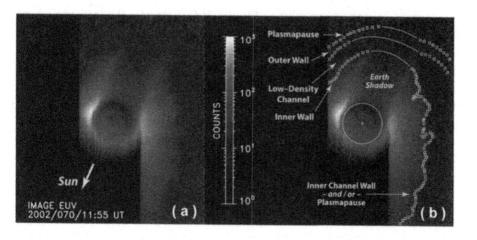

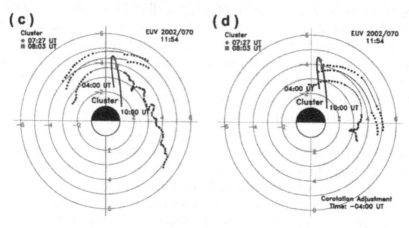

Figure 3. Data from the IMAGE EUV instrument. (a) EUV image taken at 11:55 UT on 11 March 2002 with the direction to the sun annotated with an arrow; (b) Annotation of (a) with the salient plasmaspheric features; (c) plasmaspheric features noted in (b) mapped onto the equatorial plane of the solar magnetic (SM) coordinate system and presented in relation to the projection of the Cluster orbit from 04:00–10:00 UT onto this plane; (d) remapping of observed plasmaspheric features under the assumption of 4-hours of corotation. Remapping results in Cluster immersed in the main trough and transiting the plasmapause and constrained by the plasmapause and outer wall of the low–density channel at the time of WBD–observed triggered emissions.

shadow and would suggest that the WBD–observed triggered risers/fallers had occurred as Cluster transited the outer wall of, and into, the low–density channel. However, the Cluster/IMAGE observations are not concurrent; the IMAGE EUV observation occurs ~4-hours post–Cluster. Assuming plasmapause stability, plasmaspheric feature constraint to dipole field lines, and corotation, the plasmaspheric features observed by EUV, and presented in Figure 3(c), can be remapped back in time to concurrence with the presented Cluster observations. Figure 3(d) presents the remapping of observed plasmaspheric features under the assumption of 4-hours of corotation. Remapping back in time results in Cluster immersed in the main trough and transiting the plasmapause, being constrained by the plasmapause and outer wall of the low–density channel at the time of WBD–observed triggered emissions. The remapping is consistent with density data observed *in–situ* by Cluster (Figure 2). Together with the observed *in–situ* density data, the EUV image may help explain the appearance of the triggered emissions observed by WBD in proximity to the density gradient of the plasmapause.

3.2   June 23, 2003

Next we turn to an example of some well-isolated risers and hooks observed on June 23, 2003 when the Cluster spacecraft were in the middle of their major maneuver period of going from large spacecraft separations (order of 5,000 km) to their small separations (order of 200 km). At this time the two pairs of spacecraft (1,3 and 2,4) were separated by about 35,000 km, while the distances between spacecraft 2 and 4 were less than 150 km. Because of the large distance between the two pairs, we resort to showing only the data from one pair (2 and 4) in order to observe the correlation of the triggered emissions. Figure 4 is a spectrogram in the same format at Figure 1 showing the WBD data from these two spacecraft while they were at 4.4 R_E, 12.1° magnetic latitude, 16.8 hours magnetic local time and 4.6 L-shell. Clearly, the isolated riser/hook combinations are correlated between the two spacecraft in the frequency domain, although the riser/hook combinations are considerably more intense on SC2. This suggests that the source may be closer to SC2. The frequency range of these emissions is about 2 to 3.5 kHz and their frequency drift is on the order of 1 kHz/s. In this case the riser/hook combinations appear to be triggered out of the low frequency waves (hiss) that cover the frequency range from the lowest frequency measured (about 50 Hz) up to about 2 kHz. Note also the presence of secondary triggered emissions from the risers, some of which are seen only on one spacecraft, such as the one centered at 03:45:13 UT on SC4.

In order to determine the direction of propagation of these isolated riser/hook combinations, it is necessary to go to the time domain to do the analysis since the spectrograms are created using FFTs which average over time scales much larger than the delay times. We have chosen the riser/hook centered on 03:45:13 UT for our analysis. Figure 5 shows the filtered waveforms for 0.5 seconds beginning at 03:45:13 UT. Plotted on the vertical axis are the calibrated electric field amplitudes, in mV/m, of the waveforms

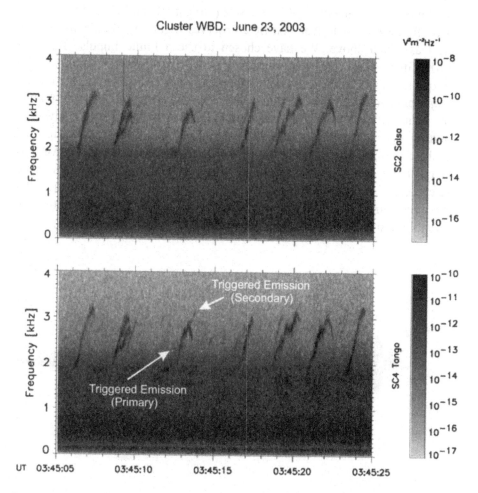

Figure 4. Cluster WBD spectrogram of riser/hook combinations observed on June 23, 2003 on two Cluster spacecraft. The riser/hook combinations appear to be triggered out of the hiss that cuts off around 2 kHz.

Table 2. Delay times associated with event times shown in Figure 5.

Event Number	Correlation Event Time		Delay Time (s)
	SC2 (UT)	SC4 (UT)	
1	03:45:13.10839	03:45:13.11284	0.00445
2	03:45:13.21639	03:45:13.21990	0.00351
3	03:45:13.36511	03:45:13.36839	0.00328
4	03:45:13.40467	03:45:13.40767	0.00300

vs. time, in s, obtained by WBD on SC2 (top panel) and SC4 (bottom panel). This figure makes it clear that SC2 observes higher amplitude waves than SC4 as stated above. We have chosen to use a Finite Impulse Response (FIR) filter over the 2-3 kHz frequency range in order to remain outside the frequency range of the intense low frequency waves and yet fully contain a major portion of the riser portion of the triggered emission. We have pointed out four different correlated wave packet intervals for further analysis by putting a grey-shaded box around them and numbering them 1 through 4 in each of the two panels. For each of the four chosen wave packets we have calculated a delay time of arrival at SC4 from SC2 using the peaks as our reference. The results of these calculations are shown in Table 2. Notice that the delay time, which varies from about 3 to 4.5 ms, decreases with time. Events 1 and 2 are from a time when the riser frequency is increasing.

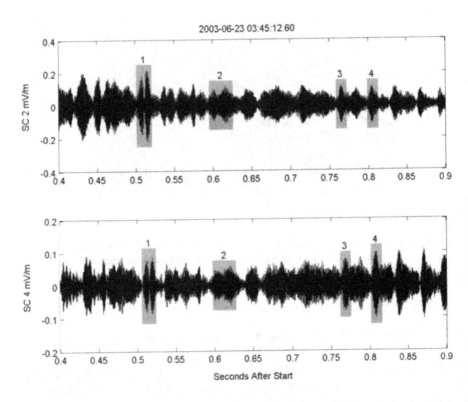

Figure 5. FIR-filtered waveform in the frequency band of 2–3 kHz of the riser/hook combination observed around 03:45:13 on June 23, 2003. Note the good correlations of the waveforms across the two spacecraft, which are used to obtain a time delay (see Table 2). In this case the riser/hook combination is seen first on SC2.

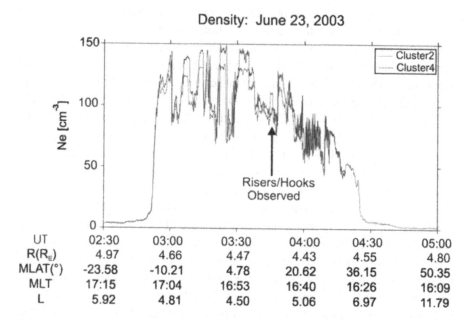

Figure 6. Electron density profile constructed using primarily EFW spacecraft potential measurements on June 23, 2003. The arrow indicates the time of the WBD measurements of triggered emissions, this being inside a broader localized density cavity in or near the plasmapause.

Events 3 and 4 are from a time after the peak of the hook when frequency is starting to decrease slightly. Although the spacecraft are getting closer to each other at this time, the closing distance is not sufficient to account for the decreasing delay times. The error in the time measurement is also not sufficient to account for the decreasing delay times. At this time SC2 is farther from the earth than SC4. SC2 is also at higher magnetic latitude than SC4. Since SC2 observes the wave packets first, the group velocity of this riser-hook combination must have a positive projection in the direction of the SC2-SC4 separation vector, i.e., in the direction toward the Earth and toward the magnetic equator. The same is true for the other risers/hooks observed in Figure 4. Based on the separation distance along the direction of the magnetic field as shown in Table 1 and the delay times shown in Table 2, and an assumption that these electromagnetic triggered emission waves are traveling along, or nearly along, the local magnetic field, we find that their group velocity is on the order of $6-9 \times 10^7$ m/s directed toward the magnetic equator. Cold plasma theory predicts for a wave of this frequency a group speed of 2×10^7 m/s (Stix, 1992).

The density profile for this encounter is shown in Figure 6 and is annotated using the ephemeris of Cluster 2. Because the density during most of this time period is above the range of Whisper (80 cm-3), EFW spacecraft potential measurements have primarily been used to obtain a density, plotted in cm^{-3} along the vertical axis, vs. time, in hours:minutes along the horizontal axis. The two traces are coded by spacecraft as shown in the upper right-hand corner. The density values clearly indicate that the Cluster spacecraft encountered a commonly observed localized region of highly–structured dense plasma in the local time sector ~1600–1700 MLT between 4–7 R$_E$ (Moldwin et al., 1994). The time of the riser/hook combination observations by WBD is indicated by an arrow, and is consistent with the spacecraft being located within one of the several broader localized density cavities within this localized region of dense plasma, even though SC2 is embedded in a narrower localized density enhancement.

The STAFF-SA data that cover the time period of the risers observed by WBD are shown in Figure 7. We show the data of SC2 where the emissions are more intense. The data of SC4 are not shown but lead to similar results. The upper two panels demonstrate that both the hiss below 2 kHz and triggered emissions between 2 and 3 kHz are electromagnetic waves. We have verified that the enhancements of the intensity of hiss between 03:45 and 03:46, and after 03:48 are directly correlated with the density enhancements seen for Cluster 2 in Figure 6. This most probably corresponds to ducting of these waves by the density enhancements. Additional analysis shows that the waves are right-hand polarized consistent with their propagation in the whistler mode (not shown). Panel (c) represents results of analysis of the wave vector direction using singular value decomposition (SVD) of the magnetic cross-spectral matrix (Santolík et al., 2003). The wave vector of the hiss and triggered emissions is found to be within 20-30 degrees of the direction of the stationary magnetic field. Panel (d) shows the electromagnetic planarity estimator (Santolík et al., 2003) indicating if the waves propagate close to one single direction (values close to 1) or if the energy is distributed between antiparallel wave normals (values close to 0). The results at frequencies above 1 kHz where the electromagnetic planarity estimator gives values of 0.7 indicate that wave vectors are rather concentrated around one single direction. Finally, panel (e) shows the component of the Poynting flux parallel to the stationary magnetic field normalized by its standard deviation (Santolík et al, 2001). Positive values obtained for the hiss and triggered emissions signify that these waves propagate approximately in the direction of the stationary magnetic field, i.e., from the magnetic equator toward higher magnetic latitudes. For some cases of short-duration triggered emissions at higher frequencies, however, we cannot exclude opposite propagation since the results are not statistically reliable.

Cluster 2   2003-06-23   STAFF-SA

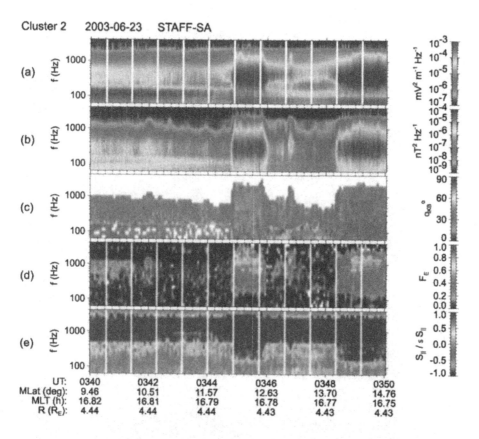

Figure 7. Time-frequency spectrograms measured by the STAFF-SA instrument onboard SC2 on June 23, 2003. (a) Sum of the power spectral densities of the two orthogonal electric components in the spin plane of the spacecraft. (b) Sum of the power spectral densities of the three orthogonal magnetic components. (c) Angle between the wave vector and the stationary magnetic field obtained from the cross-spectral analysis of the magnetic components using the SVD method. (d) Estimate of the electromagnetic planarity of wave fluctuations obtained using the SVD analysis. (e) Parallel component of the Poynting flux normalized by its standard deviation. Position of the spacecraft is given on the bottom, MLat being the magnetic dipole latitude in degrees, MLT being the magnetic local time in hours, and R being the radial distance in Earth radii.

3.3   August 31, 2003

As a final example, we show in Figure 8 a series of correlated risers, centered at 2 kHz, that were observed on August 31, 2003 at around 02:25 UT and that range in frequency from about 1.5 kHz up to about 2.5 kHz with frequency drifts on the order of 1 kHz/s. They appear to arise out of a lower frequency hiss-type emission whose upper cutoff frequency is about 1.8 kHz

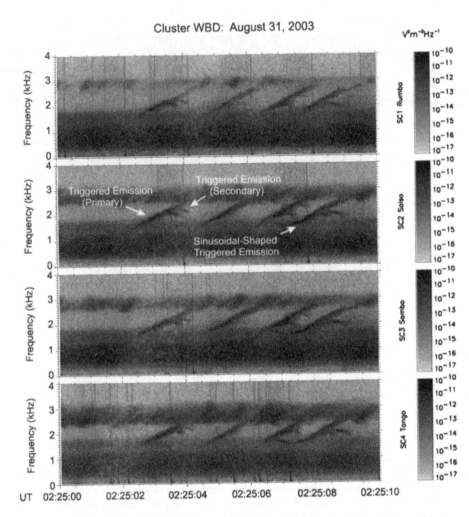

Figure 8. Cluster WBD spectrogram of risers observed on August 31, 2003 on all four Cluster spacecraft. The risers appear to be triggered out of the hiss.

and extend up into a band of more structured emission centered at about 2.9 kHz. Note also the secondary triggers coming off some of the risers, e.g., at 02:25:04 UT and the presence of a sinusoidal-shaped trigger starting around 02:25:07 UT, similar to those observed on March 11, 2002. The risers observed on August 31, 2003 appear to be triggered out of the diffuse plasma waves observed over the frequency range of 200 Hz to 1.8 kHz and be weakest on SC1 (perhaps furthest from source).

Table 1 shows that the maximum separation distance of the different spacecraft was more than 800 km both in the direction along the DC magnetic field and perpendicular to it. The maximum absolute separation is over 1100 km in this case. It is important to note that the triggered emissions are still well correlated in Figure 8, even for those relatively large maximum separations between the points of observation.

On this date the four spacecraft were located at approximately 4.6 R_E, -19.3° magnetic latitude, 12.5 hours magnetic local time and L-shell of 5.2. The density profile shown in Figure 9 (same format as Figure 6, but using the ephemeris of Cluster 1) indicates that these triggered emissions were observed during entry into a post–noon localized, highly–structured region of dense plasma (Chappell et al., 1971) with the emissions occurring in the region of steepest density gradient (see arrow). Well defined density modulations appear during that time in the Whisper data at higher time resolution (not shown).

4. DISCUSSION

The Cluster WBD instrument has obtained multi-spacecraft measurements of the various types of triggered and discrete emissions that have long been observed on the ground and in space by one spacecraft. Upward and downward hooks, or linked risers and fallers, of the type shown in Figure 1 look very similar to those obtained at Halley station, Antarctica

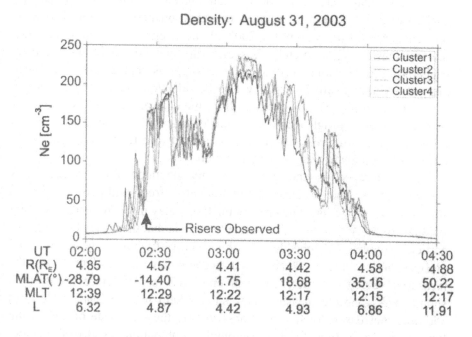

UT	02:00	02:30	03:00	03:30	04:00	04:30
R(R_E)	4.85	4.57	4.41	4.42	4.58	4.88
MLAT(°)	-28.79	-14.40	1.75	18.68	35.16	50.22
MLT	12:39	12:29	12:22	12:17	12:15	12:17
L	6.32	4.87	4.42	4.93	6.86	11.91

Figure 9. Electron density profile constructed using Cluster Whisper Sounder and EFW spacecraft potential measurements on August 31, 2003. The arrow indicates the time of the WBD measurements of the risers, this occurring on the steep density gradient on plasmasphere entry.

shown in Figure 1(e) of Smith and Nunn (1998). Likewise, the risers with downward hooks that arise out of the hiss as shown in Figure 4 look very similar to those which rise out of the hiss as shown in Figure 1(g) of Smith and Nunn (1998). Using the Cluster multispacecraft capability, we have found that the triggered and discrete emissions are often observed to be correlated across distances as great as 800 km across and along the magnetic field direction. Santolík and Gurnett (2003) found a significant correlation at distances of a few hundred km, but the correlation coefficient decreased with a characteristic scale of ~100 km across the magnetic field direction. They used WBD chorus data obtained under disturbed (high Kp) conditions in the low density region outside the plasmapause. Thus, it appears that correlation distances may be shorter well outside the plasmapause than inside based on these two studies. Statistical studies of these correlation lengths have now been started in order to determine what these correlation lengths are, both inside and outside the plasmapause, and the dependence, if any, on solar wind pressure and interplanetary magnetic field configuration.

Another characteristic of the triggered emissions discussed in this study is that they sometimes propagate toward Earth and the magnetic equator as observed on June 23, 2003. Chorus has generally been shown to propagate away from the equator (LeDocq et al., 1998), although Parrot et al. (2003) have shown that there is a reflected component of chorus that propagates back toward the equator after lower hybrid resonance reflection. Likewise, the reappearance on March 11, 2002 of the triggered emissions at much lower intensity approximately ½ s after the appearance of the principal triggered emissions observed by Cluster suggests that a series of linked upward and downward hooks may also be reflected. Supposing a constant group speed of 2×10^7 m/s obtained from the cold plasma theory, the delay of 0.5 s gives a rough estimate of the distance of ~1 Earth radius to a reflection point. Future research will attempt to identify the location of the reflection point using a ray tracing algorithm.

The difference in the group velocity obtained for the riser/hook combination on June 23, 2003 and that obtained from cold plasma theory is still under investigation. The error in the time delay measurements could not account for this difference. The possibility exists that there may be a warmer particle population embedded in this highly-structured, generally cold plasmaspheric plasma which could account for this difference in the measured velocity to the theoretical one. Analysis of future events with simultaneous particle measurements should help resolve this problem. In addition, the reason for the trend observed on the same date of decreasing time delays with time for the riser/hook combinations is also still under investigation. A Doppler shift could not account for this trend since the average phase speeds of the triggered emissions is on the order of 1.5 x 10^7 m/s.

Several studies discussed the generation mechanism of the narrowband and/or the fine structures (e.g., Lutomirski and Sudan, 1966; Melrose, 1986; Nunn et al, 1999; Willes, 2002). Regarding a generation mechanism for the triggered emissions observed on Cluster, we have briefly explored the generation mechanism proposed by Bell et al. (2000) for triggered emissions observed on Polar inside the plasmasphere. This generation mechanism involves a nonlinear gyroresonance interaction between the energetic electrons and the VLF waves, which in the Cluster case would be VLF hiss and constant-frequency emissions. A cursory analysis of the Cluster PEACE (Johnstone et al., 1997) electron data suggests that there may be enough amplification of the waves as they propagate across the magnetic equator in the presence of 15-26 keV electrons to produce the triggered emissions. However, analysis of these data have literally just begun on Cluster for these types of studies since PEACE electron data have just started (in mid 2003) to be taken consistently through Cluster perigee. Thus, there is much work still ongoing in this area.

5.   SUMMARY AND CONCLUDING REMARKS

A summary of our main findings with respect to triggered emissions observed on Cluster multi-spacecraft at or near the plasmapause using the Wideband plasma wave receiver as a detector is as follows:

- The VLF triggered emissions are observed on Cluster as electromagnetic fine-structured risers, fallers, hooks and combinations thereof in the Fourier-transformed spectrograms in the frequency range of 1.5-3.5 kHz with frequency drifts on the order of 1 kHz/s.
- Secondary triggered emissions are sometimes observed to be triggered by the primary triggered emissions.
- Occasionally the principal triggered emissions are observed less than 1 s later at greatly reduced amplitudes.
- The triggered emissions are observed within about 20 degrees north and south of the magnetic equator at around 4-5 R$_E$, all magnetic local times, and L-shells of 4-6.
- Correlation distances are as great as about 800 km both along and cross B.
- The triggered emissions are seen propagating both toward and away from the magnetic equator at group velocities less than 1 x 10^8 m/s.
- The triggers appear to be hiss and narrowband, constant-frequency type emissions.

Cluster's orbit has proven to be very advantageous for obtaining observations of triggered emissions in the form of VLF risers, fallers and hooks since its perigee around 4 R_E is nearly at the magnetic equator and often cuts through at the plasmapause, or just inside or outside of it, at L-shells around 4-6. Thus, this orbit combined with the multi-spacecraft capability promises to provide even more discoveries and a greater understanding of these intense emissions and their connection to solar activity and changes in interplanetary magnetic fields.

ACKNOWLEDGMENTS

This work was supported under NASA/Goddard Space Flight Center Grant No. NAG5-9974. We thank all of the many groups on the European side for their part in obtaining the WBD data (ESA, ESTEC, ESOC, JSOC, Sheffield University, and the Cluster Wave Experiment Consortium), as well as those on the U.S. side (JPL/DSN). We also thank ESA and NASA for valuable analysis support tools, namely, CSDSWeb, SSCWeb and CDAWeb. The authors also wish to thank B.R Sandel and the IMAGE/EUV team at the University of Arizona for their assistance in the processing and analysis of EUV data.

REFERENCES

André, M., Behlke, R., Wahlund, J.-E., Vaivads, A., and Eriksson, A.-I., 2001, Multi-spacecraft observations of broadband waves near the lower hybrid frequency at the Earthward edge of the magnetopause, *Ann. Geophys.* **19**:1471–1481.

Balogh, A., Dunlop, M. W., Cowley, S. W. H., Southwood, D. J., Thomlinson, J. G., et al, 1997, The Cluster Magnetic Field Investigation, *Space Sci. Rev.* **79**:65–92.

Bell, T. F., 1985, High amplitude VLF transmitter signals and associated sidebands observed near the magnetic equatorial plane on the ISEE 1 satellite, *J. Geophys. Res.* **90**:2792–2795.

Bell, T. F., Inan, U. S., Helliwell, R. A., and Scudder, J. D., 2000, Simultaneous triggered VLF emissions and energetic electron distributions observed on POLAR with PWI and HYDRA, *Geophys. Res. Lett.* **27**(A2):165–168.

Chappell, C.R., Harris, K. K., and Sharp, G. W., 1971, The dayside of the plasmasphere, *J. Geophys. Res.* **76**:7632–7647.

Cornilleau-Wehrlin, N., Chauveau, P., Louis, S., Meyer, A., and Nappa, J. M., 1997, The Cluster Spatio-Temporal Analysis of Field Fluctuations (Staff) experiment, *Space Sci. Rev.* **79**:107–136.

Décréau, P. M. E., Fergeau, P., Krasnosels'kikh, V., Lévêque, M., Martin, Ph., et al., 1997, WHISPER, a resonance sounder and wave analyser: performances and perspectives for the Cluster mission, *Space. Sci. Rev.* **79**:157–193.

Gurnett, D. A., Huff, R. L., and Kirchner, D. L, 1997, The Wide-band plasma wave investigation, *Space Sci. Rev.* **79**:195–208.

Gurnett, D. A., Huff, R. L., Picket, J. S., Person, A. M., Mutel, R. L., et al., 2001, First results from the Cluster Wideband plasma wave investigation, *Ann. Geophys.* **19**:1259–1272.

Gustafsson, G., Bostrom, R., Holback, B., Holmgren, G., Lundgren, A., et al., 1997, The electric field and wave experiment for the Cluster mission, *Space Sci. Rev.* **79**:137–156.

Helliwell, R. A., 1965, *Whistlers and Related Ionospheric Phenomena*, Stanford University Press, Stanford, California.

Helliwell, R. A., Katsufrakis, J. P., Bell, T. F., and Raghuram, R., 1975, VLF line radiation in the earth's magnetosphere and its association with power system radiation, *J. Geophys. Res.* **80**:4249–4258.

Johnstone, A. D., Alsop, C., Burge, S., Carter, P. J., Coates, A. J., et al., 1997, Peace: a Plasma Electron and Current Experiment, *Space Sci. Rev.* **79**:351–398.

Lauben, D. S., Inan, U. S., Bell, T. F., and Gurnett, D. A., 2002, Source characteristics of ELF/VLF chorus, *J. Geophys. Res.*, **107**(A10):1429, doi: 10.1029/2000JA003019.

LeDocq, M. J., Gurnett, D. A., and Hospodarsky, G. B., 1998, Chorus source locations from VLF Poynting flux measurements with the Polar spacecraft, *Geophys. Res. Lett.* **25**:4063–4066.

Luette, J. P., Park, C. G., and Helliwell, R. A., 1979, The control of the magnetosphere by power line radiation, *J. Geophys. Res.* **84**:2657–2660.

Lutomirski, R. F, and Sudan, R. N., 1966, Exact Nonlinear Electromagnetic Whistler Modes, *Phys. Rev.* **147**:156–165.

Matsumoto, H., 1979, Nonlinear whistler-mode interaction and triggered emissions in the magnetosphere: a review, in: *Wave Instabilities in Space Plasmas*, P. J. Palmadesso and K. Papadopoulos, eds., D. Reidel, Dordrecht, pp. 163.

Melrose, D. B., 1986, A phase-bunching mechanism for fine structures in auroral kilometric radiation and Jovian decametric radiation, *J. Geophys. Res.* **91**:7970–7980.

Meredith, N. P., Horne, R. B., and Anderson, R. R., 2001, Substorm dependence of chorus amplitudes: implications for the acceleration of electrons to relativistic energies, *J. Geophys. Res.* **106**:13,165–13,178.

Moldwin, M.B., Thomsen, M. G., Bame, S. J., McComas, D. J., and Moore, K. R., 1994, An examination of the structure and dynamics of the outer plasmasphere using multiple geosynchronous satellites, *J. Geophys. Res.* **99**:11,475–11,481.

Moldwin, M. B., Downward, L., Rassoul, K., Amin, R. and Anderson, R. R., 2002, A new model of the location of the plasmapause: CRRES results, *J. Geophys. Res.* **107**(A11):1339, doi: 10.1029/2001JA009211.

Nunn, D. and Smith, A. J., 1996, Numerical simulation of whistler-triggered VLF emissions observed in Antarctica, *J. Geophys. Res.* **101**(A3):5261–5277.

Nunn, D., Omura, Y., Matsumoto, H., Nagano, I., and Yagitani, S., 1997, The numerical simulation of VLF chorus and discrete emissions observed on the Geotail satellite using a Vlasov code, *J. Geophys. Res.* **102**(A12):27,083–27,097.

Nunn, D., Manninen, J., Turunen, T., Trakhtengerts, V., and Erokhin, N., 1999, On the nonlinear triggering of VLF emissions by power line harmonic radiation, *Ann. Geophys.* **17**:79–94.

Omura, Y., Nunn, D., Matsumoto, H. and Rycroft, M. J., 1991, A review of observational, theoretical and numerical studies of VLF triggered emissions, *J. Atmos. and Terres. Phys.* **53**(5):351–368.

Parrot, M., and Zaslavski, Z., 1996, Physical mechanisms of man made influences on the magnetosphere, *Surveys in Geophysics* **17**:67–100.

Parrot, M., O. Santolik, N. Cornilleau-Wehrlin, M. Maksimovic, and C.C. Harvey, 2003, Magnetospherically reflected chorus waves revealed by ray tracing with CLUSTER data, *Ann. Geophys.* **21**:1111–1120.

Pasmanik, D. L., Demekhov, A. G., Nunn, D., Trakhtengerts, V. Y., and Rycroft, M. J., 2002, Cyclotron amplification of whistler-mode waves: A parametric study relevant to discrete VLF emissions in the Earth's magnetosphere, *J. Geophys. Res.* **107**(A8):1162, doi: 10.1029/2001JA000256.

Pedersen, A., Decreau, P., Escoubet, C.-P., Gustafsson, G., Laakso, H., et al., 2001, Four-point high time resolution information on electron densities by the electric field experiments (EFW) on Cluster, *Ann. Geophys.* **19**:1483–1489.

Sandel, B. R., Broadfoot, A. L., Curtis, C. C., King, R. A., Stone, T. C., et al., 2000, The extreme ultraviolet imager investigation for the IMAGE mission, *Space Sci. Rev.* **91**:197.

Sandel, B.R., King, R. A., Forrester, W. T., Gallagher, D. L., Broadfoot, A. L. and Curtis, C. C., 2001, Initial results from the IMAGE extreme ultraviolet imager, *Geophys. Res. Lett.* **28**:1439.

Santolík, O., Lefeuvre, F., Parrot, M., and Rauch, J. L. 2001, Complete wave-vector directions of electromagnetic emissions: Application to INTERBALL-2 measurements in the nightside auroral zone, *J. Geophys. Res.* **106**:13,191–13,201.

Santolík, O., Parrot, M., and Lefeuvre, F., 2003, Singular value decomposition methods for wave propagation analysis, *Radio Sci.* **38**(1):1010, doi: 10.1029/2000RS002523.

Santolík, O., and Gurnett, D. A., 2003, Transverse dimensions of chorus in the source region, *Geophys. Res. Lett.*, **30**(2):1031, doi: 10.1029/2002GL016178.

Sazhin, S. S. and Hayakawa, M., 1992, Magnetospheric chorus emissions: A review, *Planet. Space Sci.*, **40**(5):681–697.

Smith, A. J., and Nunn, D., 1998, Numerical simulation of VLF risers, fallers, and hooks observed in Antarctica, *J. Geophys. Res.*, **203**(A4):6771–6784.

Stix, T. H., 1992, *Waves in Plasmas*, Am. Inst. of Phys., New York.

Trakhtengerts, V. Y., 1999, A generation mechanism for chorus emission, *Ann. Geophys.* **17**:95–100.

Trakhtengerts, V. Y., and Rycroft, M.J., 2000, Whistler-electron interactions in the magnetosphere: new results and novel approaches, *J. Atmos. and Terres. Phys.* **62**:1719–1733.

Willes, A. J., 2002, Jovian S burst drift rates and S burst/L burst interactions in a phase-bunching model, *J. Geophys. Res.* **107**(A5):1061, doi: 10.1029/2001JA000282.

Printed in the United States
by Baker & Taylor Publisher Services

Printed in the United States
by Baker & Taylor Publisher Services